Mary Beth Wilson

Impacts of Participatory Development in Afghanistan

Studien zum Modernen Orient

herausgegeben
von Gerd Winkelhane

Studien zum Modernen Orient 24

Mary Beth Wilson

Impacts of Participatory Development in Afghanistan

A Call to Reframe Expectations

The National Solidarity Programme in the Community of Shah Raheem

K S
KLAUS SCHWARZ VERLAG • BERLIN

Bibliographic Information of Deutsche Bibliothek
Deutsche Bibliothek registers this publication in the Deutsche Nationalbibliografie; detailed bibliographical data can be accessed at *http://dnb.ddb.de*.

Cover page: Research Assistant and community member walking towards the main street in Shah Raheem (photo: M.B. Wilson)

D 188

www.klaus-schwarz-verlag.com

First edition
Producer: Tim Mücke / Donatha Czichy
Cover: J2P Berlin
Printed in Germany on chlorine-free bleached paper
ISBN 978-3-87997-431-3

Tabel of Contents

Research in Review: A Brief Synopsis

Participatory development approaches have been gaining in prominence and importance in recent decades. These approaches stand in stark contrast to the more classical development approaches of modernization theory, dependency theories, and theories that stem from these two main paradigms. It is based upon the concept that development should include the opinions, understandings, and efforts of the people where a development program is to be implemented.

Increasingly, development funders, organizations, implementers, and other stakeholders have been investing their time, energy, and money into development projects and programs that include a participatory component. This is evidenced in participatory development funding figures from the World Bank and from the explosion of literature devoted to the topic, along with the increased number of participatory development projects and programs being implemented around the world.

The rhetoric surrounding participatory development approaches is very positive, based upon the expectation that it provides better development outcomes than classical development approaches. This is both supported and contradicted by the participatory development literature, indicating the complexity of the interplay of participatory approaches and development programs in real world situations.

This dissertation adds to the academic literature focused on participatory development approaches using empirical data. The research utilizes an intensive qualitatively-based case study methodology to look at the impacts of the National Solidarity Programme (NSP) on a community in Northern Afghanistan, thus presenting a new perspective on participatory development: a community level perspective.

Focusing on four areas of investigation – personal security, economic security, social security, and empowerment[1] – this research provides a historically grounded picture of the community and how the people there experienced the process and implementation of the NSP and its impacts. The fieldwork was conducted over an extended period

1 Empowerment, as defined in this research, consists of four factors: access to information, participation and inclusion, being able to hold others accountable, and the capacity to organize.

of time, providing an in-depth understanding of the community, its history, the history of the district and country, and the impact of the NSP that is not often brought to light.

Development efforts in Afghanistan have historically been based upon approaches embedded in the two classical development theories. They have mostly concentrated on large infrastructural projects that have been determined by outsiders. This top-down approach has been challenged in the post-Taliban time period, with participatory development programs such as the NSP being implemented.

The NSP, as the largest participatory development program in Afghanistan, has brought infrastructure and capacity building subprojects to over 29,000 communities in Afghanistan, resulting in over 48,000 completed subprojects (IRoA, MRRD et al. 2012b: 6). These statistics and the rhetoric that stems from them have resulted in a great deal of positive commentary about the program from government officials, development stakeholders, and journalists. This, however, only tells one side of the NSP story. This research presents another side, one that challenges the standard NSP rhetoric. This research shows that the NSP has not had a significant influence in the NSP community of Shah Raheem, and that the positive impacts felt were more incremental than transformative in nature.

While personal security was seen as good by many in the community, the NSP was not identified as playing a role in this. The NSP also was not seen as having a strong impact on economic security, with only a few people in this research indicating that it resulted in them saving money or time. The NSP had a positive impact on social security in the community of Shah Raheem, with some reporting stronger social connections, however, this impact was limited to the men, with the women seeing no shifts in their social connections due to their lack of participation in the NSP process. Small shifts in bridging social capital were found in the community, with community perceptions of the local government improving and vice versa, however, this did not manifest itself in actions or interactions between the two, only in rhetoric.

Empowerment appears to have improved due to the NSP, but in nuanced and minimal ways and only for particular populations. Some community members were better informed about development in their community (almost solely men) and were newly involved in community-level consultations and meetings about development, but many community members remained excluded from such knowledge. The NSP did not significantly alter the community members' ability to hold others accountable, nor has it actively increased the peoples' capacity to organize themselves. Each of these is seen as a vital aspect of empowerment in this thesis.

These findings indicate that the rhetoric surrounding the NSP is not necessarily representative of lived or experienced reality, implying that the rhetoric itself may be unrealistic and potentially damaging. The NSP produced only minimal changes in the community of Shah Raheem in terms of the four areas of investigation, and is not the silver bullet expected. Though only a case study community, there is evidence that Shah Raheem's experience with the participatory development program was no more or less participatory than the other NSP communities within the district of Khulm and was evaluated to be no more or less successful than the other communities in terms of the process of implementation and the usage and maintenance of the projects completed. The experience of the community was in alignment with the 74 other NSP communities in the same district. Additionally, many of the programming constraints and shortcomings noted in this research are supported by other research conducted on the NSP, though it is not often highlighted in those works.

These findings of minimal positive impacts of Afghanistan's largest participatory development program on four areas of investigation in the community of Shah Raheem, lead me to conclude that it is time to reevaluate the expectations placed upon such participatory programs. I believe they should be based in experienced realities, such as presented in this dissertation. By obtaining a deeper understanding of a development program and how it is impacting at the micro-level, new expectations and objectives for such programs can be generated. Focusing on what a development project has actually done in terms of process and the resultant impacts, versus what it theoretically could do, can help prevent participatory development from being abandoned as a failure. Utilizing such research could also bring greater honesty into the development discourse, an honestly that at times feels decidedly lacking.

The NSP has many positive aspects and has created a means to distribute development funds and projects to rural areas that were not being reached in Afghanistan. However, I would argue that it is unlikely to have more than an incremental impact over the two and a half years usually committed to a community, and I do not blame the program for this. I believe that these incremental shifts are important and can help lead to greater changes in the future, but to assume that greater shifts and changes have already been made could be dangerous for Afghanistan and for the future of participatory development. If we can temper rhetoric about participatory programs, such as the NSP, perhaps the incremental changes that are occurring can be rightfully celebrated as steps towards future change.

Forschungsbilanz und Zusammenfassung

Partizipative Entwicklungsansätze haben in den letzten Jahrzehnten an Bekanntheit und Bedeutung gewonnen. Diese Ansätze stehen im starken Kontrast zu den eher klassischen Entwicklungsansätzen der Modernisierungs- und Dependenztheorie sowie Theorien, die aus diesen beiden Paradigmen entstanden sind. Sie basieren auf dem Konzept, dass Entwicklung die Meinungen, Wahrnehmungen und Anstrengungen der Menschen einbeziehen sollte, für die ein jeweiliges Entwicklungsprogramm durchgeführt wird.

In zunehmendem Maße investieren Geldgeber, Durchführungsorganisationen und andere Akteure ihre Zeit, Energie und Finanzen in Entwicklungsprojekte und -programme, die eine partizipative Komponente beinhalten. Davon zeugen Daten der Weltbank zu diesem Thema, die Fülle an Literatur und die steigende Anzahl von partizipativen Entwicklungsprojekten und -programmen, die weltweit implementiert werden.

Was generell über partizipative Entwicklungsansätze gesagt wird, ist meist sehr positiv und basierend auf der Erwartung, dass sie bessere Entwicklungsergebnisse liefern würden als die klassischen Entwicklungsansätze. Dies wird durch die Literatur zu partizipativer Entwicklung sowohl bestätigt als auch teilweise widerlegt. Dies weist auf die Komplexität des Zusammenspiels zwischen partizipativen Ansätzen einerseits und Entwicklungsprojekten und -programmen in der Praxis andererseits hin.

Diese Dissertation leistet einen Beitrag zur wissenschaftlichen Auseinandersetzung mit partizipativen Entwicklungsansätzen auf der Basis empirischer Daten. Die Studie nutzt eine intensive, auf qualitativen Daten basierende Fallstudienmethode, um die Auswirkungen des „Nationalen Solidaritätsprogramms" (NSP) auf die Gemeinde Shah Raheem in Nordafghanistan zu untersuchen. Sie zeigt hiermit einen neuen Blickwinkel auf partizipative Entwicklung: die Perspektive der lokalen Ebene.

Die Untersuchung basiert auf vier Schwerpunkten: persönliche Sicherheit, ökonomische Sicherheit, soziale Sicherheit und Empowerment.[2] Damit zeichnet diese Forschung ein historisch fundiertes Bild der Gemeinde und wie die Menschen dort den Prozess und die Durchführung des NSP und seiner Wirkungen erlebten. Die Feldforschung wurde über einen längeren Zeitraum durchgeführt und liefert somit ein tiefes Verständnis der Gemeinde, ihrer Geschichte, der Geschichte des Distrikts und des Landes und der Auswirkungen des NSP, das ansonsten noch wenig beleuchtet ist.

Die Entwicklungsanstrengungen in Afghanistan gründeten sich in der Vergangenheit auf Ansätzen, die in die zwei klassischen Entwicklungstheorien eingebettet waren. Sie konzentrierten sich zumeist auf große Infrastrukturprojekte, vorgegeben durch Außenstehende. Dieser Top-Down-Ansatz wurde in der Post-Taliban-Zeit konterkariert durch partizipative Entwicklungsprogramme wie zum Beispiel das NSP.

NSP als das größte partizipative Entwicklungsprogramm in Afghanistan brachte Infrastruktur- und Weiterbildungsmaßnahmen durch Unterprojekte in mehr als 29.000 Gemeinden in Afghanistan und resultierte in der Fertigstellung von mehr als 48.000 Unterprojekten (IRoA, MRRD et al. 2012b: 6). Diese Statistiken und die damit verbundene Aufmerksamkeit führten zu einer außerordentlich positiven Berichterstattung über das Programm durch Regierungsbeamte, Entwicklungsakteure und Journalisten. Dieses positive Feedback ist, wie sich vorstellen lässt, nur eine Seite der Medaille. Die hier vorgestellten Forschungsergebnisse zeigen eine andere Seite – eine, die die mustergültige Bericht-

2 „Empowerment" enthält in dieser Studie vier Faktoren: Zugang zu Information, Teilhabe und die Fähigkeit, andere zur Rechenschaft zu ziehen, und zu organisieren. "Ermächtigung" als deutscher Begriff für "Empowerment" gibt nicht die volle Bedeutung des mit dem englischen Begriff verbundenen Konzeptes wieder. Daher wird bei der englischen Form verblieben.

erstattung über NSP anzweifelt. Diese Arbeit zeigt, dass NSP keinen signifikanten Einfluss auf die untersuchte Gemeinde Shah Raheem hatte und dass die positiven Wirkungen eher geringfügig ausfielen, als dass sie einen essentiellen Wandel mit sich gebracht hätten.

Während die persönliche Sicherheit von vielen Menschen in der Gemeinde als gut wahrgenommen wurde, wurde NSP keine bedeutende Rolle hierin zugeschrieben. NSP hatte keinen starken Einfluss auf die wirtschaftliche Sicherheit. Nur wenige der in dieser Studie befragten Personen gaben an, dass sie durch das Programm eine finanzielle oder zeitliche Ersparnis erlebt hätten. NSP hatte eine positive Wirkung hinsichtlich der sozialen Beziehungen zwischen Männern in Shah Raheem, aber mangels ihrer Partizipation am NSP-Prozess keinen Einfluss auf die sozialen Beziehungen zwischen Frauen. Kleinere Verschiebungen ergaben sich bei der Wahrnehmung der lokalen Verwaltung durch die Gemeinden und umgekehrt. Dies manifestierte sich jedoch nicht in Handlungen oder Interaktionen zwischen den beiden.

Das Empowerment-Niveau scheint sich durch das NSP verbessert zu haben, aber nur auf nuancierte, minimale Weise und nur bezogen auf bestimmte Bevölkerungsgruppen. Einige Gemeindemitglieder (fast ausschließlich Männer) sind besser über die Entwicklungen in ihrer Gemeinde informiert und beteiligen sich neuerdings an Beratungen und Treffen über Entwicklungen auf Gemeindeebene. Viele Gemeindemitglieder sind jedoch weiterhin von solchen Prozessen ausgeschlossen. NSP hat diese Menschen nicht wesentlich in die Lage versetzt, von anderen Akteuren verantwortliches Handeln einzufordern, noch hat es die Kapazität der Menschen erhöht, sich selbst zu organisieren. Das sind aber alles Aspekte von Empowerment, wie es in dieser Dissertation definiert wird.

Diese Erkenntnisse deuten darauf hin, dass die Rethorik die üblicherweise im Zusammenhang mit NSP kommuniziert wird, nicht zwingend der Realität entspricht, insbesondere da eben diese NSP-Gemeinde bisher gemeinhin für eine erfolgreiche Umsetzung des NSP stand. Die Ergebnisse der hier vorliegenden Forschungsarbeit sagen etwas anderes aus. Diese Tatsache sollte nicht ignoriert werden. Ich bin davon überzeugt, dass die hier untersuchte Gemeinde keine Ausnahme ist. Es gibt Belege dafür, dass die Erfahrungen in dieser Gemeinde mit den 74 anderen NSP-Gemeinden desselben Distrikts übereinstimmen. Wenn auch oft nicht explizit betont, so erwähnen oder diskutieren andere Berichte dennoch übereinstimmend einige der Restriktionen, die auch in der untersuchten Gemeinde festgestellt wurden.

Die Erkenntnis, dass das größte partizipative Entwicklungsprogramm Afghanistans nur minimale positive Auswirkungen in den vier Untersuchungsbereichen in der Gemeinde Shah Raheem hatte, führt mich zu der Schlussfolgerung, dass es an der Zeit ist, die Erwartungen, die an solche partizipativen Programme gestellt werden, neu zu formulieren. Es sollte angefangen werden, sie basierend auf erlebter Wirklichkeit neu zu bewerten, wie es in dieser Dissertation getan wird. Indem man ein tieferes Verständnis eines Entwicklungsprogrammes und dessen Wirksamkeit auf der lokalen Ebene erreicht, können neue Erwartungen und Ziele formuliert werden. Wenn sich darauf konzentriert werden würde, was ein Entwicklungsprojekt tatsächlich leistet und bewirkt im Gegensatz zu was es tun beziehungsweise bewirken sollte, könnte man verhindern, dass der partizipative Gemeindeentwicklungsansatz generell als Misserfolg abgetan wird. Dadurch würde mehr Ehrlichkeit in den entwicklungspolitischen Diskurs gebracht werden – eine Ehrlichkeit, die oft zu kurz kommt.

NSP hat viele positive Aspekte und hat zudem ein Instrument geschaffen, mit dessen Hilfe Entwicklungsgelder und -anstrengungen auf ländliche Gebiete ausgeweitet werden konnten, die vorher in Afghanistan nicht in dem Maße erreicht wurden. Ich behaupte, dass es unrealistisch ist davon auszugehen, dass NSP in zweieinhalb Jahren mehr als einen minimalen Einfluss auf eine Gemeinde hätte haben können. Die geringen Erfolge sind auf strukturelle Probleme zurückzuführen und nicht die Schuld von NSP. Ich bin dennoch der Auffassung, dass auch diese minimalen Veränderungsprozesse wichtig sind und dass sie helfen können, größere Veränderungen anzustoßen. Aber anzunehmen, dass bereits weitgehende Veränderungsprozesse angestoßen wurden, wäre für die Entwicklung Afghanistans und partizipativer Entwicklungsansätze gefährlich. Die Berichterstattung über partizipative Gemeindeentwicklungsprogramme (wie NSP) sollte dahingehend angepasst werden, dass diese kleinen, aber wichtigen anstoßbaren Veränderungen als kleine Schritte gefeiert werden, denen weitere folgen werden.

Acknowledgements

This research would not have been possible without the support, guidance, friendship, and assistance of many people and organizations.

Financial assistance stemmed from Deutsche Forschungsgemeinschaft (DFG). I appreciate their interest in conducting research on human security in Afghanistan, and applaud their efforts to continue supporting research in this complicated, challenging, and sometimes dangerous environment. I hope that their work helps to support the inclusion of myriad indicators in the development discourse and helps to increase awareness of the situation inside Afghanistan.

Following the DFG, I would like to first and foremost thank my friends and "family" in Afghanistan for their logistical, emotional, and pragmatic support. A special thank you must be made to the community members of the NSP community of Shah Raheem for sharing their insights and life with me; to Mirwais Omeri and Waheed Miakhel for helping me to more fully understand what the community was communicating and for keeping their senses of humor; to Bashir for driving us on a daily basis and letting us choose the music once in awhile; and to the CARE International Team in Afghanistan, especially the Balkh Provincial and Khulm District offices. I would also like to thank my friends who joined me for Lucky Wednesday Lunches, for all of the background and insight they provided me with, especially Haji Assadullah and Sayed Yaseen of Khulm District.

A sincere thank you is also extended to Professor Hermann Kreutzmann who brought me into the research project that made this dissertation possible and helped to guide me away from dead ends and towards a thematic thread during the writing process. I would also like to thank my colleagues at the Zentrum für geographische Entwicklungsforschung (ZELF): Andi, Andrei, Bernd, Bettina, Dörte, Henryk, Lamis, Lavinia, Lira, Matthias, Mayinur, Nurridin, Stefan, and Tobias for introducing me to the world of pastoral studies and regularly challenging my approach to this research and answering my many questions. A special thank you is also extended to Emilia for helping me deal with university bureaucracy in an unknown language. I would also like to thank my good friend Gunda for helping me with the German translation of my research synopsis.

A final thank you is extended to my family and especially to my sister Kathleen for her continual support for all of my adventures. And a special thank you to my father who read through multiple drafts of this dissertation with patience, a sense of humor, and a keen eye for typos. Any and all mistakes that remain in the text, however, are mine and mine alone.

Preface to the Research

Transliteration, Language, and Spelling

Most foreign words included in this manuscript stem from the two official languages of Afghanistan, Dari and Pashtu, with a much greater concentration of Dari due to its dominance in the research area. Dari words have been transliterated in alignment with Junker and Alavi's Persian-German Dictionary (2002), with the common transliteration in the Afghan context also provided. Transliteration of Dari into the Roman alphabet is challenging and often Dari words are transliterated in different ways, with multiple ways to spell a single word. Utilizing the transliterations of Junker and Alavi is intended to provide the reader with a standardized transliteration, the key of which can be found in Annex P: Transliteration of the Persian Alphabet.

I have also standardized the names of locations mentioned in this manuscript in alignment with the spellings used by the Afghanistan Information Management Services (AIMS), unless otherwise noted. I have included alternative spellings encountered throughout my research for locations and the proper names of people and groups in the footnotes for cross referencing with alternative sources.

Researcher Biases

As a researcher, in order to be transparent, I believe that it is very important to try to identify one's own biases and try to be as forward about them as possible. Biases cannot be eliminated, but should be analyzed in the research process (Patton 1980: 333). One of the most obvious biases that I faced is that I am not from Afghanistan and do not have Afghan roots. This has provided me with challenges, but also advantages.

Due to this outsider status, there was the threat of misunderstandings. While familiar with many of the general customs, traditions, and history of Afghanistan due to having lived there for multiple years previously, I was still not intimately acquainted with the social, cultural, and historical norms of the community in which the research was conducted. There were customs that were foreign to me and required clarification, social structures that had to be explained to me in detail, and other phrases, ideas, and ways of operating that required me to not take for granted my interpretation of a word in English or Dari, nor their usage by others.

This outsider status and the lack of indoctrination into the social environment, at the same time, provided me with a great advantage in that it allowed me to ask questions about topics that might have caused offense if they had been asked by someone from the area. It allowed me to question the reasons behind diverse beliefs in a way that would have been unacceptable by someone from Afghanistan, who is expected to have similar values and ways of viewing the world. I was able to ask people to define what they meant by a term or an expression without shame, and this sometimes revealed that multiple definitions were being used for the same expression even though the meaning was assumed to be universal.

I tried to remain as open-minded as possible, working to prevent myself from judging what I saw based upon my own education and social background, but instead tried to comprehend social norms in their specific context. This does not mean that I shied away from working to understand the deeper connections to specific social norms. Instead I tried to explore the rationales for the existence of them from different local points of view. For example, when someone said that women could not vote because it was against their culture, I asked questions to delve deeper into where this tradition had started and what was at the root of the tradition for that person and for the community. I attempted to make everyone feel comfortable expressing their beliefs, understanding, and knowledge without feeling judged.

I also tried to address the six main biases that Chambers (1981: 13-23) warned about in regards to development research: spatial bias, project bias, rural elite voices bias, dry season bias, diplomatic biases, and professional biases. The community focused on was not directly off of a paved road, and was removed from the provincial urban center by about 50 km (about 30 miles), the last few kilometers of which were dirt roads. This research only focused on a single community where a project had indeed taken place, which leaves me open to criticism regarding project bias.[3] All attempts were made to include community

3 The original research plan had included conducting multiple case studies, of

members who were not a part of the rural elite. This was a multi-step process that involved a specific sampling doctrine for the household survey and the non-randomized selection of key informant interviews with the intention of speaking with people from all sectors of the community, from the richest and most powerful, to the poorest and most vulnerable. This also helped to lower professional biases, as we were able to speak with people with many different livelihoods. The research was carried out in three different waves, which took place in both the summer and the winter, helping to eliminate seasonal bias. The length of time spent in the community helped to mitigate some of the major diplomatic biases of being too polite or timid, as I and my research team became familiar with the ways of the community and they became familiar with us, allowing a great deal of frankness. It also helped to mitigate any biases the community might have had because we had been introduced to them by the facilitating partner staff.[4]

Community Concerns

I was keenly aware that in some areas of the country regular visits to a community by a foreigner could cause problems for that community with armed opposition groups, anti-government elements (AGE), or with other communities who believe that the research community was obtaining special benefits. I spent a considerable amount of time considering my choice of community in an effort to operate in an area where a foreign visitor would not be too disruptive or be seen as an excuse for attacking a community, but also understanding that such a risk could not be mitigated completely. This concern contributed to the de-

---

lesser intensity, in three communities, one of which would have been without a project to address this, but security concerns prevented this approach.

4 After participating in the initial gathering of the elders and introducing us to the community, the facilitating partner staff left us to our own devices for the remainder of the research. I believe that this initial introduction could have provided the research team with more positive answers about the community-driven development program in the hopes of obtaining more benefits from it. While I believe this did happen to some extent, I noted that the greater the amount of time we spent in the community, the more problems or concerns were being shared with us, implying that the positive reporting bias was at least in part overcome. The Facilitating Partner for the NSP is translated as *mo'assasāt-i hamkār*.

cision to base myself in the provincial center of Mazar-i-Sharif, traveling to the case study community on a regular basis. This does mean that this research was not able to evaluate the impact of the NSP on a community actively threatened by the AGE regularly.

1 The Emergence of Participatory Development in Development Theory and Approaches

1.1 Researching Community Driven Development in Afghanistan

Classical academic theories of development utilize an analytical framework that focuses on the state as the primary actor, with economic growth as the main indicator of "development." From Rostow's modernization theory (1956) to the dependency theories of Prebisch (1950) and Frank (1969b), and including Wallerstein's world system theory (1974), these two underlying characteristics have dominated development thinking and actions in the field and stimulated development cooperation since the 1950s. Rostow's (1956: 25) modernization theory hypothesis focuses on a unilinear economic vision of development: "The process of economic growth can be usefully regarded as centering on a relatively brief time interval of two or three decades when the economy and the society of which it is a part transform themselves in such ways that economic growth is, subsequently, more or less automatic." Frank (1969a) also focuses on the economic in dependency theory. Despite bringing to the forefront a historical perspective on development, he still defined development through economic strength, "historical research demonstrates that contemporary underdevelopment is in large part the historical product of past and continuing economic and other relations between satellite underdeveloped and the now developed metropolitan countries" (1969a: 4).

Though many criticisms of these traditional theories have been voiced by other classical theorists and non-classical theorists alike (e.g., Frank 1969b; Hettne 2002; Higgott 1983; Peet and Hartwick 1999) and alternative concepts of development have been put forward such as Basic Needs and Human Security, no umbrella theory has been adopted to replace or actively compete with these classical theories. These alternative concepts of development have introduced a new way to look at development that is dependent, however, upon multiple indicators of success and multiple actors (Seers 1969: 5; UNDP 1990: 11; 1994: 25-33). While not unified under one named theory, the expanded definitions of development have impacted on the development discourse and on development practices.

Development practices have diversified to include not only programs that reference the state as the main agent of change, but increasingly include programs where non-state actors such as communities and civil society organizations (CSO) have become the main "beneficiaries." These new approaches stem in part from an expanded definition of development that stretches beyond gross domestic product (GDP) and other economic indicators to include measures of human security and social capital, with some focusing on empowerment of the local population.

Participatory development focuses on the inclusion of development project beneficiaries in the planning, implementation, and maintenance of development projects. The basic premise being that increased collaboration and involvement in the project will result in more positive outcomes and impacts. In other words, it is envisioned that participatory development approaches such as community-based and community-driven development (CBD and CDD, respectively) will succeed where classically based development approaches have failed. Though still far from having proven its success, participatory development has become a mainstream approach with a positive reputation, generally. From infrastructure projects, to education, to farming initiatives, and beyond, local people are being written into development proposals, and those proposals are receiving funding.

This shift towards a more participatory approach is reflected in the large number of participatory development programs that have been implemented in Afghanistan, especially since the fall of the Taliban. After decades of classically driven development projects, participatory approaches have begun to take on a more important role in Afghanistan. Since the end of the Taliban Regime, which has allowed for a development sector that is more open to international aid and agencies, the development landscape in Afghanistan has been inundated with programs that claim to consult with, train, empower, and utilize the local population.

Participatory development is seen by many of its academic supporters and development stakeholders in the field as a means of changing societies, empowering the beneficiaries to take responsibility for their own development (Chambers 1983; Narayanan 2003; Oakley 1991; Operations Evaluation Department 2005). Such programs in Afghanistan focus on giving voice to the marginalized and disempowered, and creating newly elected development councils that aim to be inclusive and prevent elite capture.[5] The inclusion of community members and expected increased transparency in the decision-making processes are expected to increase social capital and trust within communities, and result in greater empowerment of the community members.

These, and many other hoped-for benefits of participatory development, are not easily measured through quantitative surveys and one-time visits to communities, which are the most commonly employed research methods utilized by development funders. These benefits require a qualitative approach, where the researcher is able to spend an extended period of time learning about and observing the community, its structures, its history, its social networks, and getting to know community members through focus group discussions (FGDs) and key information interviews (KIIs).

Few studies, academic or non-academic, have focused on evaluating the benefits and the costs of the participatory development approach at the community level. Most of the research conducted has concentrated on counting the number of wells dug, the number of school walls built, or other easily measurable project indicators.

The research presented here provides a qualitatively-driven picture of how one community has been impacted by a community-driven development program that employed a participatory approach. It evaluates how community members perceive the program and how it has affected their lives, their actions, and their attitudes. It incorporates the non-classical development thinking that has taken place in recent decades, using it to frame the research and direct the collection of data.

While the rhetoric surrounding participatory development promises a great deal, what it actually delivers at the micro, or community, level is not well understood. This research looks at how rhetoric differs from experienced reality in the community of Shah Raheem in Khulm District of Northern Afghanistan.[6] This single-community focus provided

5 Some others utilize indigenously established community groups, though this is less prevalent (Mansuri and Rao 2004).

6 Khulm can also be written Kholm, Khoolm, Khooloom, and Khoulm.

the time needed to gain a multifaceted understanding of the community, the participatory program, and its implementation. From this perspective, the impacts of this particular participatory development program are analyzed.

Are these programs delivering as much as has been promised? How does community-driven development programming impact on communities? As Mansuri and Rao (2004: 51) so succinctly stated, "If the development community is going to embark on a major change in development strategy, surely it should be concerned about how well it might work." This dissertation contributes to creating a better understanding of how a participatory strategy in Afghanistan has impacted the NSP community of Shah Raheem and helps illustrate the impact it had on personal security, economic security, social security, and empowerment through the eyes of the community members and other external stakeholders.

1.2 Historical Shifts in Development Theory and Approaches

In order to understand the basis for the current revitalized enthusiasm for participatory development practices, it is important to situate the concept of participatory development into the main historical trends and paths of the theory and practices of international development. This is especially relevant in development thinking due to the non-sequential-temporal manner in which theories and approaches have evolved, "stacking up" upon one another and co-existing instead of replacing one another (Potter 2002: 63). This is supported by the idea that, "Evidence of the persistence of ideas in some quarters, years after they have been discarded elsewhere, will be encountered throughout the development literature" (Potter 2002:63).

This section explores the main trends in development thinking, while integrating into the picture the development approaches being utilized in the field. Through this historical perspective, two major shifts in thinking can be seen that have contributed to the current focus on participatory approaches. The first is a shift from the state as the focus of development to individuals, and the second is an expansion from the narrow vision of development as defined by economic growth to an expanded definition which includes a larger array of indicators, including education, poverty, life expectancy, and more.

This section will guide the reader through the development discourse in a chronological order. This will allow us to break through some of the more common assumptions made about what was happening during different time periods and can highlight the immense amount of temporal overlap between different development approaches and theories. It also allows for greater insight into the strength and pull of different theories and approaches that can sometimes be lost when they are not discussed within their historical context.[7] This will allow us to trace the more dominant thinkers, while also highlighting some smaller threads of thought, some of which gain importance later in terms of participatory approaches.

As will be seen, there is the tendency in development thinking to recycle past ideas and approaches, though often under the guise of a new creation. This has been noted by Higgott (1983: 1) in his book *Political Development Theory*, "... developments in theorizing over the last decades are largely the result of a return to older and more tried theories – or variations thereof." This is especially true for participatory approaches to development, as will be explored.

1.2.1 The Emergence of Development Studies Following World War II

The word "development" has existed for centuries, describing everything from the growth of a caterpillar into a butterfly, to the intellectual growth of human beings as they mature, to the building of a new office complex, and even to describe breaking news (Woolcock 2009: 5). Following World War II though, the word began to be used in a different realm, a setting in which development concentrated on the economic well-being or growth of a country. The focus of this section will be on development from the mid-1940s onwards, though this is not intended to indicate that prior to that no thought was put forward on what would come to be known as development studies. This time period is

7 This research defines development theory in alignment with Potter, Binns et al. (2008: 80), who say that theories are, "sets of apparently logical propositions which purport to explain how development has occurred in the past, and/or should occur in the future." Approaches are defined as the different means or strategies used to pursue development at the field level.

given as the starting point because it was only in the post-World War II period that international development studies emerged as a separate area of theorization and scholarship (Gregory, Johnston et al. 2008: 155) or as Kiely (1999: 31) states, when the "idea of development was made explicit."

A few historical notes of interest that will be relevant in the exploration of development theories and approaches should be made. The first is the creation of the League of Nations in 1919, which was the forerunner to the United Nations. The second is the creation of three institutions that were formed at the Bretton Woods Conferences held in 1944: (i) the International Monetary Fund (IMF) was established in order to help maintain currency stability and develop world trade, (ii) the World Bank Group of five agencies was established to support development through loans and other means in European and in underdeveloped nations, and (iii) the General Agreement on Tariffs and Trade (GATT) was formed to promote free trade amongst its members (Willis 2010: 36-7).[8] The UN, IMF, and the World Bank would all become prime actors in development and would be influential in the adoption of approaches to development throughout the decades (Peet and Hartwick 1999: 53). Other actors that would become influential were national development agencies, such as USAID of the United States, DFID of the United Kingdom, and JICA of Japan. These organizations, and the governments that support them, have at times weighed heavily in the rise and fall of different development theories and approaches.

The emergence of the usage of "development" as a term came about in part due to the realization of the need for outside assistance for the infrastructures and economies of Europe that had been damaged or ruined during World War II. Whole cities had been destroyed, along with a large number of industrialized areas that were the engine for much of Europe's economy. In response to this destruction, the US Government announced the Marshall Plan, officially named the European Recovery Program. This program provided economic and technical assistance worth billions of dollars to help reconstruct the European economies (Rist 2008: 69). The Marshall Plan helped to shape, and to define, the path of international development. As Hettne (2002: 7) describes, "The reconstruction of war-torn Europe provided the model for state-di-

8 The GATT evolved into the World Trade Organization (WTO) in 1995.

rected *modernization* of the 'new nations.'"[9] The objectives of the Marshall Plan were two-fold: to help rebuild the destroyed countries and their economies, and to create political ties that established alliances.[10]

Based upon Keynesian economic theory, investment in infrastructure projects in Europe was seen as a way of providing necessary reconstruction and physical infrastructure, and also as a means of providing an impetus to get the economies of those nations started again (Willis 2010: 37). Keynes believed in the concept of the "trickle-down" effect, in other words, that development benefits would spread to all sectors of society, and felt that sometimes government intervention was necessary. While this concept would fall in and out of *vogue*, the idea of one agent helping another to change has remained at the heart of development throughout the last seven decades, even as participatory or self-help approaches have increased in popularity and number. There remains a belief that outside assistance, be it technical or otherwise, is necessary and that Western countries are the natural suppliers of this. The "success" of the Marshall Plan helped to plant this idea.

Colonization and the process of de-colonization also played a strong role in development thinking. Following World War II, colonies began to gain their independence and nations with colonies began to express a sense of responsibility for the development of their former colonies.

The Marshall Plan experience, and preparing for decolonization, led to comparisons between countries, those "with" and those "without." This comparison was voiced by President Harry S. Truman in his inaugural speech of 1949. This speech, while little noticed at the time (Rist 2008: 5), came to be known as the "Four Point Speech" and is seen by some, such as Escobar (as noted in Gregory, Johnston et al. 2008: 563) as the beginning of development studies, or the "age of development" as Wolfgang Sachs (1992: 1) called it. This is because the speech contained what is considered to be one of the first introductions of the term "underdeveloped" into the political dialogue (Esteva 1992: 6).[11] Truman

9 Italics included in the original text.

10 The Marshall Plan was announced June 5, 1947 (Rist 2008: 69; Willis 2010: 38). It was seen by many as a means of addressing concerns about the Communist threat that arose following World War II.

11 It should be noted that the usage of "underdeveloped" in Truman's speech was not unique, in that academics involved in studying the world and the progress of different sectors of the world had already used the term, such as Wilfred

(1949) said, "We must embark on a bold new program for making the benefits of our scientific advances and industrial progress available for the improvement and growth of underdeveloped areas." In the speech, Truman placed a normative comparison between Western countries and non-Western countries into the public sphere. He also offered technical assistance to underdeveloped countries to help eradicate the "poverty" and "misery" in which the people of those countries lived (Truman 1949). It was a bold statement of assistance, unlike anything previously put forward by a US President. In this speech, it was made obvious that from a US perspective, Western economies and societies were viewed as the goal towards which underdeveloped countries should strive.

It also presented a believed truth at the time that has proven to be false: "For the first time in history, humanity possesses the knowledge and the skill to relieve the suffering of these people" (Truman 1949). As history has shown, this was a premature statement, as 1.345 billion people were living on less than 1.25 USD per day in 2011, with over 1.02 billion people estimated to suffer from chronic hunger (World Hunger Education Service 2012). Sixty years later, much of the world's population remains mired in poverty, notwithstanding the massive amounts of money spent, programs and projects implemented, and many good intentions. Despite all that was known then, and the many lessons and insights gained along the way, there still remains significant suffering and poverty around the world.

It is interesting to note that Truman also struck on the theme of self-help, which would not gain true acceptance and support in the development discourse until a later date. He said, "Our aim should be to help the free peoples of the world, through their own efforts, to produce more food, more clothing, more materials for housing, and more mechanical power to lighten their burdens" (Truman 1949). As can be seen, though not often highlighted, self-help, a vital part of participatory development, was an aspect of development from a very early stage, though focus on it has waxed and waned through the years.

Benson who is credited with being the first to use the word as a term to reference the status of a nation-state in his article entitled, "The Economic Advancement of Underdeveloped Areas," which was published in 1942 (Esteva 1992: 7; Rist 2008: 72). Some authors, such as Esteva (1992: 7), argue that the label of "underdeveloped" shifted the way people viewed themselves and others.

This speech also contributed to narrowing the definition of development that would dominate for years to that of economic growth. It was believed that economic growth would help establish the necessary conditions for "personal freedom and happiness for all mankind" (Truman 1949). W. Arthur Lewis proposed that economic growth should be the goal of development in 1944, and this was then supported by the United Nations Charter in 1947 (Esteva 1992: 12).[12] Influential economists on the left and right took over this idea, such as Paul Baran and Walter Rostow, who talked about development in terms of increased per capita production (Esteva 1992: 12). From these thoughts emerged the theory of modernization, which came to dominate the first UN Development Decade, lasting from 1960-1969, and is one of the classical development theories.

The difference between developed and underdeveloped nations was therefore seen as connected to their economic situation. Progress was considered to be equal to economic growth and industrialization.[13] It is interesting, however, that Truman's speech hinted at multiple layers of value: economic growth appears to be the goal of development, but still just a means to the final objective of freedom and happiness (Truman 1949). As development theories, measurements, and approaches shifted over the decades, economic growth would come to be seen as only one driver of development amongst many, and issues such as freedom and happiness would begin to receive more attention, leading to the emergence of participatory development.

---

12 Lewis was a well-known development economist.

13 Economic growth has most often been measured by gross national product (GNP) or gross domestic product (GDP), on a nationwide basis or a per capita basis. GNP is the annual total personal consumption expenditure, gross domestic investment, government purchases of goods and services, and net exports. GDP is the annual market value of all goods and services produced within a country. There are many technical difficulties in capturing such information, which is a common critique of using GNP or GDP per capita as an indicator of development. This was detailed by Peet and Hartwick (1999: 8), who point out that the accuracy of GNP or GDP data is very questionable because it leaves out the informal sector of the economy, it does not represent the distribution of wealth, and very few nations are able to accurately gather such data.

1.2.2 Different Development Approaches in the Early Years: Community Development and Modernization Theory

In the 1950s, there was confusion as to the best means to encourage development. Divergent thoughts about how to bring about economic growth in colonies were looked at, including self-help, as noted by Illich (1992: 95). During this time both top-down approaches that focused on economic development and industrialization, and participatory approaches that were more concerned with individuals or communities, were implemented.[14]

In 1955, the Bandung Conference was held in Indonesia. This conference was important to the development discourse as it was where the non-aligned movement began and the countries in the developing world voiced their own opinions about development and what should be happening (Rist 2008: 86).[15] The "third world," as a term in development, derived from this conference, with the non-aligned countries seen in contrast to the first world of capitalist politics and the second world of Communist politics (Kreutzmann 2008: 683). As will be seen, this coming together of people from the developing world was unusual in the development discourse until the 1970s when voices from the developing world began to gain greater attention and influence. The main focus of the communiqué resulting from the conference was a sharp criticism of colonialism, and a demand for the independence of all colonies. Rist (2008: 87) sees the outcomes of this conference as the first assertion of collective demands from the nations of the developing world and the discussion of "collective self-reliance." This conference brought to the forefront the idea that development experts were not the

---

14 Top-down approaches involve the imposition of a project in a specific area based upon external, "expert" ideas about what will bring development to that area. These efforts do not usually include the local population or administration in the planning or implementation stages.

15 Conferences intended to provide views from the non-dominant countries from outside of Europe and North America occurred previously on a number of occasions, but they received very little attention, which cannot be said of the Bandung Conference. Congresses for the Advancement of the Oppressed People were held in Paris in 1920, London in 1923, Moscow in 1924, and Brussels in 1927 (Rist 2008: 81). Additionally, Nehru held a conference on Asian Relations in Delhi in 1947 (AgriInfo 2011).

only ones whose voices should be heard, an idea that permeates participatory development thinking.

Development practitioners in the 1950s began to believe that it was important to take into consideration the opinions of people in the developed and the developing world, when they began to analyze failures of top-down development projects and programs in the field (Rahmena 1992: 117). The failure of top-down projects that relied on expert advice was contrasted with the success of programs such as the Etawah Project in India that was started in late 1948. Etawah is seen as one of the first successful examples of participatory development approaches (Korten 1980: 481; Ruttan 1984: 393).[16] This project was grounded in the idea that the failures of top-down projects could be attributed to a lack of involvement and input from the local communities. This was the start of the first wave of experimentation with participatory development.

This first wave of participatory development did not last long, reaching its peak in the 1960s (Mansuri and Rao 2004: 6). Two different versions of this approach were adopted: the "community development" approach of the United States and the *animation rurale* approach of France.[17] Community development (CD) focused on the need for an outside catalyst to help communities develop. Started after World War II, community development was implemented in India and the Philippines before migrating to other parts of Asia and Africa and even to the West Coast of South America (Gow and Vansant 1983: 428).[18] Over 60

16 The Etwah Project was an economic development project originally implemented in 64 villages of Uttar Pradesh, but came to include over 300 villages. It became a prototype for similar projects throughout India. The project aimed to educate villagers and train village level workers with a strong component of villager participation (AgriInfo 2011).

17 *Animation rurale* was based on a Catholic Humanist school of thought. In this approach, one person, internal to a community, was trained to identify the needs of a community and to act as the internal catalyst of change (Gow and Vansant 1983: 428; Nagar and Raju 2003: 44). It failed to produce successful development projects and while it was intended to be bottom-up, in reality it was still very much top-down, with the local *animateur* often succumbing to the pressures of involved line-ministries (Gow and Vansant 1983: 429). Even then, however, it did enhance local input. These programs were influenced by the Ghandian cooperative movement in India and Paulo Freire's work on the oppressed (Mansuri and Rao 2004: 5-6).

18 Holdcroft (1976: 3, 9) disagrees with some of these details and argues that the

nations had launched such projects by 1960 (Holdcroft 1976: 3; Ruttan 1984: 393), with UN agencies and the United States, via USAID, as the major funders (Holdcroft 1976: 3; Mansuri and Rao 2004: 6). Community development garnered much more support and money than *animation rurale.*

The basis of the US's community development approach was that a person with a high school education from outside the village would live and work with villagers in order to organize village groups, help them identify their needs, and to increase their confidence (Gow and Vansant 1983: 428). Gow and Vansant (1983: 428) claim that the main benefit of this approach was that it allowed for a local technician to be placed in every village without the large cost of placing teams of external technicians in the villages, as required for top-down approaches.

Community development gained a great deal of support. Masters programs in community development were even offered by US and British universities (Holdcroft 1976: 10). In 1954, the Community Development Division was created in the foreign aid agency of the US Government. This division supported a journal, conferences, and large projects in Asia and the Middle East (Holdcroft 1976: 20-3).[19] The US Government supported community development programs in 30 countries, spending 50 million USD on them between 1952 and 1962 (Holdcroft 1976: 24). However, community development's popularity did not last. The Community Development Division was abolished by 1963, as the community development approach lost its appeal (Holdcroft 1976: 24).

How solidly this approach was based on saving money, versus belief in a new viewpoint can be debated. What is not to be debated, however, is that this experiment in community development was, and continues to be, viewed as a failure (Cohen and Uphoff 1980: 215), and following its failure, top-down approaches became the dominant means of inter-

---

first community development projects were implemented by the UK in Africa in the early 1950s.

19 Holdcroft (1980: 482) listed a number of large US Government funded community development projects: India in 1952, Iran in 1953, Pakistan in 1953, the Philippines in 1955, Jordan in 1956, Indonesia in 1957, and Korea in 1958. Smaller projects included: Iraq in 1952, Afghanistan in 1953, Egypt in 1953, Lebanon in 1954, and Ceylon and Nepal in 1956. Despite many efforts, I was unable to find detailed information about the nature of the Afghanistan project.

national development. The CD programs did not improve the economic and social standing of the people, "elite capture" was seen as prevalent, and there were concerns that if political gains were made, it could disrupt the local political situation (Ruttan 1984: 394). Holdcroft (1976: 28) supports this with his number one criticism of community development, which was that it was inefficient in improving the economic standing of participants, the main indicator of development at the time.[20]

Khadija Haq (2002: 36) states in the *Human Development Report for South Asia* that the community development projects were simply too broad in scope and that the food shortages that emerged in the 1950s and the subsequent Green Revolution of the 1960s, "put community programmes on the backburner." She provides the example of the Village-Aid Programme in Pakistan, and what would become Bangladesh, which ran from 1952 to 1961. It is seen as having failed due to a lack of local political support as it redistributed local power in ways that did not please everyone (Haq, K. 2002: 85). In general, the community development programs in South Asia are now seen as not having taken into consideration the heterogeneity of communities and their development priorities and for excluding the landless (Haq, K. 2002: 85). The details of community development program failures in other parts of the world are not elaborated upon, though community development programs were evaluated by Fukuyama (2005: 86) to have, "degenerated into corruption, self-dealing, or, at best, a rent-seeking mentality."

Funding for community development and *animation rurale* had "dried up" by the mid-1960s (Holdcroft 1976: 3; Korten 1980: 482; Mansuri and Rao 2004: 4).[21] While it is reported that there was some level of community mobilization and an increased understanding of participation resulting from community development projects, accord-

20 This was also a main point for Korten (1980: 482).

21 The US Government was one of the main financial backers of development studies through different government entities. The Center for Research in Social Sciences and the University of Chicago's Center for Social Organization Studies were all founded by different US Government entities, and in testimony from the House of Representatives Lt. General Dick Jr. attested that the army alone had spent 8.2 million USD on development related studies in 1965 (Ish-Shalom 2006: 292, footnote 2).

ing to Gant and Vansant (1983: 428), the approach did not lead to successful actions and initiatives.

1.2.3 Modernization Theory: Laying the Foundation of Mainstream Development Approaches

Throughout the rise and fall of community development, other approaches were also being explored, ones that were top-down in nature grounded in Modernization Theory. Modernization theory, one of the most impactful theories of development, is based upon the concept that economic growth is the answer to underdevelopment. Though it has its roots in the 1950s, modernization theory became widely debated and began to be adopted by development thinkers and practitioners in the 1960s (Peet and Hartwick 1999: 83), though it was not highly regarded in academic circles (Streeten 1981: 105). It was in 1960 that Walt Rostow's tome *Stages of Economic Growth: A Non-communist Manifesto* (1960) was published.[22] Based on an essay he wrote in 1956, the book laid out the theory that growth and industrialization were at the heart of development, following the examples of the United States and Europe, and that all states were on a linear path towards this kind of development (Rostow 1956). "It was widely supposed that it was the business of the US to reconstruct the world in its own image" (Preston 2002: 168).

Rostow's theory stemmed from a historical comparison of the evolution of different countries in Europe, North America, and Asia (Rist 2008: 95).[23] Rostow determined that all nations could be found along a spectrum that went from underdeveloped to developed. He established five different steps along that spectrum in which every country could be identified, as shown in Figure 1 below. "It is possible to identify all societies, in their economic dimensions, as lying within one of five cate-

22 Rostow was a US economist, political theorist, professor, and an advisor to both US Presidents Kennedy and Johnson. He served as the Chairman of the Policy Planning Council at the State Department and also as the Special Assistant to the President for Security Affairs (Ish-Shalom 2006: 294-5).

23 Walt Rostow's book stood as a response to Karl Marx's *The Communist Manifesto* (Marx and Engels 2002). Both books based their analysis on a comparison of England and India, which makes their different understandings of "progress" and "development" even more interesting (personal communication, Hermann Kreutzmann, May 3, 2011).

gories: the traditional society, the preconditions for take-off, take-off, the drive to maturity, and the age of high mass-consumption" (Rostow 1991: 4). Development was therefore defined as the stage at which the general population of a nation was able to purchase consumer products regularly (Willis 2010: 39), also termed as self-sustained growth (Ish-Shalom 2006: 298; Latham 1998: 199). Nations at this stage were expected to be largely urban, with an economy that was not based on agriculture (Willis 2010: 39).

Figure 1: Rostow's Unilinear Five-Stage Modernization Theory Model

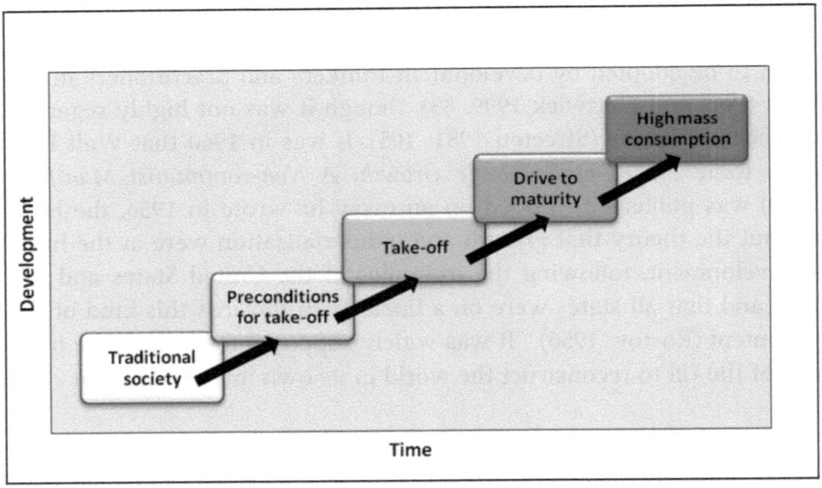

Source: own design based on Rostow (1956)

Development was presented in terms of a dichotomy, or dualism, between the modern and the traditional, with the West representing the modern and all other countries representing the traditional (Preston 2002: 172). Modernization theory is grounded in the idea that a country's development is based on its internal structures and society, and that underdeveloped countries can follow in the same footsteps as developed countries with developed countries' assistance. This theory does not look at fundamental differences in societies, in resources, or in allotments. It also does not take into consideration the histories of different countries, which will be seen as one of its largest faults by later development theorists. It limits the paths of development to only one, and it puts forward foreign aid, in the model of the Marshall Plan, as the key to movement through the stages of development (Preston 2002: 162). It's focus on the state as the main actor and/or agent of development means there is a lack of focus on communities and individuals, leaving no room for them to be active. They remain passive recipients. This stands in stark contrast to community development, and other participatory development approaches.

Preston (2002: 177) claims that this time was the height of interventionism. This way of thinking built upon the Enlightenment vision of non-Europeans as "noble savages," which then shifted to "uncivilized savages" who needed to be "civilized" (Preston 2002: 143). Modernization theory was widely adopted and supported during its period of popularity. Samuel Huntington, though most famous for his theorization on the clash of civilizations (Huntington 1993), was one of the strongest proponents of modernization theory (Kreutzmann 1998: 255). Leys (1982: 342) even credits Huntington with having, "revitalized the ideology of modernization." Huntington's work in development was very influential, and his 1968 book *Political Order in Changing Societies* (Huntington 1969) was found to be the most quoted book of its time amongst American university professors in political science (Kenski and Kenski 1975: 571).

Modernization theory was also heavily criticized, and by the mid-1960s was starting to lose some of its support, though themes from this theory can be found in many of the development theories and approaches that arose in the following decades. Some of the major criticisms of modernization theory include the inability of the theory to explain income gaps within nations, the Eurocentric nature of the theory and its focus on a unilinear development in the image of the West, a lack of historical perspective for underdeveloped nations, and its simplicity in terms of analysis in general (Higgott 1983: 20). Others, in agreement, found modernization theory's unilinear, path-bound approach to be too constricting, leaving no room for alternative paths outside of the European and North-American models (e.g., Kreutzmann 1998: 256). Development scholars have also criticized modernization theory for being too focused on internal or endogenous factors, while skimming over external influences (Hettne 2002: 7; Kreutzmann 1998: 257).

This brief sketch of the basic tenets and beliefs of modernization theory helps to illuminate the primary approaches to development that were based on its premise. The belief that a model for development already existed, led to development policies and approaches that were top-down in nature (Willis 2010: 46), where Western countries and international organizations would provide development funds or loans for development projects planned, implemented, and monitored by Western experts. Those providing the development funds or loans were the experts with all of the answers to the question of why a country was not developed. In other words, these projects utilized what Korten (1980: 496) described as the "blueprint model." External funding and knowledge was exported to a country and acted upon the local environment, with little, if any, consultation and without understanding the context in which they were working. Development was supposed to be done in the same way in all parts of the world, independent of things such as culture, religion, geography, etc. This style of plan was highly controlled, as problems and projects were defined and expected outcomes were visualized (Korten 1980: 496).

In the 1950s and 1960s, foreign aid was mostly focused on industrial development, with the implementation of projects utilizing technology to improve agricultural outputs and large infrastructure projects, such as the building of damns and roads (Willis 2010: 46). One of the largest of these projects was the Alliance for Progress (AFP). With support from Walt Rostow, who was advising President Kennedy on development matters, the US established the Alliance for Progress in 1961.[24] Focused on Latin America, the AFP consisted of 20 billion USD worth of investment over a ten year period (Willis 2010: 68). The AFP was supposed to help countries of Latin America go from "undeveloped" to "self-sustained growth" (Latham 1998: 199). Rostow advised Kennedy that with US Government foreign assistance and aid, 80% of Latin American countries could make this jump (Latham 1998: 199). There was the belief that the US was superior to Latin America and that it, therefore, could serve as a model for the countries of Latin America to

24 The AFP was officially started in August of 1961 at the Punta del Este Conference in Uruguay (Ish-Shalom 2006: 294-5).

emulate. This supports Ish-Shalom's (2006: 288) argument that the very roots of the AFP were enmeshed with modernization theory.

The Alliance for Progress was intended to ". . . [touch] every aspect of Latin American life and promised to alter a traditional society within a decade" (Smetherman and Smetherman 1972: 79). Economic growth, land reforms, better housing, increased education opportunities, and tax reforms were all part of the package. The US was going to bring development to Latin America. However, the experiment of the AFP did not end in success. While there was some evidence of positive gains in the areas of focus, such gains were small and incredibly limited in nature (Willis 2010: 68). By the 1970s, the AFP had lost its allure and was generally thought to have been a failure (Willis 2010: 68), especially as dictators had taken over many of the AFP countries. Slow growth rates and lack of visual progress were noted in the first annual report on the program, and things only got worse in the following years (Latham 1998: 224).

> *In the case of the Alliance for Progress, the ideology of modernization shaped and legitimated an ambitious, highly publicized effort to combine the promotion of Latin American development with the containment of Communism. It also forced analyses of social change into a particularly rigid template, presented sweeping economic and political reform as matters of simple volition, and reiterated a profound sense of historical destiny for the United States, a nation claiming the power, knowledge, and right to transform a struggling world. (Latham 1998: 200)*

Another example of a modernization theory based development program was the "green revolution," which took place in the 1950s and 1960s. This revolution was focused on improving the agricultural production of developing countries. It had mixed results as some nations gained self-sufficiency or even an excess in products, while in others inequality increased and the programs were seen as being very detrimental environmentally (Willis 2010: 152). Modernization, with its heavy focus on industrialization, did not always focus on the environmental impacts of different projects or policies. This became more and more problematic as different scholars began to focus on environmental issues and some even started to predict the limits of growth, such as the

Club of Rome's 1972 report, *The Limits to Growth* (Meadows, Meadows et al. 1972).

In the 1960s and 1970s, modernization theory was the foundation, or touch-stone, of development thinking, and it continues to remain that today in many ways. It shaped, and continues to shape, perceptions, approaches, and measurements of development. Despite the many valid critiques that have been made, the idea that modernization leads towards functioning like countries in Europe or in Northern America continues to persist both within the world of development academics and development professionals, and also amongst individuals in developing countries. The economic strength of these countries and the many perceived benefits are still a focus of development thinking.

Modernization theory's fundamental reliance on a universal and unilinear development model stands in marked contrast to participatory development approaches. This is especially true in that modernization theory supports the concept of development experts and the need for external funds and personnel to bring about changes in a state. Projects that are grounded in the theory of modernization and their failure to produce the expected results have helped to bring forward participatory development approaches. As such participatory approaches can be seen as a response to such failures, and are supposed to address some of the perceived weaknesses of modernization theory based approaches. Some scholars, such as post-development thinkers, however, see participatory development approaches as a façade, hiding top-down approaches based on modernization theory.

Despite a lack of popular support, alternatives to the modernist approach to development were still quietly being put forward in alignment with community development and *animation rurale* into the first UN development decade. For example, John F. Kennedy's inaugural address in 1961 pledged American help for those living in mass misery to help themselves (Illich 1992: 92), thus supporting the idea of participatory development, while also reinforcing some of the main components of modernization theory, such as the necessity of outside assistance. Participation was also still a part of the USAID rhetoric, with the Foreign Assistance Act of 1966 calling for maximum participation (Cohen and Uphoff 1980: 214; Shrestha 1994: 45). The "operational effectiveness" of this, however, was minimal at best, having "no evident impact on AID programming" (Cohen and Uphoff 1980: 217).

Despite the murmur of these alternative perspectives, modernization theory dominated the first UN development decade (Esteva 1992: 13), hitting its peak of popularity before the decade's end in 1969 (Preston 2002: 171). Simply put, this meant that, "The goal of development was growth; the agent of development was the state..." (Leys 2005: 110). In alignment with this concept, the World Bank established a "poverty line" based upon a country's GDP in 1962.[25] This indicator operationalized poverty, making it measurable and comparable across countries (Illich 1992: 92), feeding into the concept of the linear path of development. This helped to more fully ingrain the concept of economic growth as the equivalent of development, and despite later successful and unsuccessful efforts to increase the range of indicators used to identify development, GDP per capita continues to be a very influential measure. It is used by the World Bank (Kreutzmann 2008: 679) and other organizations despite many criticisms voiced over the past few decades.

1.2.4 Dependency Theory Rises in Response to the Failures of Modernization Theory

Modernization theory, though dominant in the 1960s, was most notably criticized by scholars and development thinkers in Latin America. Dependency and world system theories evolved from these development thinkers. This school of thought was labeled "structural-internationalists" by Todaro (1977: 51), though other scholars find this label to be misleading, believing that modernization theory is also structuralist in nature (personal communication, Herman Kreutzmann, June 2011). These new theories emerged from the perceived lack of progress in developing underdeveloped nations using top-down development techniques grounded in the theory of modernization. They were also deeply indebted to Marxism.

25 As of 1990, the World Bank determined that any person living on a salary below 1 USD per day at purchasing power parity was living in absolute poverty (Thirlwall 2002: 42). This was altered to 1.25 USD per day in 2008 (Ravallion, Chen et al. 2009: 165).

Dependency theory is one of the classical or traditional development theories. It approaches the study of development from a historical perspective, in contrast to modernization theory; looking to history to understand the sources of underdevelopment in developing nations (Shrestha 1994: 14). Frank (1969a: 3) wrote, "We cannot hope to formulate adequate development theory and policy for the majority of the world's population who suffer from underdevelopment without first learning how their past economic and social history gave rise to their present underdevelopment." A country's level of development is, therefore, seen as being related to its history and its resulting place in the international economy (Hunt 1989: 66). Dependency theory looks at underdevelopment as being based on the relationship between developed and underdeveloped countries, with developed countries having risen to their state of economic strength and power based upon the exploitation of non-industrialized, underdeveloped countries (Willis 2010: 69). This theory sees the success of the developed states as stemming from their industrialization, which relies on cheap raw materials from underdeveloped states. The developed nations are seen as the center, while the underdeveloped states are labeled as periphery, with a strong sense of responsibility being placed upon the center, and a focus on class control over resources and surplus (Hunt 1989: 67).

Dependency theories evolved over the decades through the work of scholars such as Prebisch (1950), Baran (1957),[26] Baran and Sweezy (1966), Furtado (1964, 1965, 1969, 1978), and A.G. Frank (1969b, 1991). Dependency theory, as an economic and development theory, was presented in Latin America in the 1950s, with Raúl Prebisch being given a great deal of the credit for its promotion.[27] Higgott (1983: 48), however, argues that it did not gain popular acceptance until the mid-1960s, when A. G. Frank entered the picture.

Prebisch was the Executive Secretary of the Economic Commission for Latin America (ECLA), which was established in 1948. He, in association with other members of the ECLA, including Celso Furtado, theo-

26 While Baran's work (1957: 1-21) centers on economic surplus, he does emphasize the dependence of developing countries on underdeveloped countries. He talks about how underdeveloped countries lack access to the same resources that developed countries have and also emphasizes the underdeveloped world as the source for raw imports.

27 Raúl Prebisch was an Argentine economist.

rized that it would not be possible for underdeveloped nations to become developed while they were still a part of the global economic structure (Willis 2010: 66). This contrasted sharply with modernization theory. In alignment with modernization theory, however, dependency theory defined development in terms of industrialization and urbanization (Willis 2010: 66). The means to reach those indices of modernity, and therefore development, however, differed greatly between the two classical development theories.

Although A. G. Frank has become noted for his involvement in the establishment of dependency theory (Shrestha 1994: 14), it has been argued that no leader for this school of thought ever emerged. It is evident in reading about the topic that within the greater school of dependency theory there were many disagreements, some of which were fundamental (Smith 1979: 247). Debates about theoretical differences amongst those supporting dependency theory would last well into the 1970s (Rist 2008: 117).

While dependency theory left the economic focus of modernization theory unchallenged, it stressed history and causation in new ways. As Frank said, "The entire approach to economic development and cultural change attributes a history to the developed countries but denies all history to the under-developed ones" (Frank 1969b; as quoted in Preston 2002: 174). From the viewpoint of dependency theorists, trade, as practiced in the existing economic structure, was a preventative to development for Latin American countries. Due to the global environment in which Latin American countries were situated, they would not be able to follow the same development path as European and North American countries. Dependency theorists saw the world market as divided into powerful economies at the center or core and weaker economies in the periphery (Preston 2002: 181), as diagramed in Figure 2. From this perspective the powerful economies control the system, and as they accumulate technical expertise, they maintain their dominance and further ensure their position in the global economy, thus making it impossible for periphery countries to compete. Many Latin American countries exemplified this, with economies that were dependent upon the exportation of primary products to the markets of Europe (Peet and Hartwick 1999: 42; Preston 2002: 180). This was recognized as a liability during the Great Depression of the 1930s when demand and then prices for primary goods dropped drastically, leaving the economies of such exporting countries very vulnerable (Rist 2008: 112).

Figure 2: Diagram of the Extraction Process in Dependency Theory, from Periphery to Core

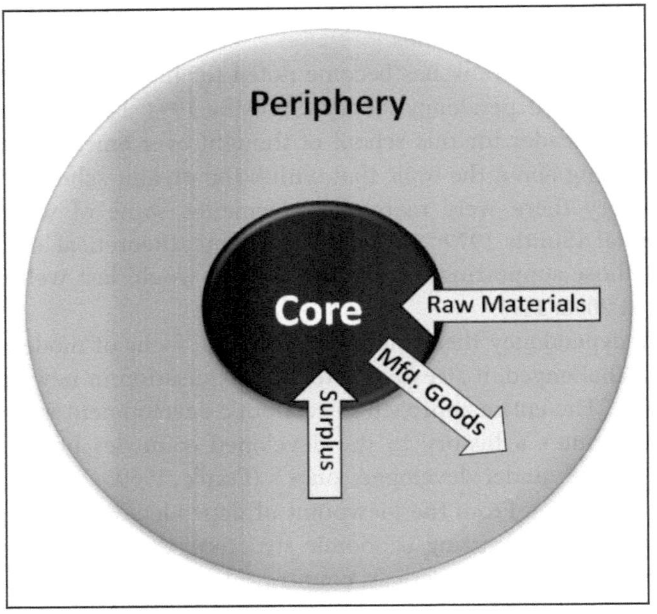

Surplus refers to material quantity of goods produced and economic surplus.
Source: own design based upon Willis (2010: 69-72)

The world economic system is seen as surviving based on relationships of exploitation that not only exist between countries, but also within countries. "Frank's predominant image was that of the now ubiquitous metropolis-satellite relationship in which the satellite was kept dependent by a sucking out of surplus to the metropole" (Higgott 1983: 49). Peasants are exploited by land-owners, land-owners are exploited by merchants and the "chain" continues until the profits are eventually extracted from the periphery and enter the core (Willis 2010: 70). There is a strong focus on economic surplus and how it is distributed (Hunt

1989: 65). This type of dependency was not only seen as economic in nature, but also political and cultural (Preston 2002: 195).

The cause of underdevelopment, therefore, was considered to be exogenous to a country, with little to no focus on possible endogenous factors that might also have contributed to the underdevelopment. This was pointed out as a flaw in dependency theory by Smith (1979) in a *World Politics* article. He said, "...dependency theory has systematically underestimated the real influence of the South over its own affairs" (Smith 1979: 249).

From this perspective, development approaches that were based upon modernization theory were seen as restricting developing countries to a permanent status of underdevelopment. The open market was seen by dependency theorists as unbalanced and not fair or open. The already existing inequalities between countries were not eliminated through the implementation of modernization influenced development policies and programs, but in fact were reinforced. Developing countries lacked investment funds to build up their economies and had to export raw goods and materials to already industrialized countries, leaving developing countries at the lower end of the value chain: exporting raw materials while having to pay for more expensive processed materials from industrialized countries. It was understood that foreign competition made it impossible for national industries to be established and thrive, and that, therefore, government intervention to protect state companies and development would be needed.[28]

Frank, and other dependency theorists, were strongly opposed to Rostow's theory of modernization (Rist 2008: 110-1). Frank felt that it had no empirical validity and that it was theoretically inadequate (Shrestha 1994: 11). Additionally, he believed that it was politically ineffective. He was a strong proponent of overthrowing the capitalist system, in contrast to other scholars, though less aligned with Marx, in contrast to Furtado, who felt that the system just needed to be reformed (Willis 2010: 70).

While the theories of dependency are similar in their understanding of the causes of underdevelopment, their viewpoints on the feasibility

28 Additionally, dependency theorists called for land reform, in opposition to the large land-holding patterns that remained from the colonial period (Hunt 1989: 51). They also advocated for reforms in existing economic policy to address these issues (Willis 2010: 66).

of successful development for underdeveloped nations differ. The two main schools that emerged were the "reformists," who believed that the capitalist trade system could be reformed with greater state intervention to benefit the periphery, and the "Marxists," who felt that the only answer was the complete overthrow of the capitalist system (Willis 2010: 70).

The dominant policy reform or development approach that stemmed from dependency theories came from the Marxist school – import substitution industrialization (ISI) – which required the government to create tariff barriers to manufactured imports in order to protect internal infant industries and internal production of manufactured products within the underdeveloped country.[29] These tariff barriers would increase the costs of foreign imports, allowing competing national industries to develop. In support of this were the historical experiences of the US, Russia, Germany, and Japan, all of which had practiced industrialization behind protective trade barriers in their own development histories (Preston 2002: 183). The idea was that, given enough time, these national industries would satisfy the local demand for products that had formerly been imported and would help the nation industrialize and modernize. A number of Latin American countries adopted this strategy, including Brazil, which practiced it from the 1940s onwards (Willis 2010: 67) and Argentina which started practicing it in 1942 (Hunt 1989: 47). As Preston (2002: 186) states, "These policies met with early success and there was extensive industrialization and urbanization. Following WWII the economies of Latin America grew rapidly. However, there were unanticipated problems . . ."

One of the main problems was that the process of industrialization, whether done behind trade tariffs or not, was dependent on foreign technology and the resulting production was absorbed by only a relatively small local market (Preston 2002: 186; Willis 2010: 69). The import substitution industrialization model was on shaky ground by the 1960s (Preston 2002: 187; Willis 2010: 69),[30] and with the onset of the oil

29 Import substitution industrialization was, according to Willis (2010: 66), built upon the "infant industry" approach developed by the German economist Friedrich List. However, Don Marshall (2002: 103) states that it stems from Arthur Lewis' model from 1955.

30 One author cited the exhaustion of the ISI experiment as having occurred by the 1950s (Smith 1979).

crises of the 1970s, people began to think that national development strategies in Latin America needed to be reviewed. Despite the increased dependence on foreign imports of technology and the widening gap between industrial workers and non industrial workers, import substitution industrialization was still practiced in some nations into the 1980s. "Over time, however, import substitution was associated with high-cost, low-quality industrial output, the economically damaging neglect of agriculture, and entrenched positions for foreign capital" (Peet and Hartwick 1999: 42-3). It was not long before the perceived remedy to the problem began to be seen as the cause of the problem (Peet and Hartwick 1999: 43; Preston 2002: 188). Dependency theory was also criticized, even by some of its own creators who began to perceive weaknesses in the theory and its applicability.

Frank was eventually seen as turning his back on dependency theory in the early 1990s, as he moved towards a world system view. He said that he found the theory of dependency no longer convincing and that national development was not possible in the current global system (Leys 2005: 115). He felt, therefore, that the focus should shift from national development to world development, with the beneficiary focus being on "particular groups or classes" (Frank 1991 as quoted in; Leys 2005: 119). This shift towards a non-state agent as the development focus is something that Frank (1992) continued to support, and something that is now a part of participatory approaches to development. While he continued to believe in dependency, he also felt that the theory of dependency never managed to provide answers for how to eliminate dependence and experience independent growth (Frank 1992: 136). Cardoso also eventually became very critical of the dependency theory he had contributed to, finding that the theorists could not offer solutions to the problems that they presented (Rist 2008: 120). "If one follows dependency theories, then, *one may know the origins of 'underdevelopment' but not much about 'development' itself*" (Rist 2008: 120).[31] In this vein, dependency theory managed to open up a new angle to the discourse of development: ". . . the *dependistas* shed greater light upon the mechanisms of underdevelopment and its links with class interests and the international system as a whole" (Rist 2008: 121).

---

31 Italics included in the original.

The Latin American economist Max-Neef (1991) blamed the failure of dependency theory on its singular focus on economics. He argued that by focusing only on economic growth, social and political processes were ignored. This argument would begin to garner more attention in the mid-1980s and 1990s, as wider definitions for development began to take shape. These expanded definitions coincided with and supported participatory development approaches.

Dependency theory was also challenged by the success of the newly-industrialized countries in Asia that came about in the 1970s, as these were peripheral countries that had been able to develop under the capitalist system, going against the fundamental understanding of some dependency theorists that capitalism had to be abolished or reformed before development could come to the periphery.[32] The theory was also criticized for its concentration on economics, without understanding more about the context of development, such as the social, political, or cultural situation (Willis 2010: 72). It did not take into consideration original endowments, such as climate, natural resources, and population (Higgott 1983: 57). Dependency theory was also seen as failing to take into consideration the fact that the characteristics of dependent countries also existed in other non-dependent countries (Conway and Heynen 2002: 100).

Dependency hit its peak of influence in the 1970s, though according to a number of development scholars, it had little actual influence on policy-making (Rist 2008: 110; Willis 2010: 72). It never gained the large scale institutional support that modernization theory gained in the 1950s and 1960s (Higgott 1983: 76), though it was supported by the "third worldists." It did, however, have a large impact on the development discourse. It helped to bring forward the notion that development and underdevelopment were in fact linked, in opposition to modernization theory, which saw the two as unrelated (Higgott 1983: 49). It also expanded the different factors on which it was believed development was dependent, and thus made room for even more indicators of development to be identified. This contributed to creating room for the establishment of the concept of human security, the use of social capital,

32 Rist (2008: 143) argues that the development of the Newly Industrialized Countries (NICs) disproves dependency theory.

and the creation of alternative development approaches such as basic needs and participatory development in the future.

Dependency theory shifted the paradigm of development thinking, firstly, by providing an alternative understanding of the source of underdevelopment. It looks at development from a more holistic viewpoint that involves history and the interactions of states with one another. This helped to pave the way for the current revival of participatory development approaches, as it inserted more reflexivity into development thinking.

By criticizing the dominant development paradigm of modernization theory, dependency theory also helped to deflate the theory of modernization and make room for alternative ways of thinking about development to evolve. This also involved the re-conceptualization of development as a situation of imbalance.

Even though dependency theorists did not shift the definition of development outside the sphere of economic growth and increased industrialization, it did introduce the concept of internal self-help and the idea that a state could possibly benefit from removing itself from the global economy, through the example of import substitution industrialization. This made room for different understandings and conceptualizations of development to emerge over the past four decades. It, however, still used other countries' historical modernization as a development model, keeping the concept of development unilinear, with a clearly defined path.

Additionally, dependency theorists looked upon the population of the State, and especially the rural population, as units of workforce. Their worth was in being a workforce, and nothing else. The population was not recognized as a development actor, this role remained at the State level. This would change dramatically as participatory approaches gained favor.

As has been shown, though seen as fundamentally opposing theories, modernization theory and dependency theory are based upon many of the same assumptions: there is a path towards development that leads towards imitation of North American and European countries, the State is the main actor, and economic development is the main indicator of success. These two theories, and the forthcoming world system theory, which is based upon the tenets of dependency theory, can be categorized as classical approaches to development. Though the theories of course differ regarding some of the understandings of why a country is undeveloped, they still evaluate this based upon economic indicators and hope to resolve it through the implementation of top-down development approaches that engage the State, but not the people of the State. These classical approaches continue to inform many development projects and programs worldwide today, but they now share time and money with alternative development approaches that have emerged in response to the failures of these classical approaches.

1.2.5 Expanding the Definition of Development beyond Economics: Shifts towards Participation

These alternative development approaches begin with the emergence of different understandings of the indicators used to measure development. Towards the end of the 1960s, Dudley Seers (1969) began to advocate for a set of indicators to measure development, instead of the single economic indicator used up until that point. He found the use of GDP as the sole indicator of development, inadequate and put forward the idea of including inequality, poverty, and unemployment in development analysis (Bins 2002: 79; Parnwell 2002: 113). This was a radical shift that acknowledged some of the perceived shortcomings recognized by dependency theorists in terms of inequalities within national economies and not just between them. This helped to pave the way for the basic needs approach, which would gain prominence in the 1970s.

As the second UN Development Decade approached (1970 to 1979), it became evident that many underdeveloped countries that had followed the programs of modernization theory remained ensconced in poverty, with inequality not subsiding, but actually increasing (Esteva 1992: 13; Moseley and Pandit 2007: 351). The assumed trickle-down effects of modernization were not in evidence, and it was the richer countries and people that were benefitting (Willis 2010: 93). A shift in focus to the poor of the world took place in some circles in the beginning of the 1970s. This helped lead to the idea that in order for rural people to take control and manage their own lives, participation in development was needed (Gow and Vansant 1983: 427). While it did not lead to strong changes in the approaches used to address development concerns at the time, it did lead to an expansion of the development debate and the inclusion of participation as a theme of the United Nations Research Institute of Social Development (UNRISD) (Chambers 2005: 99). It was in this decade that the interconnections between economic and social development began to be seen. No unified approach or theory, however, was established.

Some development scholars began, around this time, to elevate the value of local knowledge and a people-centered view of development, both of which are highly valued in participatory approaches. Brinkerhoff and Klauss (1985: 145) said,

> *This people-centered conception does not ignore the need, for example, to raise agricultural productivity, provide employment opportunities in rural or urban areas, or build infrastructure. However, it takes as its starting point local people's knowledge, capacities, aspirations and desires, and places development activities in the service of bettering people's lives in ways that best suit them. People become both the means and the ends of development.*

Supporters of the value of local knowledge, sometimes called indigenous knowledge, could be found in the creators of participatory rural appraisal (PRA), which focuses on involving the local population, or community, to participate in research that informs development planning and the conceptualization of development projects and programs. PRA arose from theorists such as Orlando Fals-Borda and Anisur Rahman, amongst others, in the 1970s and was popularized most noteably by Robert Chambers. It gained a great deal of popularity and support, but many were concerned that while people used the terminology of PRA, their techniques were actually not as participatory as conceptualized.

The expansion of the definition of development has been integral to the evolution of participatory development approaches, which involve the people in order to create multiple levels of impact on a community. By putting forward more means of measuring development, Seers helped to create a more holistic picture of underdevelopment and development, a picture that is utilized in participatory development approaches today. Many different factors are seen as having an impact on the development of a community, and there is a greater understanding of how these different indicators are interconnected. Seers was just one of the first to rethink what development meant, others followed and continued to identify new indicators of importance.

The increased valorization of local knowledge, as opposed to a complete reliance on expert knowledge, also marked a big shift in development thinking. Although it did not penetrate all development projects, it did create room for the involvement of people in development projects that affected them. This was vital for the growth of participatory development approaches.

1.2.6 NIEO and Efforts to Shift Power to the Periphery

The 1970s also represent a time period when the developing world began to gain power and a voice. From the OPEC oil price hikes of the early 1970s to the Vietnam War, developing countries' impact on developed countries began to gain attention (Rist 2008: 142). This led to multiple conferences that focused on "collective self-reliance," problems relating to raw materials and development, and other topics of importance to underdeveloped nations.[33] The most impactful was the United Nations Conference on Trade and Development held in April and May of 1974. The Declaration on the Establishment of a New International Economic Order (NIEO) was produced at it, along with a program of action. The NIEO, drafted by members of the developing world, laid out a series of proposals meant to shift the balance of economic power, which was in the hands of the developing countries. The declaration proclaimed the developing countries' determination to establish a new international economic order to, "correct inequalities and redress existing injustices, mak[ing] it possible to eliminate the widening gap between the developed and the developing countries" (General Assembly of the United Nations 1974: n.p.). The tools for achieving this included trade reforms, monetary reforms, debt relief, technology transfers, and regional cooperation (Preston 2002: 242). The understanding was that by changing the rules of the game, developing countries could catch up to the developed countries.

33 In 1970 a conference was held in Lusaka where "collective self-reliance" was an adopted principal and a conference on raw materials was held in 1972 in Algiers by heads of State and the governments of the non-aligned countries (Rist 2008: 143).

The NIEO was still focused on economic growth, expansion of trade, and increased aid from industrial countries, so it did not break from the vision of what development looked like from modernization theory. It simply attempted to change the rules of the game to provide developing nations with a chance to play a more dominant role (Rist 2008: 148-9). This understanding of the unbalanced nature of the global economy was grounded in dependency theories, as was argued by Higgott (1983: 91). Rist (2008: 154) sees the NIEO as the last stand of modernization thinking and the belief that industrialization and economic growth were equivalent to development. This is in strong contrast to the dependency theorists who encouraged disassociation from the global market (Rist 2008: 150).

The NIEO was a strong statement of solidarity, but in spite of aspects of its program being included in, "the various organization structures of the United Nations" (Preston 2002: 244), it had very little practical impact. "Despite its rhetorical list of demands, it proposes nothing new – or anyway, nothing that sheds fresh light on how to improve living conditions for the peoples of the South" (Rist 2008: 151). Rist (2008: 151, 153) sees the NIEO as an effort by national elites to gain a bigger share of the economic benefits of development and believes it is fortunate that it was never truly integrated into development policy.

The NIEO represents one of the first times that developing nations organized themselves to put forward what they saw as an alternative *modus operandi* for the world economic order. While, not directly related to the rise of participatory development, I believe the introduction of the NIEO did provide the opportunity for an alternative concept of development to emerge. It allowed voices from the developing world to emerge, though these voices still stemmed from elite circles, leaving the general population out. Unequal distribution of wealth and development were seen as something that required fundamental changes in the rules of the game in the minds of those who put the NIEO together. It also represents a high point for third worldism, and it received a good deal of attention in international development circles (Rist 2008: 5).

Some see the lack of success of the NIEO as related to a breakdown in the solidarity of the developing countries in the mid-1970s (Preston 2002: 244; Rist 2008: 153). Rist (2008: 153) highlights three different types of developing countries: the rich countries living off of oil revenues, the extremely poor countries, and the Newly Industrialized Countries (NIC) in order to highlight that their objectives were no longer in alignment. It should be noted, however, that the issue of dependence and interdependence in the global structure has not been lost from the development discourse.

1.2.7 Adding to Seers: Support for More Expansive Development Indicators

The lack of success of approaches stemming from modernization theory also led some to continue to question the strong economic focus of development thinking and practice, not just the structure of the system. Robert S. McNamara, as President of the World Bank, tried to emphasize that development in the second development decade should focus on more than just an increase in GNP (Esteva 1992: 13), as was suggested by Seers towards the end of the 1960s. McNamara's 1973 speech focused on the need to address what he termed as "absolute poverty" and to look at more than GDP. He stated that there was a need:

> *... to redefine the objectives and measurement of development in more operational terms. While most countries have broadened the statements of their development goals to include references to reducing unemployment and increasing the income of the poor ... they still measure progress toward these complex objectives with a single measuring rod: the growth of GNP. (McNamara 1973: n.p.)*

McNamara suggested measuring increases in the income of the poorest, and not just the well-off. He also talked about the elimination of malnutrition and illiteracy, the reduction of infant mortality, and increasing life-expectancy rates. This, therefore, allowed for the expansion of the measures of development greatly, in alignment with similar themes that run through approaches presented later, such as basic needs, human development, and human security. This expansion was supported by the US Congress in its 1973 directive that USAID's efforts be dedicated to meeting the needs of the poorest of the poor (Ruttan 1984: 397) through the use of participation and inclusion (Cohen and Uphoff 1980: 213; Finsterbusch and Van Wicklin III 1989: 573; Korten 1980: 482), known as the "New Direction mandate." Escobar (1992: 138) notes that McNamara's speech was the beginning of the era of "poverty oriented" approaches to development. The World Bank, UNESCO, and DFID all soon followed suit and began to focus more directly on poverty (Cohen and Uphoff 1980: 213, footnote 3).

McNamara's speech also supported the revival of rural work programs that, "make extensive use of local labor" (McNamara 1973: n.p.), a participatory approach. Though he noted that such programs in Bangladesh, Tunisia, Indonesia, and other countries faced limited success in the past; the blame for this was placed on inadequate management and limited scale (McNamara 1973: n.p.). He provided no evidence to back up this statement, but said that, "Experience shows that there is a greater chance of success if the institutions provide for popular participation, local leadership, and decentralization of authority" (McNamara 1973: n.p.). The participatory programs he referred to were the same community development programs from the 1950s and 1960s discussed previously.

The focus on self-help and participation continued with the Third World Forum of 1973, held in Karachi, and the New Directions Mandate mentioned above. Additionally, the World Health Organization (WHO) conference in Almaty, Kazakhstan in 1978 emphasized the importance of participation for the extension of primary health care (Shrestha 1994: 45-6). Participatory approaches, however, would not become mainstream until the 1990s.

1.2.8 Basic Needs Approach: Introduction of Individuals into the Development Discourse

Calls to redefine development as something that went beyond economic growth were supported by the Declaration of Cocoyoc of 1974. This declaration, which was written by a group of intellectuals, many of whom were members of the Third World Forum (Rist 2008: 165), encouraged a new approach in which basic needs were the focus.[34] "Our

34 The Cocoyoc Declaration put forward some ideas that were, and are still, considered to be radical (Rist 2008: 156-7), including an international taxation system that would automatically transfer resources to development assistance. Development was seen as a problem to be solved both within developed and developing countries, by raising consumption in the former and lowering it in the

first concern is to redefine the whole purpose of development. This should not be to develop things but to develop man" (UNEP and UNCTAD 1975: 896). This declaration also rejected the linear nature of classical development theories, arguing that there were many different paths to development and that imitating the historical model countries that "happen to be rich today" was not the answer (UNEP and UNCTAD 1975: 897).

In 1975 the Dag Hammarskjöld Foundation's report, "What Now," was published. It called for human-centered development (Esteva 1992: 15).[35] It also called for development to be seen as much more than just an economic process. It found that development should be based on the satisfaction of needs in a sustainable way, and that this would require the North to change its consumption patterns (Rist 2008: 156).[36] "What Now" was not adopted and had very little impact, but it helped to lay the foundation for the introduction of the basic needs approach, which was put forward by the International Labour Organization (ILO) in its declaration issued from the World Conference on Employment in 1976 (Esteva 1992: 15).

These reports were in reaction to the understanding that top-down approaches had not been benefiting all segments of society and by the desire to expand the notion of development beyond economic growth to include food, health, education, and other factors. The basic needs approach gained the most support, becoming the approach of choice with international development organizations (Peet and Hartwick 1999: 54). Basic needs represented an important shift from macro-focused development policies that focused on the State as its main beneficiary to a focus on people; the poorest people (Willis 2010: 93). It was adopted by the ILO, UNEP, and the World Bank (Potter, Binns et al. 2008: 115) and many NGOs (Peet and Hartwick 1999: 3; Preston 2002: 237; Rist 2008: 173).[37] "Without a doubt, it signaled a turn in the operations of the

---

later (cf. UNEP and UNCTAD 1975: 899, 897).

35 "What Now" was an initiative of both the Hammarskjöld Foundation and the UNEP.

36 The Dag Hammarskjöld Foundation report recommended setting limits on meat and oil consumption, it proposed to extend the life of consumer goods, and proposed the replacement of privately owned cars with public transportation (Rist 2008: 156-7).

37 Basic needs' popularity was not long lasting, according to Esteva (1992: 15), who said that it gained a following for only two or three years.

World Bank, which suddenly became more anxious to intervene at the grassroots, without ceasing to fund large-scale projects" (Rist 2008: 163).

It also called for more than just economics to be relevant in the development debate (Preston 2002: 245); outlining a number of basic needs, including: basic personal consumption, which includes food, shelter, and clothing; access to essential services, such as water, sanitation, education, healthcare, and transportation; access to paid employment; and what was termed as qualitative needs, which included the ability to participate in decision-making and the establishment of a healthy and safe environment in which to live (Hunt 1989: 265-6), as outlined below in Figure 3. Other individuals and organizations defined their own sets of basic needs, many of which concentrated most strongly on what Hunt considers basic personal consumption.

Figure 3: Outline of the Definitions of Hunt's Basic Needs Indicators

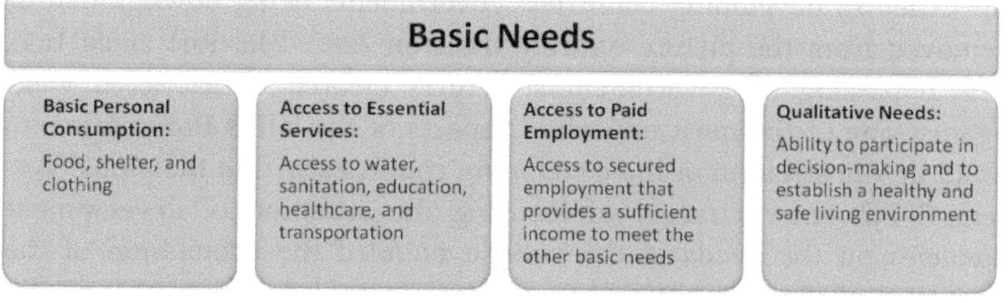

Source: own design based upon Hunt (1989: 265-6)

As implemented, the basic needs approach resulted in programs that looked at health, education, farming, and more, focusing on the weakest or most vulnerable groups in society. Planning revolved around district and regional entities, instead of the State as a whole, and specific social groups were targeted (Elliott 2002: 46), which was a shift from classical development theory inspired programs. Greater focus was placed on small-scale activities and on the poor, both ideas that would also be utilized in participatory development approaches.

Despite all of these differences from other theories of development, Preston (2002: 246) argues that basic needs was approached through the same type of "blueprint planning" of the past and that it was guided by modernization theory. "It is clear that the basic needs approach is far from being a radically new departure in theorizing and action" (Preston 2002: 246). Basic needs raised concerns that it would result in a movement away from modernization theory, with its concentration on both agriculture and the urban informal sector (Willis 2010: 96). Willis (2010: 95) argues that it did not contradict modernization theory, but was an approach to be used in conjunction with modernization theory to address those who were not benefitting from projects or programs based only on modernization theory.

Paul Streeten (1981: 332), one of the proponents of the basic needs approach, saw it as a return to the original intentions of development: "In one sense, this is a home-coming. For when we embarked on development 30 years ago, it was surely the needs of the poor of the world that we had in mind." [38]

In contrast to previous efforts to involve state governments in development, the World Bank's focus on efficiency in the delivery of basic needs led to developing countries' governments being downgraded or removed from the picture entirely (Preston 2002: 246; Rist 2008: 164). The bypassing of government structures created critics, as it went against one of the most important aspects of the NIEO Programme, to build up the strength of the developing countries within the global system. "[T]he new strategy of focusing the attention of development agencies on the needs of the poorest entailed the submission of the countries of the Third World to greater external direction of their development efforts" (Preston 2002: 244). However, there were some proponents who argued that the NIEO and basic needs were compliments and not in competition with one another. Streeten was one such advocate. He saw the NIEO as trying to formulate "a framework of institutions, processes and rules" to correct the bias of the international system to ensure that basic needs would become a program from self-reliant governments and not "global charity" (Streeten 1981: 357-8).

38 Paul Streeten is a well known development economist who has been very influential in the study of development. He has worked most notably with the UNDP on the Human Development Report, UNESCO on its World Culture Reports, and with the World Bank regarding its basic needs policies. He has also been involved with the Ministry of Overseas Development in the UK, the University of Sussex (where he founded the Development unit), Boston University, and Oxford (Ralph Bunche Institute for International Studies 2011).

As with other theories and approaches, the basic needs approach has been criticized. There were problems in defining "basic needs" in clear quantitative terms that were standardized and easily comparable, which are the preferred indicators of multilateral and bilateral development agencies and organizations. The complicated nature of determining how much an individual needs in terms of food to survive and other basics revealed that defining basic needs was much more complicated in reality than theoretically (Rist 2008: 167). Nevertheless, the seeming simplicity of the basic needs message made it an easy approach for NGOs to adopt, as it registered with the populations of the developed world (Preston 2002: 248), in a similar way to participatory approaches today.

The Brandt Commission, also known as the Independent Commission on International Development Issues, produced a report in 1980 called *North-South: A Programme for Survival* that also supported the basic needs approach.[39] In addition, it introduced into the development lexicon the binary of North and South (Potter, Binns et al. 2008: 28), with the North representing developed countries, be they in the Northern half of the globe or not, and the South representing underdeveloped countries. The author of the report was Willy Brandt, whose celebrity for being a Nobel Prizewinner and the ex-chancellor of West Germany brought attention to it. The overall objective of the report was to eliminate dependence, oppression, hunger, and distress (Rist 2008: 161). It was not viewed as adding a great deal new to the development discourse, but it did advocate for basic needs, supporting the ILO (Brandt 1980: 54-59, 63), and it gained the attention of people concerned with poverty (Lal 1983: ix).

The basic needs approach was different in that it stopped addressing development only from the perspective of a State and a State's well-be-

39 The Brandt Commission was made up of economists and politicians, acting in an independent capacity, representing different countries from both developed and underdeveloped countries. It's members included: Abdlatif Y. Al-Hamad (Kuwait), Rodrigo Botero Montoya (Columbia), Antoine Kipsa Dakoure (Upper Volta), Eduardo Frei Montalva (Chile), Katherine Graham (US), Edward Heath (UK), Amir H. Jamal (Tanzania), Lakshmi Kant Jha (India), Khatijah Ahmad (Malaysia), Adam Malik (Indonesia), Haruki Mori (Japan), Joe Morris (Canada), Olof Palme (Sweden), Peter G. Peterson (US), Edgard Pisani (France), Shridath Ramphal (Guyana), and Layachi Yaker (Algeria) (Centre for Global Negotiations 2012).

ing, shifting the focus from the State to individuals. This was important for the future rise of participatory approaches, which also focuses on the local population instead of the State. The extensive use of multiple indicators of development seen in the basic needs approach will later be reflected in the belief that participatory approaches address multiple sources and effects of underdevelopment.

1.2.9 World System Theory: Introducing More Nuance into Dependency Theory and Presenting a Multi-directional Model of Development

Also theoretically important in the 1970s was the world system theory of development that was an expansion on the ideas embedded in dependency theory. It was presented by Immanuel Wallerstein (1974), in collaboration with his wife, Terence Hopkins, and has been categorized as an adaption of dependency theories (Conway and Heynen 2002: 100). Instead of there only being the two areas of the core and the periphery, as outlined in dependency theory, this theory adds a third area, the "semi-periphery." This third category acts as the periphery to the core, while at the same time it exploits countries in the periphery, as diagramed below in Figure 4. A country's membership in a single category is not fixed in the world system theory, but is movable depending on the country's fortunes (Klak 2002: 110). This theory was therefore multidirectional, with a country able to shift towards the core or away from it (Willis 2010: 73). This represents a movement away from dualist thinking, which was characteristic of the classical development theories of modernization and dependency. Another major difference from dependency theory is the placement of the world system at the center of the analysis (Wallerstein 1974: 7).

The world system theory led to very few development policies. Some development scholars find the theory to be good at describing certain patterns, but unable to explain them (Willis 2010: 75), which makes the writing of policy very difficult. One such thinker is quoted as saying, "Perhaps it is better to think of a world-system *analysis* or *perspective* rather than a fully-fledged world-system theory" (Klak 2002: 111).[40] The presentation of a multi-directional theory of development

40 Italics included in the original.

cut into one of the foundations of modernization theory: development is a path on which the movement is always forward, never backwards.

Figure 4: World System Theory:
Relationship Between Core, Periphery, and Semi-Periphery

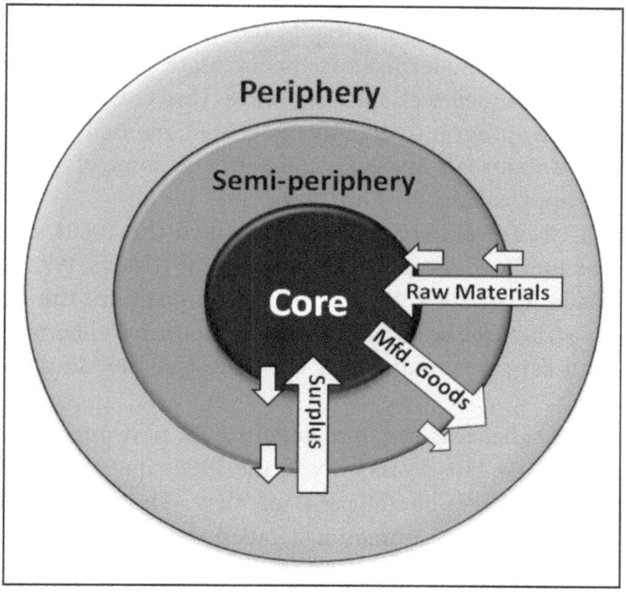

Surplus refers to material quantity of goods produced and economic surplus.
Source: own design based upon Preston (2002: 219-21)

The theory still presents states as the primary focus of development, and it does not include the expanded definitions of development explored above. Therefore, it did not help pave the way for participatory approaches in a large way, but it helped placed more nuance into the discourse, contributing to breaking down the expectations that one approach or one theory holds all of the answers.

1.2.10 Belief in the Forces of the Market: Neo-liberal Approaches Dominate

A sense of pessimism entered the development scene towards the end of the 1970s and moving into the 1980s, as fears about instability, debt, and a lack of development success in the post WWII period began to be recognized (Preston 2002: 288). Despite the questioning of development approaches based on modernization theory, state aid programs inspired by the theory continued well into the 1980s. However, this was accompanied by smaller programs such as those implemented by NGOs, with a focus on the small-scale and on empowerment, some of which utilized participatory approaches (Preston 2002: 289).

The third UN Development Decade covered the 1980s and is often referred to as the "lost decade" for development (Esteva 1992: 16) or the "wasted decade" (Rist 2008: 178). It is known mostly for the new policy approach of structural adjustments, based upon neo-liberal economic theory, which became dominant at the behest of the World Bank and the IMF.

Development theorists began to advocate for non-interference in the market in the 1970s, but it was only in the 1980s that the policy debate tipped towards the neo-liberal point of view. Those who believed in neo-liberalism, or what is sometimes called the "New Right," felt that the "modern free-market capitalist system [was] maximally effective in producing and equitably distributing the economic, social, political, and intellectual necessities of civilized life" (Preston 2002: 253).

Inspired by such theorists as Deepak Lal (1983) and Bela Balassa (1971, 1981), neo-liberalism was built upon the economic theories of Adam Smith and his successors, such as David Ricardo (Gregory, Johnston et al. 2008: 497; Willis 2010: 47), and is strongly connected to modernization theory. The main tenets of the approach were that the public sector was too involved in development, that infrastructure was not the best place to focus, and that economic controls were distorting prices (Willis 2010: 47). Neo-liberalism stands in opposition to Keynesianism, which included government assistance and interference in the market and was popular when modernization theory was first developed.

Structural adjustments were seen as a response to the debt crisis that exploded, starting with Mexico defaulting on its loans in 1982. It was believed this defaulting indicated that governments and individuals were living beyond their means, and that restructuring would bring them back to equilibrium (Rist 2008: 172). Therefore, the answer to underdevelopment required all companies to be privatized; all tariffs, quotas, and subsidies to be eliminated; the economy had to be opened to foreign investment; reforms had to occur in the tax system; and stabilization measures had to be implemented, such as cutting back on government expenditures and wages (Preston 2002: 258; Willis 2010: 52).

These adjustments reflected what came to be known as the "Washington Consensus," a term coined by Williamson in 1990 (Peet and Hartwick 2009: 92; Willis 2010: 52). This term is used to refer to neo-liberal policies that were proposed or required by the IMF and World Bank to receive loans in the late 1980s (Willis 2010: 52). The adjustments were included in a country's structural adjustment program (SAP), which was required by the World Bank and the IMF in order to receive financial assistance from them, something many countries were seeking to do as the debt crisis expanded.[41] In essence, SAPs were intended to lower the interference of the state in the economy, giving the market more power (Willis 2010: 52). It was assumed by neo-liberal thinkers that implementing stabilization and adjustment measures together would increase government income and would lead to greater efficiency, and therefore economic growth. Structural adjustment programs became dominant in the 1980s, with 228 SAPs introduced in sub-Saharan Africa by 1996 (Simon 2002: 89). Preston (2002: 269) said, "In development work it led to a stress on outward-directed development strategies and the encouragement of unregulated markets."

Neo-liberalism in development resulted in a renewed focus on modernization theory. "Alas, the right has had nothing better or even new to offer; *neo*liberalist monetarism and supply-side theory were no more than warmed over 'neo-'classical theory, disinterred from its half-century old grave" (Frank 1992: 135). It acts strongly in opposition to many of the shifts that were leading towards a decentralized approach to development, with the people becoming the focus, and development being

41 Potter et al. (2008: 93) indicate that USAID also required SAPs as a condition for obtaining loans.

measured by many different indicators. Neo-liberalism pushed these approaches to the side and saw a return to a State-centered, economic focused form of development. During the 1980s neo-liberalism dominated the field, as nations adjusted their policies in order to obtain needed development funding from the IMF, World Bank, etc.

As with so many other development approaches, neo-liberalism or market fundamentalism, did not result in the economic growth expected. Poverty levels often increased as the cost of living rose and real wages fell, along with increases in the rates of unemployment (Schuurman 2002: 13; Willis 2010: 52). Preston (2002: 260) describes it as having done "untold damage" to the economies and societies of the "third world." Additionally, public services saw a drastic cut in many nations, including health and education benefits (Rist 2008: 173). While these negative realities cannot be blamed on the adjustments directly, as there were many other factors involved at the time, it can be seen that the SAPs did not lead to poverty reduction, as was pointed out by McIlwaine and Willis (2002: 99).

Neo-liberalism's negative impacts were met with new ideas regarding participatory development, and those voices rose throughout the decade. While neo-liberal policies still exist, they are now often dressed up in the terminology of participatory development, with more emphasis on the people and their problems. The negative fall-out from such programs has likely contributed to the increased attention being paid to participatory development approaches, as people were searching for alternative development approaches that would produce better results.

Following this return to the classical theories of development, the arrival and acceptance of alternative approaches entered the development sphere in a way that had not been accomplished prior. While the classic or "grand old" theories of development are still relevant to any discussion on development and can often be discovered as guiding inspiration for development programs and projects that cover all development sectors, these alternative approaches, especially participatory approaches began to gain more respect and more attention.

1.3 Movement towards Participatory Development

1.3.1 Participatory Approaches: People as Agents of Change

Small adjustments to approaches informed by the classical development theories began to evolve in the 1980s, making room for increased participation of different actors and the decentralization of development work. While the third decade of development was dominated by neoliberal thinking and structural adjustment programs, there continued to be voices that called for development to be defined more broadly than just along economic lines, echoing Seers and those who followed after him.

It was in the 1980s that a "populist approach" to development began to emerge (Shrestha 1994: 23), and development organizations started to utilize participatory approaches (Cooke and Kothari 2001a: 5).[42] The participatory approach builds upon the idea that local communities are not "backwards" but are actually efficient (Korten 1980: 498; Shrestha 1994: 23). It also calls into question the generally accepted principle that industrialization and economic growth should be the aims of development (Shrestha 1994: 23).

Participatory development was supported by authors such as Chambers (1983, 1994, 1997, 2005) and Cernea (1985), and is considered by Chambers (2005: 98) to have really been developed mostly by people working for NGOs in the field, not just academics. These authors, together, put forward the positive aspects of local development that is participatory, in contrast to the top-down approaches that by the mid 1980s were seen as unsuccessful (Mansuri and Rao 2004: 7). Programs that utilized this participatory approach, such as the Self-Employed Women's Association in India, the Orangi Pilot Project in Pakistan, and the Iringa Nutrition project in Tanzania, were seen as models of participatory development that were working (Mansuri and Rao 2004: 7). This approach would gain increased attention in the 1990s.

The fall of the Soviet Union ended the Cold War between the USSR and the US. This political shift meant that the development proxy war

42 This approach has also been called variously: another development, alternative development, people-centered development, counter development, and participatory development (Oakley 1991: 2).

between the two countries slowly ground to a halt, and there was the hope that development would become a true priority that was not mired in political motives, but was valued in its own right (Rist 2008: 197; UNDP 1994: 71). It also cleared the way for a continued focus on market-oriented answers. Depending on one's viewpoint, development still remains highly political today.

In the 1990s, it became evident that the neo-liberal structural adjustment plans of the 1980s were not leading to an eradication of poverty, and were in some cases even contributing to increased poverty, e.g., in Jamaica (Willis 2010: 53). Such disappointing results made people begin to re-evaluate their understanding of development yet again, leading to movement away from classical approaches, towards an exploration of grassroots and participatory approaches. It was in the 1990s that the theorization of development seemed to stall, with no overarching theory rising in the same manner as the classical development theories. The focus instead shifted to the development agent and development indicators.

1.3.2 Human Development at Center Stage and a New Development Index

The main development paradigms and approaches brought forward in the 1990s were characterized by the continued expansion of the definition and measurement of development beyond economic indicators and an increased focus on people as the actors of development. The United Nations Development Programme (UNDP) highlighted both of these shifts in its first *Human Development Report* in 1990, written by Mahbub ul Haq, who Amartya Sen describes as "the pioneering leader of the human development approach" (UNDP 2010b: vi).

The authors of the report say that development is about creating an "environment for people to enjoy long, healthy and creative lives," and they see this as a "simple truth" that has gotten lost in the focus on the accumulation of economic wealth and growth (UNDP 1990: 9). Human development, as defined by the authors of the first UNDP report, is "a process of enlarging people's choices. The most critical ones are to lead a long and healthy life, to be educated and to enjoy a decent standard of living. Additional choices include political freedom, guaranteed human rights and self respect ..." (UNDP 1990: 10). The authors of the 1990 report see the introduction of human development as a refocusing of development on the ends and not the means (UNDP 1990: 9).

The human development approach diverges from classical development theories in that economic growth is seen as "not sufficient for human development" (UNDP 1990: 11). Its supporters believe that it is wrong to think that "high economic growth will automatically translate into higher levels of human development" (Haq, K. 2002: 9). Therefore, economic growth is not seen as a good proxy indicator of development. Economic growth is seen as only one factor amongst many that contribute to human development. In this way, it is similar to the basic needs approach, which came before it, and the human security approach which followed, in that it identifies multiple indicators that are necessary for human development. Human development is applicable to both developing and developed countries (UNDP 1990: 11), which allows for problems and flaws in developed countries, unlike classical theories, which can be used to address inequalities within a country.

Mahbub ul Haq, however emphasizes, that human development is a new perspective and not just a twist or repeat of previous development theories:

> *The human dimensions of development is not just another addition to the development dialogue. It is an entirely new perspective, a revolutionary way to recast our conventional approach to development. With this transition in thinking, human civilization and democracy may reach yet another milestone. (Haq, K. 1999: 2)*

The UNDP *Human Development Report* was first published in 1990. It became an annual report that looks at different factors of development. It introduced human development as a concept and a new index of development – the Human Development Index (HDI) – which incorporates the main ideas of human development into a simplified index.[43] HDI added two new indicators to the common development indicator dominant since the 1950s, Gross National Product (GNP) (Streeten 1981: 374). In addition to measuring material well-being using the purchasing power adjusted GDP (or command over resources), it also measures ed-

43 The HDI was developed by the UNDP in the late 1980s (Willis 2010: 5). It is based upon Amartya Sen's "capabilities" approach (Stanton 2007: 3).

ucational attainment, and the life expectancy of a population. From these, it creates a single statistic to evaluate both social and economic development in one index.[44] Peet and Hartwick (1999: 7) called HDI a redefinition of development and Rist (2008: 209) called it a, "break with the sacred cow of economic growth."

Some development thinkers complained, however, that this new index was nothing more than a revision of previous economic measures (Esteva 1992: 17). However, it has become a highly influential index and is still a part of the development discourse today, and has even inspired other alternative indices of development.[45] Sen, one of the group members to have developed the HDI, believes that it is useful in that it does not ignore everything beyond economics, though he still sees it as crude (UNDP 2010b: vi). While many countries' development standing is similar to its HDI standing, this is not always the case. For example, many oil producing countries have a high GDP, but not a high HDI (UNDP 2010b: vi). Afghanistan has an HDI value that is near to that of Zimbabwe (.376), Malawi (.400), and Cote d'Ivoire (.400). If measured by Gross National Income per capita (constant 2005 PPP$), Afghanistan would rank at the top of this group, followed by Cote d'Ivoire, Malawi, and then Zimbabwe. However, if development is measured by the human development index the order would be different, with Cote d'Ivoire and Malawi ranking above Afghanistan, all of which would rank above Zimbabwe, as displayed in Figure 5.

44 Educational attainment is calculated as a combination of the adult literacy rate and the gross enrollment ratio in primary, secondary, and tertiary education. This, along with life expectancy at birth and the GDP adjusted for purchasing power are then combined into one composite index. The average of these three figures is the HDI (UNDP 1990: 9-16).

45 These alternative development indexes look at particular populations or at specific concepts, such as the Gender Development Index (GDI) and the Gender Empowerment Measure (GEM), which both look at development through a gendered lens and the Human Poverty Index (HPI), which focuses on deprivation. The HPI measures the percentage of the population who are not expected to reach 40 years old, the adult illiteracy rate, and a deprivation index based on access to safe drinking water, access to health services, and childhood weight measurements (Thirlwall 2002: 44). GDI and GEM were introduced in 1995 and HPI was introduced in 1997 (Elliott 2002: 47).

Figure 5: Highlighting Countries with a Similar Human Development Index Values and Different Gross National Income per Capita Values

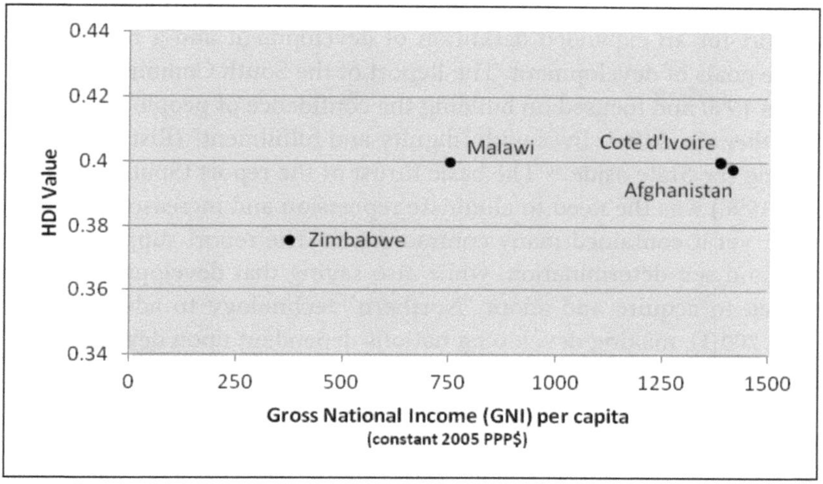

Source: own design based on data from UNDP (2011: 127-130)

The development of the HDI and its usage represents the recognition of a general shift towards seeing development as enveloping multiple factors and indicators, as was originally explored by Seers. This expansion has been widely embraced by academics and scholars who support participatory development. In the last twenty years, HDI has become a standard measurement that is used to discuss the development state of a country, joining the economic indicators, but not replacing them.

Some scholars, such as Rist (2008: 205), a post-development thinker, see the presentation of human development as a ruse. He questions the idea that development was anything but human-centered from the beginning, seeing the UNDP Human Development Reports more as a public relations campaign to make it appear as though development stakeholders have found a new answer to the problem of underdevelopment. Rist (2008: 205) said, "But the point, of course, was to rehabilitate a largely discredited concept by giving it a spiritual boost that it would be in bad taste to refute." The authors of the 1990 Human Development Report admitted that their way of looking at development as human was not really new, saying that it was in alignment with the "human good" discussed by Aristotle (UNDP 1990: 9).

Additional reports have come out that have helped to garner wider support for an expanded definition of development and a restatement of the goals of development. The Report of the South Commission came out in 1990 and focused on building the confidence of people in order to help them lead their lives with "dignity and fulfillment" (Rist 2008: 202), leaving the State aside.[46] The basic thrust of the report (South Commission 1990) was the need to eliminate repression and increase empowerment, yet it contained many contradictions. The report supported self-help and self-determination, while also saying that developing nations needed to acquire and adopt "Northern" technology to advance (Rist 2008: 200-1), making developing nations dependent upon developed nations for their development.

The World Bank began a series of Participatory Poverty Assessments in the mid-1990s, and began research that asked the poor what they wanted. The *UNDP Human Development Report of 1991* focused on the enlargement of people's choices as being the foundation of development (UNDP 1991), a theme that participatory development is also based upon. The definition of development here focused on democratic development, participatory development, economic opportunities, education and health, and a clean and safe physical environment. It also supported the involvement of all community members in decision-making, and the idea that all people needed to enjoy political and economic freedoms (Rist 2008: 208).

All of these reports and the new indices created support for the underlying themes of participatory development. The HDI incorporates multiple indicators of development and participatory development hopes to address multiple indicators of development through the participatory process.

46 The South Commission was made up of important and distinguished individuals who participated not in their official capacities as politicians and such, but "serving in their personal capacities." Established in 1987, it was an independent body that was aimed to provide findings and recommendations related to development. (South Centre 2011).

1.3.3 Human Security: Development Creating Security

The UNDP *Human Development Report* continued to be influential and introduce new ways of looking at development moving forward. The 1994 report (UNDP 1994) introduced a new framework for development that incorporated some of the previous efforts to expand the definition of development beyond the economic. This report, authored by Mahbub ul Haq, laid out the concept of human security. While the roots of human security theory run deeper into history than 1994, this report was the first from a large international body to introduce a paradigm that incorporated the intellectual branches of development and security together.

In opposition to dominant development and security theories, which historically focused on the State as the main actor, human security firmly placed people in the center, as the main actor of concern. In human Security, "Development must enable all individuals to enlarge their human capabilities to the fullest and to put those capabilities to the best use in all fields – economic, social, cultural and political" (UNDP 1994: 13). Human security connects both security and development through a multi-disciplinary understanding of both, going beyond economic measures of growth and development and beyond physical measures of security to envelope these two factors and incorporate other indicators to measure development.

The *Human Development Report* of 1994 drew a picture of a symbiotic relationship between security and development: "The peace agenda and the development agenda must finally be integrated. Without peace, there may be no development. But without development, peace is threatened" (UNDP 1994: iii). This coming together of two disparate fields of study involved a more holistic view of world interactions. Its main focus was to create "freedom from fear" and "freedom from want" for all people (UNDP 1994: 24). It was theorized that freedom from these two threats would address the problems of global insecurity and underdevelopment. Haq wrote,

> *Human security is not a concern with weapons. It is a concern with human dignity. In the last analysis, it is a child who did not die, a disease that did not spread, an ethnic tension that did not explode,*

a dissident who was not silenced, a human spirit that was not crushed. (Haq, M. u. 1995: 116)

The UNDP report outlines seven categories that are seen as vital to the establishment of human security, shown in Figure 6. These seven different categories address not only physical security but also different concepts of development including the traditional development goal of economic security. The categories include: economic security, food security, health security, environmental security, personal security, community security, and political security (UNDP 1994: 24-5).

Figure 6: The Seven Development Indicators of Mahbub ul Haq's Human Security Concept

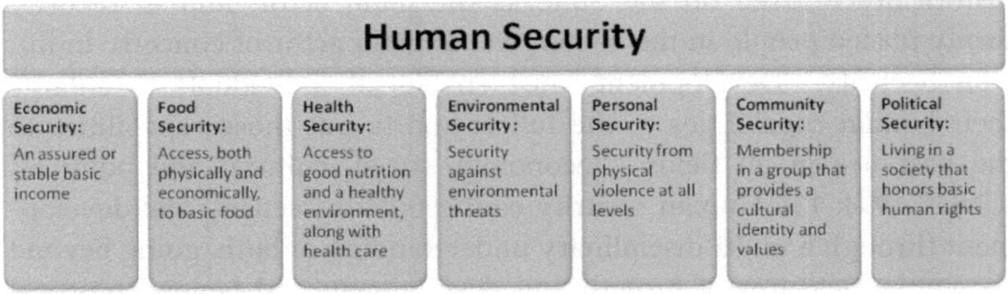

Source: own design based on the UNDP Human Development Report (1994: 25-33)

The report was used as evidence of the need to expand the basic understanding of global security to include threats to the seven named categories at the UN's World Summit on Social Development held in Copenhagen. Since then, human security has gained attention and adherents.

The governments of Canada, Japan, and Norway have adopted human security as their guiding theory, or lens, from which to view development efforts (Alkire 2003: 3; Kanda and Kuwajima 2006: 36).[47] They hope that by orienting their foreign aid policies towards the enhancement of human security that they will convince other countries of its importance and improve development efforts.

47 Canada has budget lines committed to human security efforts, Norway established a human security network that meets regularly to discuss human security threats, and Japan prioritizes human security in its diplomatic "Blue Book" (Alkire 2003: 21).

Even with some governments and the UN supporting it, human security has its critics. The main criticism put forward is that human security is not useful as a functioning theory because the many different threats are never prioritized or clearly defined in the literature (Alkire 2003: 22).

Human security is central to participatory development approaches. While not all participatory development proponents would agree, I believe participatory development approaches are aimed at addressing a number of the human security categories at once. The categories being addressed differ depending on the exact project being implemented, but the very heart of participatory development is that it is holistic in nature, intended to empower, and meant to provide greater human securities to the participants involved. Increased usage and support for participatory development approaches has gone hand in hand with an increased belief in the importance of human security.

1.3.4 Participatory Development: Development Centered Around the Agent

A greater focus on grassroots efforts and a shift in concern from the State to the people can be seen in the emergence of the buzzwords empowerment, participation, civil society, and social capital in the development dialogue in the 1990s. These concepts are embedded, to different degrees, in the ideas of basic needs, human development, and human security (Willis 2010: 102). They are also supported by an increased interest in local knowledge and its role in producing better development results:

> *In simple terms, it is suggested that as development plans are always translated into practice at the local level then a knowledge of local-level cultural patterns is seen as a logical precondition of the success of planning. (Preston 2002: 296)*

This shift towards an agent-centered approach was very dramatic, though not highly focused on, in the conceptualization of the roles of different actors in development. The agent-centered approach is, "grounded in the everyday life experiences and understandings of men and women, be they poor peasants, entrepreneurs, government bureaucrats, or researchers" (Long and Long 1992: 5). The knowledge that evolves from this perspective is an awareness that ideas, interests, and understanding of States and other bureaucratic entities do not always represent local interests and ideas (Preston 2002: 300). This was often emphasized by scholars who focused on gender issues, as they felt that females were not represented or included in most development projects.

Predominantly seen as passive recipients of development historically, "beneficiaries" were now seen as having a role to play in their own development. Local knowledge was given a platform, questioning the previously held belief that development experts had all of the answers. Participatory research techniques, such as participatory rural appraisal (PRA) and the participatory learning approach (PLA), are just two examples of methods that were introduced in the 1970s and were supported heavily in the literature in the 1980s (e.g., Chambers 1983). These research techniques were introduced in part because they were seen as being able to help in "uncovering the realities" of those living in poverty (Cleaver 2001: 38).

The slow build up of so many shifts in thinking appears to have inspired a new wave of support for participatory development approaches. The understanding of development as moving beyond the economic has created a more expansive understanding of development that requires different indicators of "success" and some believe it also requires a different conceptualization of project delivery approaches. While many in the field and in academia discuss participatory development as if it is a new idea, some, such as Korten see, instead, a recycling of past ideas. This returned focus to participatory approaches is seen by Korten (1980: 483) as a pendulum swing back to the approaches of community development and *animation rurale* of the 1950s and 1960s, with little to no critical thinking or analysis of the challenges noted from that time period.

Participatory development approaches such as community-based and community-driven development are theorized to provide many benefits that traditional development approaches had proven unable to produce, from creating more sustainable projects that are increasingly economical, to the more intangible objectives of empowering communities and creating greater social cohesion. By engaging the local population, participatory development projects are expected to increase effectiveness, have a stronger impact, and to directly serve the needs of the local population, instead of their externally perceived needs (Dongier, Van Domelen et al. 2003: 3; Mansuri and Rao 2004: 2).

Projects steeped in participatory approaches are expected to address some of the major perceived faults of top-down approaches (Chambers 1983). This more locally based concept of development, and belief in its multi-layered transformative capabilities, has made participation of the people a standard requirement of many development projects today. The idea being that people have a right to be involved in decision-making that impacts their lives, and that people can improve development efforts by sharing their local knowledge and perspectives (Desai 2002: 117). Additionally, peoples' participation is expected to lead to their "empowerment," a complicated term that will be discussed in more detail later, but something that was rarely, if ever, a consciously considered outcome of top-down development approaches. Khadija Haq (2002: 83) discusses empowerment as a new focus that arose in the second of two broad periods of development, dating from 1985 to the present.

Another characteristic of participatory development that sets it apart is its perceived flexibility. Participatory projects are supposed to be "process planning" projects where the design and implementation occur simultaneously, with adaptations and adjustments made in the field as different observations are made. It is based upon the idea that development actors do not know what will be successful prior to implementation because they do not know enough before they begin (Dale 2004: 44). This stands in stark contrast to projects stemming from classical development theories.

USAID, one of the largest providers of development funds worldwide, began to shift its management structure towards one that allows for participatory development projects in 1993 (La Voy, Charles et al. 1998: 1). The USAID Administrator at that time declared that participatory approaches were the means to achieving sustainable development (La Voy, Charles et al. 1998: 1). A new directive that outlined all of the procedures for the development organization was issued in 1995, which "clearly mandates participation" (La Voy, Charles et al. 1998: 6). USAID was not alone, as the inclusion of participation in development activities also took place in the Asian Development Bank (ADB), the UN, and the World Bank (Zakhilawal and Thomas 2008: 9).

In the late 1990s, the World Bank and the IMF presented the concept of Poverty Reduction Strategy Papers (PRSPs), which were grounded in the same approach as the structural adjustment program (SAP) from the 1980s and the neo-liberal agenda. The Poverty Reduction Strategy Papers were required to qualify for continued funding from both organizations. Though the PRSPs continued to include many of the same adjustment policies contained in the SAPs, they were more concerned with the internal national situation than the SAPs and focused more intensely on the poorest of the poor (Willis 2010: 54). Another difference was that the PRSP had to be written by the national government in consultation with civil society organizations and donors of funds (World Bank 2011e: n.p.). This shifted the knowledge base, making the process more participatory in nature than the SAPs, at least on paper. There are scholars, however, who question whether this required participation actually increased the diversity of inputs into the process, or if marginalized voices remained marginalized (Willis 2010: 54).

In alignment with an increase in attention to grassroots development practices and ideas, non-governmental organizations (NGOs) became increasingly important as players in the distribution of development aid in the 1990s. While prior development funding was mostly conferred through international organizations or via the local government, the amount of funding going through NGOs expanded greatly, with an estimated seven billion USD of aid flowing directly through NGOs in 2002, instead of going through national governments (UNDP 2002: 5). NGOs are seen by some as being more closely connected to local communities and being able to react more quickly to local events, problems, and demands than other development actors (Willis 2010: 99), though this is also disputed, including by President Karzai of Afghanistan who often accuses NGOs of being corrupt (Gizabi 2006). While some have put forward the idea that NGOs could be the missing link in development, otherwise known as the magic bullet syndrome (Vivian 2008: 167; Willis 2010: 98), others believe the praise of NGOs is overdone: "However, the claims regarding NGOs have been revealed as being overstated" (Willis 2010: 100).

This increased amount of funding for NGOs coincided with a large increase in the number of NGOs operating worldwide. According to the UN, in 1914 there were just over 1,000 international NGOs, but by 2000, this had expanded to 37,000 functioning international NGOs (UNDP 2002: 5). The vast majority of this growth, almost one fifth, occurred in the 1990s (UNDP 2002: 5).

Participatory development approaches have entered into the mainstream development discourse, with NGOs, unilateral and multilateral funders, and international organizations all supporting development programs and projects with a participatory angle. According to Desai (2002: 120), NGOs have played a major role in community participation. This has shifted the development environment, making room for unique cases and for case-specific answers instead of the "blueprint planning" of the top-down approaches supported by modernization theory.

1.3.5 Post-Development: Problematizing "Development"

Another concept that entered the development discourse in the 1990s was the emergence of the post-development movement. This school of thought sees development, as it has been practiced, as Eurocentric in orientation, relating to the work of Edward Said, especially his book *Orientalism* (1979). Theorists such as Arturo Escobar, Majid Rahmena, and Wolfgang Sachs see the development discourse as non-neutral, with the Eurocentric vision of development being presented as the only "correct" option (Willis 2010: 29). They generally argue that a lack of development stems from an outside perspective and that it was only through the adoption of this outside perspective that development became a "problem" (Willis 2010: 28-9).

> *Development' has helped incorporate large areas of the globe into a Northern-dominated economic and political system which has destroyed indigenous cultures, threatened the sustainability of natural environments and has created feelings of inferiority among people of the south. (Willis 2010: 28)*

For post-development thinkers, the whole concept of development needs to be "problematized and/or rejected" (Sidaway 2002: 17). "The idea of development stands like a ruin in the intellectual landscape. Delusions and disappointment, failures and crimes have been the steady companions of development and they tell a common story: it did not work," according to Sachs (1992: 1). Esteva theorizes that the establishment of the word underdeveloped in Truman's Four Points speech created a label filled with assumptions and judgments (1992: 7). Those who were underdeveloped or undeveloped were defined in opposition to those that were seen as developed, who were Western. Peet and Hartwick (1999: 15) see post-development as a complete rejection of modern development. Post-development encourages a return to the local, advocates for simplistic living in terms of materialism, and calls for a greater focus on spirituality (Peet and Hartwick 1999: 15).

Post-development also raises the question of what kind of development people want, allowing room for the local vision of development to differ from the "Western" view of development and its connection to modernization. Willis (2010: 207) notes, "[P]ost-development theorists have argued that 'development' should be focused on what local communities want and should not be a response to a Northern-imposed model of what is a correct form of development." This is highly criticized by those who say that post-development theorists do not pay enough attention to the many different forms of development that have taken place, that they are simply critical without offering alternatives, and that they do not recognize the positive changes development has brought about, focusing only on the negative (Sidaway 2002: 18; Willis 2010: 207).

While the post-development thinkers present their ideas as a break from all other ways of thinking about development, others argue that post-development thinking is an extension of previous criticisms of classical development theories (Desai and Potter 2002: 1; Sidaway 2002: 17). Post-development is also rejected by some as being irrelevant, seeing it as an intellectual fad or fashion that has little relevance to poor people throughout the world (Sidaway 2002: 18). Others criticize it for its romanticism of the local population and indigenous means of living (Corbridge 1998: 144; Kiely 1999: 31).

At its core, post-development thinking is a rejection of previous development theories and approaches, including modernization theory, dependency theory, structural adjustment programs, etc. It views these efforts and ways of thinking as confining. These other theories and approaches limit the possibilities, while a post-development viewpoint, by removing the Eurocentric and culturally imperialist framework, allows for a much larger array of development options (Kiely 1999: 34).

For some post-development thinkers, this would mean that participatory development approaches such as community-driven or community-based development are not acceptable, as they are still seen as being within the Eurocentric framework, with Eurocentric goals. Yet, despite the abject rejection of development in general, it is easy to see a common thread between post-development thinking and participatory development approaches: both support the idea of listening to the people who are directly concerned, the people in the community or the village, and allowing them to hold the decision-making power (Rist 2008: 259). This reverence for indigenous knowledge or local knowledge might be more rhetoric than reality in the practice of participatory development, but, theoretically, it is still one of the main tenets of the approach.

For less dogmatic people who support aspects of the critique of post-development thinking, it is possible that participatory approaches can be seen as a viable means of shifting development practices towards a more post-developmentalist perspective, without abandoning the concept of development fully. This allows people to believe in the critique, but take away from it what they believe provides a different option to help bring about better living conditions for people.

1.3.6 Extracting Meaning: Concluding Remarks on the History of Development Theory and Approaches

The first decade of the 21st Century has not abounded with many new theories or approaches, but has instead maintained the momentum of the 1990s, with shifts towards local actors and participation and an enlarged understanding of what contributes to development. While these themes are gaining in strength, there are still large, ongoing State focused projects cast in the mold of modernization theory. However, even some of these projects are taking into account the idea that local input is important to their success. While a shift towards multiple development objectives has played a role in increasing the focus on participatory approaches, it is really the shift away from the State as the main development actor of importance, towards the individual and the community that appears to have pushed development stakeholders to support participatory development efforts so enthusiastically. This has required the re-evaluation of individuals living in poverty as being simply victims, towards seeing them as agents of change with capacities and knowledge that can positively contribute to development planning, implementation, and sustainability. This shift has paved the way for a movement towards the grassroots, towards the participatory. Even though classical development theory based programming with its focus on infrastructure and large scale projects.

The new development environment is one in which classical development theories are questioned, though not abandoned. These types of projects have not brought about the shiny future they promised, whether measured in purely economic terms or in terms of basic needs, human development, or human security. Within thirteen years of intensive international development activities, between 1990 and 2003, 54 countries became poorer and 21 countries saw their HDI scores drop (Rist 2008: 235-6). This overall failure does not mean that all growth focused development efforts have had negative consequences, but it does mean that overall they are not producing the hoped for results in a satisfactory manner. This has made room for alternative approaches to development, such as participatory development. Participatory development is now being enthusiastically embraced within the academic discourse and in the field. One can judge its popularity through the publication of many books and articles on the topic and the increased amount of development funds being committed to participatory development projects and programs.

What remains to be seen is whether participatory development will face the same travails of previous development approaches. Will it too fail to bring about changes that have been deemed positive by development scholars and practitioners? Will the parameters of development shift? Will different indicators of development be cultivated that will change the understanding of development completely? These are questions that can only be answered in time, but it can be said that participatory development is an important approach that needs to be evaluated from many different angles and perspectives to verify its strengths and weaknesses, and determine if it is an improvement over previous theories and approaches. The literature reviewed below begins to address some of these questions, but thus far there are few clear answers.

1.4 Participatory Development's Rise in the Wake of Failed Top-Down Approaches

As can be seen from the historical analysis of development theories and approaches above, the allure of participation as a means of improving development has been a part of the development dialogue for some time, sometimes moving to the forefront and sometimes receding. From community development approaches in the 1950s and 1960s, to calls for self-help and participatory rural appraisal research approaches in the 1970s, to the current push for community-based and community-driven development programs (CBD and CDD respectively); participation has been a piece of the development puzzle. While the concept of inclusive development is not new, Grow and Vansant (1983: 428) point out that the increasing emphasis on it is, and that there is a new faith being placed in the approach by governments and development stakeholders. Oakley (1991: vii) calls it the "age of participation."

Participatory approaches have been gaining popularity in recent decades as the failures of classical theories have come into greater focus. These inclusive approaches are seen as addressing one of the major failings of classical, top-down approaches: the lack of consultation and inclusion of the people. There is the hope that by including people, development projects and programs will be more successful in bringing about better lives for individuals (Chambers 1983: 140; Mansuri and Rao 2004: 2). Bilateral and multilateral aid agencies, NGOs, grassroots organizations, development management consultants, and others have all joined the "chorus" of support for participatory approaches (Finsterbusch and Van Wicklin III 1989: 573). Most development focused authors note that participatory development has been gaining importance since the 1970s and 1980s (Cohen and Uphoff 1980: 213; Khwaja 2005: 267; Michener 1998: 2105; Operations Evaluation Department 2005: 1; Parnwell 2002: 112). Desai (2002: 117) calls community participation today, "an indispensable part of any programme or project," while Mansuri and Rao (2004: 2) call community-based and community-driven development two of the fastest growing approaches to development. As of 2005, an estimated 74% of all World Bank country assistance strategies included community-based and community-driven strategies as an important element (Operations Evaluation Department 2005: 11-2).

The approach has become so influential that Cooke and Kothari published a book entitled *Participation: The New Tyranny?* in which they call the spread of participation in development "inexorable," noting that it is supported by all of the major development institutions (Cooke and Kothari 2001a: 3). Gow and Vansant (1983: 427) believed, already in 1983, that a consensus had been reached that participation was necessary for rural development. This consensus has spread since the 1980s to all realms of development, not just rural development. "It is now almost reactionary seriously to propose a development strategy which is not participatory..." according to Oakley (1991: 6). All major bilateral development agencies were emphasizing participatory policies by the 1990s, with some even calling it the "new orthodoxy" (Henkel and Stirrat 2001: 168).

Francis Fukuyama (2005: 85) points out that community-driven development has been gaining support since 1995. He indicates that community-driven development provides greater ownership, though he is not naïve about the failures of this approach in the 1950s and 1960s (Fukuyama 2005: 86). In the current literature, he is one of the few to refer to these earlier participatory development efforts.

Voices calling for participatory approaches began to be heard in the 1980s. Starting in the 1990s, development programs that included a participatory aspect grew in number, in popularity, and in the amount of development funds and aid that was allocated to them. Chambers (2005: 101) emphasized that though participation was around in various forms prior, it was only in the 1990s that it "became a preoccupation on a global scale," and, ". . . is now part of almost every development activity."

Participatory development has been receiving increased attention in academic circles too, with the number of scholarly journal articles and books rising drastically between the early 1990s and 2010.[48] A peak

48 I conducted multiple researches on JStor and Google Scholar and found that while there were only around 100 articles published on participatory development in 1992, there were over 13,000 readily available by 2010. These searches were conducted in March of 2011.

in academic literature production was seen between 2004 and 2008 with more than 1,000 new articles published in each of those years.

Development funders have been investing in participatory development monetarily. The World Bank has proven to be very influential in supporting and funding participatory approaches. According to conservative estimates, the World Bank was already committing $325 million USD to participatory development initiatives by 1996 (Mansuri and Rao 2004: 2). This amount would approach $2 billion USD only seven years later in 2003, a significant increase (Mansuri and Rao 2004: 2), and increase to 7.8 billion seven years later (Mansuri and Rao 2013: 44).[49] The percentage of World Bank projects that included participatory approaches increased from 2% in 1989 to 25% by 2003 (Operations Evaluation Department 2005: ix). The World Bank, though one of the main supporters of participatory approaches, was not alone in its shift towards participatory approaches. USAID also increased its funding for such programs, as did DFID, the UN, and other bilateral and multilateral development organizations and agencies. This increase in attention, usage, and funding of development programs involving the participatory approach has led to its acceptance as a dominant force in the development discourse and in the field.

As other development approaches and theories have floundered and been called into question by an ever increasing number of academics, practitioners, and funders who have lost their faith in "development," the one light of hope appears to be helping people help themselves, which requires the participation of local people: "The only help which, examined critically, did not prove disreputable or counterproductive, and which seemed to point a way out of the dilemma, was *help for self-help*" (Gronemeyer 1992: 66).[50]

The question that arises is where does this new faith in the powers of participation stem from? Rahmena (1992: 121) sees the new advocates for participation as seeking to, "regenerate the development dis-

---

49 Mansuri and Rao note that there is no "authoritative estimate" of the World Bank's community-based and -driven development program portfolio, and say that there, "is considerable potential for measurement error in these numbers." They have, therefore, put forward what they call a conservative calculation. They also note that the 2003 lending is seven billion USD when, "lending for an enabling environment for such projects is included" (cf. Mansuri and Rao 2004: 2).

50 Italics included in the original.

course and its practices." Others believe that it can improve development efforts significantly (Cernea 1985; Chambers 1983; Gran 1983; Sillitoe 2000; Woolcock 2009). This dissertation contributes to the academic research that examines the participatory approach to development and attempts to critically understand the impacts such participatory programming has had in one community in Afghanistan.

1.5 Exploration of Participatory Development in the Literature

In an attempt to better understand participatory development, and capture the full range of discourse that addresses participatory development, I used a number of different research approaches. The most influential journal articles and books were noted based upon their repeated inclusion in the more relevant literature and the number of times they were cited within different articles, books, reports, guidelines, etc. Additionally, a thorough review of journal articles that had the key terms "participatory development," "community driven development," "community-based development," and "community development" was conducted using JStor and Google Scholar.

This resulted in a selection of over 150 directly relevant articles, books, reports, etc., shown in Figure 7. Each of these was reviewed and the main topics and issues addressed in each were documented in a database that was used to analyze the main trends and themes of participatory development literature. This process revealed an internal discourse of participatory development where much of the literature referenced was built upon the basis of a limited set of books and articles, such as the work of Chambers (1983, 1997), Cooke and Kothari (2001b), and others. The literature of participatory development is an ongoing conversation where authors cite one another, build upon each other's work, and often provide evidence of conflicts and contradictions of previous interpretations or implementation of participatory development projects. This makes for an interesting dialogue that is documented below.

Figure 7: Major Participatory Development Print Resources Reviewed (1975-2009)

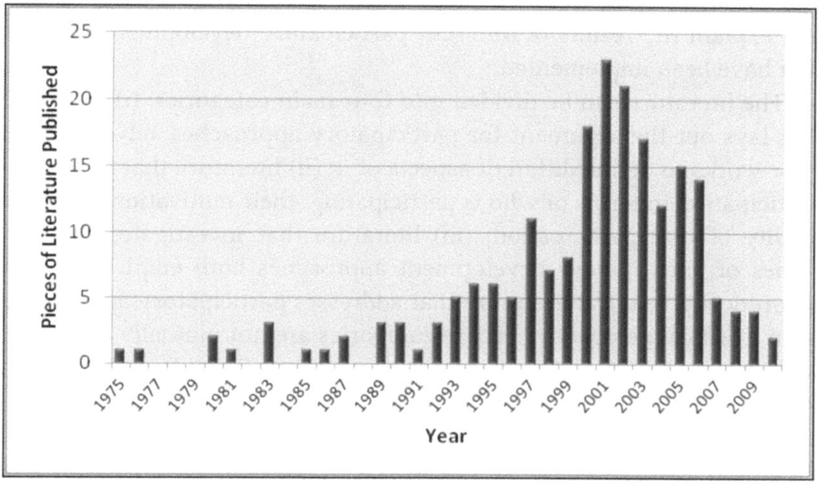

Source: own design based upon self-created database

The literature that addresses participatory development, while not vast, is substantial. Some authors, however, state that there has been little written about the topic.[51] This is supported by Hayward et al. (2004: 95-6) who said, "Participatory approaches to research and development have had relatively little academic or practitioner critique, resulting in a mythologizing of the power of participatory methodologies to accomplish problem solving, emancipation or empowerment." The vast majority of literature on the topic stems from the second wave of support for participatory approaches that started in the 1980s and continues to the present.[52] The literature does not come to many, if any, unified con-

51 Despite the existence of a good deal of literature on participatory development approaches, authors continue to state that there is a "dearth" of participatory development literature (Mansuri and Rao 2004: 1; Michener 1998: 2105; Owen 2006: 10).

52 Some authors, such as Howard White believe that there have been three different waves, with the second occurring in the 1970s and 1980s with support from the UN, but as is noted by Mansuri and Rao (2004: 10), this second wave never really caught on with funders or in practical work in the field. Therefore, this dissertation focuses on only the two main waves of participatory development, the one in the 1950s and 1960s and the current one.

clusions. It does, however, show that both the concept of participatory development and its actual implementation are each receiving increasingly critical analysis, as scholars and practitioners begin to analyze and explain the results of different participatory development projects that have been implemented.

The literature can be divided into four main categories: (i) literature that lays out the argument for participatory approaches, advocates for it, or works to define different aspects of it; (ii) literature that looks into participation in terms of who is participating, their motivations, and the quality of that participation; (iii) literature that investigates the outcomes of participatory development approaches both empirically and theoretically; and (iv) literature that addresses participatory approaches from a critical perspective. These categories are not mutually exclusive, with some overlap and interplay between them. They have been used to help frame this literature review.

1.5.1 Definitions of Participatory Development

One of the leading proponents of the second wave of participatory development has been Robert Chambers. Chambers' (1983) book entitled *Rural Development: Putting the First Last* has had a lasting impact on the development discourse and is often referred to in participatory development literature.[53] Chambers (1983: 2) starts by calling the extremes of rural poverty in the Third World an "outrage." He argues that addressing this outrage requires a set of reversals: in ways of thinking, in learning, in attitudes, and in perceptions (Chambers 1983: viii-x). The most important of these reversals is "putting the last first." For Chambers (1983: 75) this means recognizing the inherent worth of rural peoples' knowledge, which he sees as being often superior to outside experts. He focuses on the need to better understand rural poverty and the fact that this has been made difficult due to the perceived superiority of educated outsiders and the six different biases of research that prevent closer contact with the rural poor (Chambers 1983: 1-27, 76).[54]

53 Chambers' (1983) book was based upon an article he published in 1981 entitled "Rural Poverty Unperceived: Problems and Remedies" (Chambers 1981).

54 The six original biases include: spatial, project, person, dry-season (seasonal), politeness or timidity (diplomatic), and professional bias (Chambers 1981: 5-11; 1983: 1-27). The research underlying this dissertation attempted to limit these

He advocates for greater integration of the rural poor into development practices to utilize the knowledge of each of the three cultures he identifies – academic, practitioner, and rural poor – to reduce rural poverty. Chambers (1994, 1997, 2002, 2005) continues to produce literature relating to participatory approaches to research and development, supporting it and trying to improve participatory practices.

Chambers' early work was supported by Michael Cernea who also believed in "putting people first" (1985). Guy Gran (1983) also advocated for the use of participatory approaches, critiquing dominant development paradigms of the time, including neo-liberalism, and presenting participatory development as an alternative strategy for human development. Participatory development was seen as a way to reverse power dynamics by giving a "voice" to the poor and by including them in the development process (Mansuri and Rao 2007: 1). This results in taking away the vision of the poor as "victims." The World Bank also advocates the importance of listening to local voices, albeit from a more consumerist perspective: "As consumers, community members are the most legitimate, informed, and reliable source of information about their own priorities" (Dongier, Van Domelen et al. 2003: 305).

Paul Sillitoe (2000: 4) advocates for participatory development in his article, highlighting the wealth and importance of indigenous knowledge that had previously been pushed aside as inefficient or outmoded. He also focuses on how local knowledge can increase efficiency and the idea that local people are very capable of defining their own problems (Sillitoe 2000: 4). Michael Woolcock (2009: 8), a well-established development scholar, puts local knowledge and decision-making at the center of his vision of the future of development, advocating for its adoption on an even larger scale than seen today.

The World Bank started to commit to participatory development approaches and advocate for them in the 1990s, which was a big conceptual shift from its more standard practice of utilizing an "external expert stance" (Bhuvana, Kearns et al. 1996: 3). The World Bank highlighted the many perceived benefits that involving people in their own development could produce. Former World Bank President James D. Wolfensohn, in his preface to the *World Bank Participatory Hand-*

biases as much as possible as discussed previously in the section on Researcher Biases.

book of 1996, said that participation could make the World Bank's development efforts "more effective and sustainable" and that participation would help create a sense of ownership amongst the people and build their confidence in the value of the World Bank's programs (Bhuvana, Kearns et al. 1996: ix). These thoughts were echoed in the *Poverty Reduction Strategy Sourcebook* published in 2003 by the World Bank: "Experience demonstrates that by directly relying on poor people to drive development activities, CDD [community-driven development] has the potential to make poverty reduction efforts more responsive to demands, more inclusive, more sustainable, and more cost-effective than traditional centrally led programs" (Dongier, Van Domelen et al. 2003: 303).

This literature asks development planners to recognize the worth and importance of the people they are working with and tries to de-emphasize the dichotomous paradigm of "us vs. them" that exists in other development theories and approaches. It puts forward the perceived benefits of including local people in the development process and advocates that top-down approaches be overtaken or replaced by bottom-up approaches that are participatory by nature.

Following the re-emergence of participatory approaches into the development discourse, a selection of the literature produced looked at how to implement participatory approaches such as community-based development (CBD), community-driven development (CDD), participatory rural appraisal (PRA), and more. This literature stems both from a theoretical and practical perspective. The theoretical literature was produced in the earlier years, and focuses on different tools, such as PRA, and techniques to create more participation (Chambers 1994, 1997, 2002; Freire 1996). Guidelines for participatory development began to emerge in the 1980s (Oakley 1991: 174).

The World Bank has been a strong contributor to the production of papers and reports that provide guidance to participatory development practitioners (Bhuvana, Kearns et al. 1996; Dongier, Van Domelen et al. 2003; Doniger, Van Domelen et al. 2001; Khwaja 2005 ; Narayan 2005; Operations Evaluation Department 2005). This literature has looked at the bureaucratic/institutional level and the community level. For example, the *World Bank Participation Sourcebook* provides a chapter to guide practitioners through the different steps of using a participatory approach and provides a menu of options for each of the different stages (Bhuvana, Kearns et al. 1996: 121-144). This guide also put forward different methods that can be used to encourage participation, along with different tools (Bhuvana, Kearns et al. 1996: 181-204).

Guidelines were also produced outside the World Bank setting. *Participation Works!*, which was published by the New Economics Foundation in 1998, provides insights into how to conduct participatory development (Lewis and Walker 1998). Guijt and Veldhuizen (1998) provided a list of participatory methodologies, Keough (1998) provided advice on how to conduct participatory development in an effective way, and Mikkelsen (2005) covered participatory development in her guide for practitioners. CARE International, an NGO, has also published guidelines (Shah, M. K., Kambou et al. 1999). It is important, however, to note that there is not a single, clear vision of participatory development, how it should be defined, and how it should be practiced.

Some of the literature focuses on specific aspects of the participatory development process, such as Dongier et al. (2003), which explores how to include the government in participatory approaches. Others, such as Platteau and Gaspart (2003) focus on how to prevent elite capture, providing their own framework of analysis. Mansuri and Rao (2007: 51-3) provide advice on scaling up.

The more recent guidelines extract lessons learned from case study research of projects, programs, and personal experiences in the field. This literature often tries to understand how to extract the "success" of a specific participatory program and make it applicable to all locations. Botes and van Rensburg (2000) did this and provided nine common pitfalls and twelve musts for participatory development, providing more guidance to practitioners. These more empirical guidelines tend to be a bit less idealistic about the impact of participation and more realistic about the challenges that are faced in the field.

Despite the production of different guidelines, handbooks, etc. it is apparent that a few terms utilized in the participatory approach have not been clearly or critically defined in the literature (Mansuri and Rao 2007: 10). Following the revival of participatory approaches and their perceived benefits, further clarification of the terms utilized was needed and a string of literature was published that concentrated on doing this. The two terms most commonly used included participation and community, and this literature looked to define these terms and break them down. Despite these efforts and clarifications, no universal definition for participation or community has emerged in development studies (Mohan 2001b: 2), which continues to increase the challenge of understanding participatory development in theory and in practice today. As Desai (2002: 120) noted, "Participation is often treated as if it could be implemented in a vacuum, with all actors agreeing on its precise meaning and goals."

As part of my effort to evaluate a participatory development project, this dissertation takes a closer look at some of the different ways in which participation has been defined or categorized. Most authors break the term participation into multiple components, as did Cohen and Uphoff (1980: 218-9), who put forward three *dimensions* of participation: the kind of participation, who is participating, and how the participation happens. They continue on to list four *kinds* of participation: participation in decision-making, participation in implementation, participation in benefits, and participation in evaluation (Cohen and Uphoff 1980: 220-221).[55] Who is participating refers to who is included and who feels able to actively present their thoughts and opinions. How participation is happening refers to such issues as whether the participation is coerced or voluntary, amongst other issues. Oakley (1991: 8-9) builds upon their work and labels three broad interpretations of participation: contribution, empowerment, and organization. The World Bank identifies four separate means of participation: sharing information, consulting, collaborating, and empowering (Operations Evaluation Department 2005: 1). This dissection of participation has increased people's awareness that not all participation is equal and not all participation is empowering, as is sometimes assumed.

The World Bank defines participation as, "a process through which stakeholder's influence and share control over development initiatives and the decisions and resources which affect them" (Bhuvana, Kearns et al. 1996: xi). The UNDP defines participation as a development strat-

55 Cohen and Uphoff (1980: 220-1) further detail three different types of decisions: initial decisions (planning), ongoing decisions (during implementation), and operational decisions. Implementation is broken down into resource contributions, administration and coordination efforts, and program enlistment activities. Benefits are broken down to include material benefits, social benefits, and personal benefits.

egy in which, "people are closely involved in the economic, social, cultural and political processes that affect their lives," expressed in its Human Development Report of 1993 that focused specifically on participation (UNDP 1993: 21).

Another significant debate has existed around whether participation should be seen as a means or an end. As a means, it is seen as a way to achieve more efficient projects or better living standards, and as an end, it is seen as important in itself for equity and empowerment (Cleaver 2001: 37; Guijt and Shah 1998b: 9). Mikkelsen and Boyce both discuss participation as a means, a tool that is necessary to achieve the desired outcomes and results (Hayward, Simpson et al. 2004: 101). Some authors avoid the debate between the two and view participation as both a means and an end (Cohen and Uphoff 1980: 220, 227; Hayward, Simpson et al. 2004: 103; Nelson and Wright 1995: 1). This understanding of it being simultaneously a means and an end colored the research conducted for this dissertation. For those who view participation as a means, it results in projects where the goal for the project is the creation of infrastructure with less concentration on the participation of the people. This can lead to greater elite capture and other concerns about the legitimacy of participation and people's access to participating, as will be discussed below. When participation is treated as an end, it can create a completely different type of program, where a great deal of attention is paid to who is participating, how they are participating, etc. and the physical outputs of the development project become secondary in focus.

Another term that needs clarification is "community." Is a community defined by geographic borders? Is it defined by administrative borders? Can it be defined as a social group? Is it a group of individuals and/or households that join together for common interests in specific settings or contexts? There are many different working definitions that exist. The World Bank defines a community as a group of people who share development goals and form solidarity based upon social norms (Operations Evaluation Department 2005: 55). Those who do not belong are excluded. Most participatory development literature uses community to, "denote a culturally and politically homogeneous social system," that is cohesive (Mansuri and Rao 2007: 13).

This vision of community that is homogeneous in nature has been called into question as the realities of actual communities emerge. Pain and Francis (2003: 51) discuss the fact that many groups involved in participatory activities are not always "bound by geographical area or a shared sense of identity" and that within one "community" there can be many divisions and multiple "senses of community." This is also supported by Cleaver (2001: 44), Mosse (2001: 32), Guijt and Shah (1998b: 7-9), and Strand (2003: 1) who warn against assuming that communities are homogeneous and exist without differences. Communities often are heterogeneous in nature and are characterized by power differentials that can interfere with the idealized implementation of participatory development. This perspective was integrated into the research methodology for this dissertation, and diligent attempts were made to gather information from as many different people of the community of Shah Raheem as was possible, to explain differences of experience and perspective.

Community is important to participatory approaches as it is a defining factor for most participatory projects. Community-based and community-driven development projects are two types of participatory development. While the two terms were not clearly differentiated originally, recent reports have defined community-based development as a less intensive form of participatory development in which the development project is done in collaboration with a community, its people are consulted, and information is shared between the community and the external organizers of the intervention (Operations Evaluation Department 2005: 2). While this is more participatory than purely top-down approaches where communities are not consulted, it is not as participatory as community-driven development. Community-driven development refers to development projects or programs in which the control of the project resources, its design, and its implementation are in the hands of the community members (Operations Evaluation Department 2005: 116).

Despite the lack of a unified understanding, the definitional literature has been impactful, as it has made practitioners and other stakeholders question their own conceptualization of specific words, and has helped bring forward a greater understanding of the many different ways that participation can occur and the many different ways community can be defined.

1.5.2 The Process of Participation

As the first pilot programs that utilized participatory approaches in the second wave began to mature in the 1990s and 2000s, the literature started to look at the empirical results of implementation. Areas of success and areas of concern were simultaneously highlighted. Criticism and evidence of problems have gained attention in the last decade, becoming more prominent than literature uncritically supportive of participatory approaches, such as that from the 1980s. Korten (1980: 480) describes the 1980s as a time of "wishful thinking," in which many individuals and international organizations were writing or talking about implementing participatory development projects, but few were actively implementing projects in a truly participatory manner.

One of the first areas to be looked at was who is participating in participatory development projects. While the basic premise of participatory development is that everyone is able to participate or be represented, this underlying belief has been called into question based upon empirical research and experiences in the field. Individuals and teams looked at different development programs and evaluated who was and who was not actually participating, with a specific focus on women and other marginalized groups.

Some participatory development researchers have found that marginalized people and groups were able to participate in participatory programs. Timsina (2003: 239-40) found that increased numbers of women were participating in a community-based forestry project in Nepal, though he does raise concerns about the quality of that participation. Others have also found positive empirical evidence that participation of marginalized people increased due to participatory approaches (e.g., Besley, Pande et al. 2005: 656; Haq, K. 2002: 5).

However, literature based on empirical evidence has also shown that even in participatory programs, marginalized people in a community can still be excluded, be it physically or in terms of being able to actively participate in decision-making. Gender focused authors have looked at the exclusion of women in participatory projects (Agarwal

2001; Guijt and Shah 1998a; Hildyard, Hegde et al. 2001). An investigation into gender and participation in a water resource project in Thailand showed that gender and power differences were resulting in women being excluded from the project and from its benefits (Resurreccion, Real et al. 2004: 531). Fonchingong (2006: 144-146) also found that women were excluded in Cameroon grassland projects, and identified a number of barriers to participation that included a lack of literacy, culture, and infrastructure. Narayanan (2003: 2485) found that social norms could prevent marginalized individuals, especially women, from participating in projects.

Barriers to participation became an increased focus in the literature in the beginning of the 21st century. Fonchingong (2006: 144-5) sees low social status, workloads, and schedules as contributing to women being unable to participate in development projects, along with financial difficulties, poor means of transportation, low education levels, vulnerable health situations, and corruption. Additionally, Mansuri and Rao (2004: 11) discuss how the time necessary to participate in a development project has a cost for the poor that should be calculated into program planning.

Additionally, most participatory development literature acts as though the desire to participate is natural and is desired by those who live in poverty. It sees participation as a means for people to voice their desires and obtain empowerment (Narayan, Chambers et al. 2000). Less idealistic authors explore what might be considered less noble reasons for people to participate or to not participate. The OED (2005: 38) reports that some villagers see participation only as a requirement for a project, but do not connect it to a sense of empowerment. Mercer (2002) sees participation as motivated, at least in part, by the desire to raise one's own social and financial status, as evidenced by her research in Tanzania, something that was also explored by Platteau and Gaspart (2003). Rahmena (1992: 116) highlights, that often times, participation is not voluntary, that people are pushed, coerced, or asked to participate. Freire (1996: 29) also talks about people's reluctance to participate. Oakley (1991: 237) believes that one must carefully plan for participation, and not just expect it to happen naturally.

Hayward et al. (2004: 100) explore the idea that there are legitimate and valid reasons for community members not to participate, and how this does not automatically mean that these people are excluded. They do not see participation as always positive and believe that it does not always lead to empowerment (Hayward, Simpson et al. 2004: 96). This is supported by Cleaver (2001: 51) who says that non-participation can be a "rational" strategy, highlighting that participation can have benefits, but it can also have costs. This can be particularly true in an environment like Afghanistan where sometimes associating with outsiders can put one's life or family in danger.

Mancur Olson's *The Logic of Collective Action* of 1965 and Hardin's "The Tragedy of the Commons" of 1968 both put forward the notion that self-interest can result in free riding in cooperative situations, which can be seen as a rational choice (Uphoff 1992: 326-356). This means that some people will not contribute or participate under the assumption that the other community members will, and therefore they can still receive the same benefits without sacrificing any time or energy. This is a concern that has been raised with participatory development approaches in the past (Mansuri and Rao 2004: 6) based upon the concept of the "tragedy of the commons" outlined in Hardin (1968: 1244).[56] Uphoff (1992: 326-356) argues that despite this threat, not everyone is motivated by their own interests and that free riding can be avoided if organizations are small enough to deter such behavior, or if the benefits can be in some way limited only to those who participate.

The validity of the participation has also been questioned, asking *how* participation is taking place. Authors who focus on this do not believe that simply understanding who attends meetings or physically participates provides enough information on actual participation. They focus on questions such as whether participation is real or symbolic, if it is voluntary or coerced, if it is direct or indirect, etc. (Cohen and Uphoff 1980: 223-4). Another way of putting this, is asking if participation is real or if it is just another development fad or "mantra" or "tokenism" (Hailey 2001: 97). Krefetz and Goodman (1973: 373) believe there is a separation between "genuine participation," in which participants have an influence on decision making, and symbolic "participation." In other words, what is the quality of the participation in such projects?

56 Hardin (1968: 1244) attributes the concept to, "a mathematical amateur named William Forester Lloyd and a pamphlet he wrote in 1833."

The possibility, some would say reality, of symbolic or tokenistic participation has gained more attention in the last ten years, in alignment with the recognition that simply being in attendance is not equivalent to meaningful participation. Traditionally, participation has been measured by attendance, but Guijt and Shah (1998b: 3) point out that though women attend community gatherings, this does not mean that their concerns are being voiced or addressed. While knowing the number of attendees is informative, it does not tell us about who felt they had the right to contribute to the meeting and whose voices were really being heard. Oftentimes attendance is high, but active participation is low, with only a few people making the decisions. Kapoor (2002: 106) points out that not all participation is equal: "What ensures that PRA discussions and interactions are coercion free? What prevents women and disadvantaged people, whom Chambers takes pains to bring into the PRA meeting, from feeling intimidated by authority figures also present?"

Symbolic participation is sometimes viewed as stemming from donor driven timelines and expected outcomes, with the timeframes being too short to allow for real participation. It is also seen as sometimes stemming from specific power relations within a community in which certain individuals might be in attendance but do not have a say in what is transpiring, as was found to be the case in Botswana (Mompati and Prinsen 2000: 629). Pain and Francis (2003: 51) explore the idea of participation "only in name" and how this can be influenced by internal hierarchies, which is also discussed by Mansuri and Rao (2004: 11). Sillitoe (2000: 7) presents anecdotal evidence of how power hierarchies in communities can be strengthened, instead of broken down, through the process of implementing a participatory development program, as he witnessed in Bangladesh where marginalized voices gained no increased notice or power.

Some authors focus on the process level, investigating how power differentials within a community impact on project selection or how it impacts the selection of community members to spearhead the project. Platteau and Gaspart (2003: 1689-90) show how power structures can determine who is selected to represent a community and how this is often done to support already existing power structures, even if the people know that the individuals selected will extract more than their fair share of the benefits from the program. The OED (2005: 24) put forward concerns that go beyond the local power structures and focus on political interference with community selection from the local government or central government. One can also extrapolate that internal power structures can prevent the young from voicing their opinions in societies where age is highly valued or for women in a patriarchal society, such as the one in Afghanistan. Desai (2002: 119) notes that the only people who are active in participatory development, "are perhaps the richest amongst the poor," bringing economic wealth into the discussion also.

Power structures outside of a community can also impact on real participation in a project. Mosse's work (2001: 19) provides evidence that "local knowledge" is often skewed to local understandings of project possibilities, and not necessarily of actual needs. In this situation the participation is again symbolic in that the actual voices of the community are still not being heard.

As you can see, the concept of participation is complex and needs to be evaluated on multiple levels to really understand what kind of participation is taking place. Taking these different perspectives into consideration, this research provides real insights into physical participation, tokenism, and active decision making in an Afghan Community. While the number of participants in meetings is noted, this research dug deeper into the situation and tried to determine if there was any connection between attendance and participation, and participation and empowerment.

1.5.3 Connecting Participatory Approaches and Effectiveness

After participation, the most written about aspect of participatory approaches is the effectiveness of these types of projects. This can be broken down into a number of categories, including effectiveness in producing the intended project, efficiency or cost-effectiveness, whether it was empowering and built social capital, whether the project was maintained or was sustainable, and if the project contributed to conflict management. Additionally, the literature looks at how the effectiveness of participatory development programs is affected by the scale of the project and attempts to discern how these normally smaller scale projects can be successfully scaled up.

Isham et al. focused their analysis of 121 different participatory water projects on understanding whether participation resulted in projects that were more effective generally, finding that it did (Isham, Narayan et al. 1995: 175). Finsterbusch and Van Wicklin (1989: 576), found that participation increased effectiveness in their meta analysis of 52 USAID projects, albeit modestly. Other authors also found that participation increased the effectiveness of development projects (Akhtar Hameed Khan 1996: 36; Jimenez and Sawada 1999: 415; Khwaja 2004: 428). Isham et al. (2002: 667) found water projects in India and Nepal were more successful when they included participation. An evaluation of the World Bank's community-based and community-driven development activities by the Operations Evaluation Department (2005: 11) looked at World Bank programs in six different countries, and found that projects that utilized a participatory approach had more success, measured quantitatively, based upon planned goals.[57]

Participatory development projects are seen as being more efficient, with projects built for less per capita than top-down approaches that do not involve the people (Dongier, Van Domelen et al. 2003: 306; Mansuri and Rao 2004: 2). Dongier et al. (2003: 306) note that Adato and others found in 1999 that projects implemented by local organizations had a lower cost per "beneficiary" than did projects where the local people were not making the decision. The OED (2005: 27) found that efficiency was increased as costs to governments for infrastructure projects had decreased, though it also notes that participatory projects are more expensive to prepare and supervise than traditional top-down approaches.

Development researchers have also focused on alternative forms of effectiveness, such as whether improved beneficiary targeting was seen in participatory development programs. A problem with top-down approaches was that the benefits were not reaching the people most in need, as exemplified by the push for the basic needs approach discussed previously. A number of authors have alluded to the belief that participatory development does in fact improve targeting (Besley, Pande et al. 2005: 648; Mansuri and Rao 2004: 3; Oakley 1991: 18) and some research

57 The countries of the study included: Benin, Brazil, Egypt, India, Nepal, and Vietnam.

has shown empirical evidence of this (e.g., Besley, Pande et al. 2005: 654-655). Research into whether participatory development successfully helps reach the poor, as the main targets of the approach, began to appear in the late 1980s and has continued since then (Operations Evaluation Department 2005: xi).

Morris et al. looked at an antipoverty program in Honduras and found that the program was not reaching those who were most in need, but instead was benefiting some of the better off, as noted in Hoddinott et al. (2000: 18). Mansuri and Rao (2004: 25) also found that using participatory approaches did not necessarily result in better community targeting or the targeting of individuals. Therefore, there appears to be conflicting findings on whether participatory approaches result in better "beneficiary" targeting and what conditions might be necessary for a participatory development approach to result in better targeting.

Project targeting, also known as preference targeting (Mansuri and Rao 2004: 23), was also discussed in the literature. Differentiating this from beneficiary targeting, project targeting focuses on the appropriateness and usefulness of the project to the people who it is intended to benefit. In one survey, the vast majority of respondents indicated that the development projects that had been implemented in their area were not the community's priority, but at the same time 80% of the survey participants were happy with the project in the end (Mansuri and Rao 2004: 24; Rao and Ibáñez 2005: 788). This misalignment of the projects implemented from those the community would have prioritized could have been due to some of the constraints on participation mentioned above, resulting in the participants' voices and preferences not truly being taken into consideration. Another set of researchers working in China found that the participatory aspects of the program they were looking at also did not improve program targeting, as noted in Mansuri and Rao (2007: 6-7).

Researchers concerned with beneficiary targeting also explored the idea that participatory projects were open to elite capture of the benefits by local power holders. For example, in some villages of Afghanistan, the local leaders have extracted more water from irrigation canals built through participatory development projects than they were allowed (KIIs 2010). Agarwal (2001) looked at this issue in forestry projects in India and Nepal. Other development researchers focused on participatory development also explored elite capture and its connection to participatory development approaches (Katz and Sara 1997; Rao and Ibáñez 2005). Robert Chambers (2005: 94) recognized this as a problem, and said, "The tendency for local elites to capture projects and programmes and use them for their own benefit should indeed be recognized as a fact of life." He then went on to discuss how this was not always a bad thing, and that sometimes local leaders were better able to manage projects and that this sometimes gave them more incentive to spread the benefits further, including to the poorest of the poor (Chambers 2005: 94).

Some participatory development studies support this conclusion, though not all. While the elite, or those who traditionally held power in a community, had more control over a project, this did not always mean that there was elite "capture" (Rao and Ibáñez 2005: 809). This was supported by work done on community-driven development in Indonesia by Dasgupta and Beard (2007: 229) who found that groups in which there was a smaller power differential amongst the community members tended to benefit the poor less than those with elites, in their analysis of urban community-driven development in Indonesia.

In the 1990s, literature that focused on effectiveness of participatory development approaches bringing about the empowerment of the people began to gain in number, accelerating in the 2000s. The premise being that participation should increase empowerment by giving a voice to the people. Gloekler and Seeley (2006: 142), point out that some Women's Organization members in Pakistan gained self-confidence through their participation in a participatory development program, and that this led them to become local politicians. Whole books have been devoted to this topic (Craig and Mayo 1995; Narayan 2005). Echavez (2010: 5) says:

> *However, it is often assumed that the best way to achieve the empowerment of marginalized groups is through collective organizing and group work, or building 'power with.' Consequently, empowerment has become very closely linked with strategies of participation.*

However, fieldwork oriented participatory development researchers are skeptical because lived realities are not aligning with the theory that empowerment and participation are so strongly connected. Dill (2009: 736) looks at community-based programs in Dar es Salaam, Tanzania and finds that despite their best intentions, some were disempowering and produced greater exclusion, especially for the less educated and poor.

Mansuri and Rao (2004: 57) believe that it is important, "to realize that community-based development is not necessarily 'empowering' in practice." Narayanan (2003: 2486) says that, ". . . instead of providing a challenge to the existing power structure, the restricted participation actually paves the way for a legal and suitable medium for enhanced domination of the existing elites." A point that also is supported by Hildyard et al. (2001: 70). Hayward and Simpson et al. write that participatory approaches can act to reinforce and reproduce existing socio-political structures instead of changing them (Hayward, Simpson et al. 2004: 104).

Henkel and Stirrat (2001: 182) question what the people are being empowered for. They suggest that the people are being empowered to be part of a "modern" project and that this kind of empowerment aligns with what Foucault labels as "subjection" (Henkel and Stirrat 2001: 182). This argument leans towards post-development thinking. Rahmena sees the idea of helping others to help themselves as disempowering. He points out that, " When A considers it essential for B to be empowered, A assumes not only that B has no power– or does not have the right kind of power– but also that A has the secret formula or a power to which B has not be initiated" (Rahmena 1992: 123). He concludes by saying that many participatory development advocates and supporters actually contribute to devaluing traditional forms of power, and therefore disempowering the people (Rahmena 1992: 123). Gugerty and Kremer (2008: 4) argue that the outside funding of many participatory development projects can be disempowering, weakening the role of the "disadvantaged."

Social capital, entered into the popular discourse in the 1990s with the publication of Robert Putnam's book *Bowling Alone: The Collapse and Revival of American Community* (Putnam 2000), though its roots stem back further to other authors such as Pierre Bourdieu (1984). It was quickly discussed in connection with participatory development, with some authors putting forward the idea that participatory development approaches would increase social capital, therefore identifying another effectiveness of the approach. Social capital broadly refers to social connections within networks and between networks that can provide an individual or a community with rewards.

Though the rhetoric is strong in connecting social capital to participatory approaches, the empirical evidence in the literature is weak. The World Bank supports the idea that participatory development builds social capital in an area, and puts forward evidence that increased social capital has a positive impact on household welfare (Dongier, Van Domelen et al. 2003: 308), as did Rao and Ibáñez (2005: 792) in their research on social investment funds in Jamaica. The OED (2005: 18), however, found that the impact on social capital was ambiguous, with some participation programs having a positive impact on social capital according to participants, and in other areas there being no recorded impact. Wood et al. (2006: 28) warn that participatory approaches can antagonize intra group dynamics and inflame pre-existing divisions, resulting in a breakdown of social capital.

Some authors also focus on the importance of understanding the social capital and power relations within a community prior to the planning and implementation of any development project. This connects to the literature focused on the heterogeneity within communities discussed previously. Chase et al. (2005: n.p.) found that the relationship between participation and social capital was often dependent on the preexisting social structures and relationships in a community, with high levels of social capital contributing to greater participation and better outcomes. Mosse (1997: 498-499), however, found that good social capital did not always result in a good project. Social capital and empowerment were both seen as hindered by patron-client systems within communities by the World Bank (Operations Evaluation Department 2005: 20).

An increasing amount of literature has emerged in the last decade that focuses on how participatory development can play an effective role in conflict management. It is assumed that areas in conflict or post-conflict have lost their social capital or their sense of community (Owen 2006: 8; Strand, Toje et al. 2003: 9). Therefore, the World Bank, along with others, have hypothesized that community-driven development and other participatory approaches can help to increase social cohesion (Owen 2006: 9; Strand, Toje et al. 2003: 12). The World Bank found that participatory development programs performed better in conflict countries as compared to non-conflict countries (Operations Evaluation Department 2005: xiii).

Gibson and Woolcock (2005: 9) wrote about how the social environment created through participatory approaches can create spaces for conflict management and how the program can help create "the capacity to engage." Strand et al. (2003: 6) address the topic through a desk study of 14 different community-driven development projects in areas of conflict and found that, "While the establishment of participatory community processes may constitute a useful framework for negotiations and dispute mediation, and even contribute to build trust locally, community-level reconciliation need to be linked to comparable processes on a national level." They also provide a section on lessons learned regarding community-driven development in conflict areas (Strand, Toje et al. 2003: Section 3).

Another measurement of effectiveness is the sustainability of a project or its impacts. This must be differentiated from environmental sustainability, which has also arisen as a factor in the general development discourse in the last two decades. Participatory approaches and their impact on maintenance have been touched on by a few authors. Some found that community led management was associated with better maintenance and sustainability of participatory projects in various countries (e.g., Isham and Kähkönen 2002:684-685; Katz and Sara 1997: 2), though they did not establish causality. Khwaja's research (2004: 427) into the Aga Khan Rural Support Programme (AKRSP) in Pakistan found that non-technical projects that were community-based were more sustainable and better maintained than similar projects without participatory aspects. Nel et al. (2003: 11) also looked at sustainability in their research on community-based development in South Africa and found that the people there sustained and maintained their cooperative, even though it required finding outside assistance to do so.

A number of authors, using empirical evidence from different programs, found this positive connection between participation and maintenance and sustainability to not exist. Kleemeier (2000: 941-942) looked at a Malawi water project and found that there was weak sustainability due to a lack of external support, especially for larger projects. Nkansa and Chapman (2006: 509) found that sustainability was dependent upon leadership and social cohesion in their study of six different communities that had participated in a community education program in Ghana. The OED (2005: 37) found that community-based and community-driven development projects were less likely to be sustainable as compared to traditional top-down projects. Clever (1999: 604) found that communities sometimes could not sustain development projects because they lacked the material resources to do so after the implementing partner had left.

World Bank Staff told the OED (2005: 19) that community groups would need six to ten years of support in order to become sustainable. The belief in the need for longer time frames for participatory approaches to be really effective is something that is often discussed amongst practitioners in the field.

After participatory development became more mainstream, with both the UN and World Bank placing a strong emphasis on its usage, along with government aid organizations, a new focus in the literature emerged. The question of how such programming could be scaled up from its original micro-scale to a macro-scale began to be discussed and debated. Advice and analysis regarding the best ways and means to scale up participatory development projects began to appear towards the end of the 1990s.

Two examples of successful attempts to scale up participatory development programs have been written about: The Self-Employed Women's Association by Rose (1993) and the Anand Milk Union Ltd. (AMUL) dairy farmer cooperative in India by Kurien (1997). The dairy is one of the biggest producers of milk and butter in India and still functions today.

Concerns about scaling up participatory projects are common, especially as so much of the success of participatory development projects is often associated with charismatic leaders (Mansuri and Rao 2004: 48), which are not readily available in all locations. Scott (1998) is quite negative about the prospect of successfully scaling up participatory development, as he worries about implementers making participation routine and no longer a real tool, as shown in his book *Seeing Like a State: How Certain Schemes to Improve the Human Condition Have Failed*, as was noted by Mansuri and Rao (2004: 49-50), who believe that scaling up needs to be gradual and not rushed. Binswanger and Aiyer's (2003) Policy Research Working Paper for the World Bank explores some of the many challenges of scaling up community-driven development and tries to advise on the necessary factors needed to scale up such programs.

As has been shown, there are many different indicators of effectiveness regarding participatory development. Sifting through the literature helps to illuminate the very complex nature of trying to evaluate any participatory development program and its effectiveness. Effectiveness must be clearly defined and relevant indicators must be selected. While a project might have succeeded in producing a piece of infrastructure, it might have failed to involve the whole of the community, or while every member of the community might have attended a project selection meeting, the most powerful people in the community might have unduly influenced participants to vote for the project that the powerful individual would benefit from the most. Effectiveness is determined by the criteria selected to look at.

The review of this literature highlights the idea that participatory development might have advantages over classical development approaches in some areas. This does not mean that it is able to address the multiple indicators of success that are outlined in theoretical debates simultaneously. The approach itself does not eliminate some of the more fundamental issues that have arisen in previous development projects, but perhaps these expectations are unrealistic. The complexity of the situation, as will be shown later, is supported by the empirical findings of this research. A participatory approach is highly dependent on field conditions and personalities, and while a program and approach can be very effective in one area, this does not guarantee its effectiveness in all areas. These are important things to keep in mind when reading any participatory development literature, including this dissertation.

1.5.4 Criticisms of Participatory Development

As with all development approaches, participatory development has been critically analyzed and criticized. Even supporters of the approach have cautioned that a participatory approach is not a panacea (Cohen and Uphoff 1980: 213, 228). Cohen and Uphoff see participatory development as an important aspect of development projects, but highlight that participation, "can be difficult to promote and the results are not always predictable" (1980: 228).

Pain and Francis (2003: 48) identify three areas of critique, the first two of which have been addressed in brief above: the adoption of participation in name only and a critique of techniques and methodologies used to implement participatory programs. The third critique is more challenging because it questions the basic concepts of participatory development thinking (Pain, R. and Francis 2003: 48), and is what Mansuri and Rao (2004: 1) categorize as "fundamental critiques." The following literature explores some of these more fundamental critiques, which have been put forward in the last decade. These fundamental critiques question the positive connection between participation and empowerment, ownership, and the positive nature of participation in general. It should be noted that while there is a clear deviation between different types of critiques theoretically, this clarity is not as visible in the literature itself. Instead, different types of critiques overlap and merge as many authors do not separate their criticism into the same categories.

Some of the most prominent critics of the underpinnings of participatory development are contained in Cooke and Kothari's edited book *Participation: The New Tyranny?* (2001b). This book brings together fourteen development scholars and practitioners to analyze participatory development in a manner that allows the authors to individually question and analyze different aspects of the approach, and to voice some of the cynicism they cultivated due to their experiences as practitioners. The book, based upon a conference held in Manchester, breaks down some of the underlying precepts of participatory approaches and questions the capacity of these approaches to work in the real world. The authors take apart the reasoning of participatory development to determine how much truth lies within the assumptions made that support the approach, and bring to the forefront new ways of understanding participatory development and development thinking in general.

Cooke (2001:102-121) focused on looking at whether participation leads to the best possible decisions, as is often assumed, in his chapter on the psychological limits of participation. He puts forward four examples of when group decision-making can lead to less than optimal decisions that are not in the best interest of the group. He believes that group decision-making can lead to risky shift, the Abilene paradox, groupthink, and coercive persuasion (Cooke 2001: 103). He defines risky shift as occurring when group members take more risky decisions together than they would have individually (Cooke 2001: 106). The Abilene paradox is when people agree to something because they believe other people in the group agree with it, producing an "unconscious collusion to produce a false statement," and groupthink is when group dynamics lead to bad or wrong decisions (Cooke 2001: 108, 112). Coercive persuasion is when decisions are based on "brainwashing" and other intentional manipulations (Cooke 2001: 116). Cooke (2001: 121), however, does not recommend abandoning the approach completely, but instead recommends utilizing participatory approaches wisely and with a full understanding of when they are most appropriate. This is a common conclusion shared amongst the fourteen contributors to the book.

Hildyard et al. (2001: 69), question the assumption that simply gathering different stakeholders together will mean that the decisions reached are fair to everyone. They believe that the power differentials among the local population make this assumption unreliable, and that it therefore raises questions about participatory approaches on the whole.

Kothari (2001: 147) focuses on how participatory knowledge is gathered, emphasizing that, "The use of participatory techniques often requires the taking out of anything complicated, making people's lives and their social interactions linear and sterile as they fit into charts, diagrams and tables and conform to the boundaries and limitations of the methodological tools." Therefore, participatory approaches cannot achieve their intended purpose of representing the complexity of situations, but instead result in cleansed or static snapshots, which simplify the situation too much. Arguing that, even when all actors actively participate in a project, vital information might still be missing. She provides the example of weddings, and how they can impact on many decisions about a person's life, including education, work habits, etc., but that because it is a rarely occurring activity in an individual's life, it can be missed by the participatory rural appraisal "snapshot" (Kothari 2001: 147).

From a different perspective, Mosse (2001: 32) questions the assumption that the utilization of local knowledge will translate into social change and a shift in power structures. Gronemeyer (1992: 66) also questions this and implies that participatory development practices actually dis-empower people. "As development help, it must first of all destroy what it professes to save– the capacity of a community to shape and maintain its way of life by its own forces" (Gronemeyer 1992: 66). Henkel and Stirrat (2001: 171) discuss that some critiques of participatory development approaches question the value of local knowledge, with some finding that the experts and the state actually do know better.[58]

Mohan (2001a: 162) discusses another aspect of local knowledge and the belief of some development scholars and stakeholders that this kind of knowledge is the only knowledge that is valid, denying validity to all outside perspectives and understandings. "This homogenizing and demonizing of Western discourses repeats earlier criticisms of Orientalism" (Mohan 2001a: 162). It seems quite possible that relying solely on one source of knowledge, be it from outside development experts or the local population, is not the most productive of decisions, as it ignores the worth of the others' knowledge. It can also lead to localism, which is viewed as a problem of participatory development by some authors, such as Mohan and Stokke (2000). They imply that this concentration on the local leaves the larger structural situation that is seen as contributing to the marginalization and poverty of people unaddressed.

Kapoor believes that participatory development needs to be questioned on a deeper level. "When something goes amiss in practice, Chambers' response is to find better ways of doing it: his 'and then asking why' is not about asking 'why are we doing what we are doing?' but 'why does it not work?'" He sees Chamber's approach as lacking in procedures and safeguards with the result that participatory approaches are not as effective in reality as proposed in the theoretical literature. He also believes that Chambers does not pay enough attention to power structures and how that plays into a participatory development experience (cf. Kapoor 2002: 102).

Henkel and Stirrat argue from an opposite point of view, noting that while Robert Chambers (2001: 178) left a great deal of room for his participatory methods to be changed and adapted, that many current practitioners of participatory development are less flexible, and have "routinized" participatory approaches. This, thus, makes the approach

58 Henkel and Stirrat (2001: 171) categorize this type of criticism as "conservative."

formulaic in nature, which defeats one of the main differentiating characteristics of such programming: making participatory approaches similar to top-down approaches that evolved from classical development theories.

Other critics question the reasons international organizations and agencies, and NGOs have embraced participatory approaches so intensely. They ask, what are the real purposes and motivations for these stakeholders? Mohan (2001b: 9) voices the opinion that participatory approaches are being used "uncritically" as a "rubber stamp," and labels this tokenism. This is supported by Botes and van Rensburg (2000: 43), who wrote that participation is being used by governments, not to empower communities or community members, but to get approval for funding from donors who are looking to fund participatory programs. This does not just apply to governments, but equally to all development stakeholders. Hildyard et al. (2001: 59) call this "cosmetic" participation, "visible, heard even, but ultimately only there because their involvement lends credibility and legitimacy to decisions that have already been made."

Rahmena is also cynical about the motivations of different stakeholders' advocacy for, and use of, participatory development approaches. He believes that participation is used for fund-raising, for gaining legitimacy, and because it is economically appealing, etc. (Rahmena 1992: 119). In some ways Rahmena (1992: 119) believes that the large organizations and agencies have tried to imitate NGOs, which he believes are seen as closer to the people, less bureaucratic, and more efficient. Governmental and inter-governmental organizations in the 1980s and 1990s have worked to present themselves as being professional, but also connected (Rahmena 1992: 119), more in alignment with the NGOs who were gaining influence at that time. Some governments, are also viewed by Rahmena (1992: 119) as jumping on the bandwagon of participation and supporting it in the hopes of ensuring their allotment of the foreign aid pie.

Henkel and Stirrat (2001: 183) believe that the development industry supports participatory development, at least in part, because it "shifts the responsibility for the consequences of projects away from the agencies and the development workers," and places it into the hands of the people participating. Fukuyama (2005: 86) also looks as this aspect:

"The current emphasis on 'community-driven development' puts the burden on the aid recipients to assess their own needs and then take responsibility for implementation of a project that they themselves have designed."

Critical authors have begun to explore the concept that governments are using participatory approaches not just to gain approval, but also to legitimize their control, both politically and socially, in order to support existing power structures (Botes and van Rensburg 2000: 45). Governments using participatory development to enforce their views or to stabilize their control and power can be seen in Ribot's article about Sahelian forestry (1995: 1587-8), with local chiefs being responsible only up the command structure and not downwards. This same relationship was shown in other reports focusing on India and Indonesia (Mansuri and Rao 2004: 47). Nelson and Wright (1995:2) were concerned that participatory approaches will be used to co-opt or monitor groups that are not liked by the government. Mosse also discusses how participatory approaches can serve the top-down planning systems (Willis 2010: 106).

Participatory development authors with a critical eye have put forward the possibility that participatory approaches to development are being used to shift the costs of development from the government, or state, to the community (Mosse 2001: 468; Ribot 1995: 10). Participatory approaches, such as community-driven development, put the burden on the poor to improve their conditions, therefore relieving governments and other actors from taking responsibility (Rahmena 1992: 119). This view is partially supported by government respondents who contributed to the OED report (Operations Evaluation Department 2005). When asked about the benefits of participatory development approaches, they focused on the resource savings that participation gave to them (2005: 30).

One of the primary criticisms of participatory development programs are that they are not in fact "bottom-up" but that these programs continue to be "top-down" (Hildyard, Hegde et al. 2001: 59; Narayanan 2003: 2484). This is often blamed on the institutional framework at hand (Hildyard, Hegde et al. 2001: 59), with development implementers responsible to their funders and not to the people of the community. This upward responsibility has been shown to result in donor driven development projects, more often aligned with donor goals than community desires (Dill 2009: 731; Platteau and Gaspart 2003: 1687). Finsterbusch and Van Wicklin (1989: 589) found that funders are judged by how much funding they have distributed and not on the actual outcomes related to those funds, thus distorting motivations at the field level.

Donor driven timeframes and timelines are also seen as driving participatory development approaches, with negative effects. Mansuri and Rao (2004: 3-4) highlight the problem of funder timelines preventing the building up of institutions and the institutional capacity in the communities. In their eyes, this can lead to the establishment of dependency, in complete opposition to the idea that participation creates agency and empowerment. Hulme and Edwards also believe this to be true and see the drivers of participatory approaches as being based on the whims of the donors (as noted in Willis 2010: 106).

The OED (2005: 22) blames timeline constraints for the benefits of some programs being captured by the elites. Mohan (2001b: 52) sees participatory development as leaving no room for deep participation, as the implementing organizations are "time-bound and accountable to funders further up the line." These supply-led approaches do not provide enough time for communities to evolve and for institutions to be strengthened, which are seen as necessary for participatory approaches to have a larger impact on, "economic growth, democratic governance, sustainability, equity, and protection of the poor" (Platteau and Gaspart 2003: 1687).

It has been pointed out that participatory development can create parallel or simply alternative structures to both central and local governments (Mansuri and Rao 2004: 1), though this can be said of all international development that is not channeled through the government. This can create competition between community structures and the government, instead of creating coordination and cooperation between the two. Thomass-Slayter (1994: 1479) highlights how participatory approaches can set up a competitive environment between the communities and local government. This competition is not viewed as encouraging, but rather as problematic.

All of this literature will be considered throughout this dissertation, and serves as a warning that while logical connections can be deduced theoretically, it is important to learn more before prophesying that specific actions are causally related to others. It is important to problematize and question one's own hypothesizes and those of the informants used in gathering data and information.

1.6 Extracting Meaning: The Historical Review

As discussed above, participatory development has a deep history that is firmly embedded in alternative thinking to classical or traditional development theories. It focuses on the people and on the usage of their knowledge and understanding of a situation, in contrast to relying on advice from external development experts whose theories are universally applied without a strong understanding of the local physical, social, cultural, or economic environment. In opposition to classical development theories, participatory development focuses not just on creating economic growth, but also on supporting the establishment of other development indicators that interrelate to create better lives for the people. It is the people, not the state, who are the focus. Participatory development is meant to be flexible and responsive, where classical approaches are traditionally prescribed and defined.

The story of the rise of participatory development is also the story of classical development approaches, as participatory development approaches were developed, in part, in response to the repeated disappoints stemming from classical development projects and programs. The expectations for participatory development approaches are very high, especially for community-driven development, which is seen as the most intensive kind of participatory development. As outlined in the literature, participatory development is theorized to produce many different and almost entirely positive outcomes. These outcomes range from better targeting as compared to top-down approaches, to increased quality, more efficiency, and better sustainability. Participatory projects and programs are also expected to impact on the social structures of the societies in which they operate, culminating in the increase of social capital and empowerment of the marginalized. In essence, the impact of participatory development is expected to be holistic in nature.

Utilizing the critical participatory development literature reviewed, this research focuses on looking for both positive and negative impacts stemming from the implementation of a participatory development program. Additionally, critiques of participatory development have led me to be very cautions about assuming connections and causality.

Many scholars continue to support and advocate for participatory development approaches, but they are now beginning to proceed with greater caution and with more ambiguity (Cohen and Uphoff 1980: 213; Cooke and Kothari 2001b: 15). This is true even for some of the foremost names in participatory development, such as Robert Chambers (2002, 2005). As academic evidence continues to produce complex and sometimes contradictory results, the challenges of participatory approaches have begun to emerge.

Funders, however, appear to be un-phased by the increasing evidence that participatory development projects and programs are only partially delivering on the expectations placed upon them. They continue to funnel money into participatory development projects and programs with high expectations of what the results will be. It is for these funders, and others, who strongly believe in the wholly positive rhetoric of participatory development that this research was conducted. It is meant to paint a more comprehensive picture of participatory development projects and their impacts from a community perspective, and to highlight where rhetoric and expectations differ from lived experiences on the ground, and where they align.

Scientific literature indicates that research is being conducted to determine if participatory development approaches are producing the expected outcomes. Thus far, however, the results have revealed that the impact of participatory development varies greatly overall, and by indicator. While one journal article might highlight improved empowerment in one community, another article might contradict this showing how certain groups were excluded and therefore disempowered. The debate continues as researchers work to understand the mechanisms underlying participatory development and the different impacts it can have on distinctive environments and in different social systems and settings.

These mixed results do not mean that participatory development is a failure, but it does indicate the need for a deeper understanding of how this approach to development programming impacts on different communities in different temporal, geographical, cultural, and societal settings. What these mixed results seem to point to, is that there are many influencing factors that determine the level of success or failure of a participatory development approach and that replicable research and evaluations of the different possible outcomes is necessary to help realign enthusiasm for this approach more closely with the practical outcomes that have been seen in communities.

Much of the research conducted thus far on participatory approaches has been large geographically, but small in thematic scope. In other words, the methodology has most often looked at participatory development from the perspective of a whole district, project, or State, with some utilizing a meta-narrative perspective, gathering data from different projects and nations to draw conclusions. The majority of this research has been conducted from a quantitative methodological approach, with only a few ethnographic case studies (Mansuri and Rao 2004: 4).[59] The focus of the research has also been disjointed in nature, focusing on only one or two indicators of development at a time, rarely employing a more inclusive or broad analysis.

This research addresses this limited approach to understanding the impacts of participatory development. The methodology was established in order to help provide a micro-level perspective of how a participatory development program impacted on the lives of the people living in a single community. Did the program increase active participation and was it more inclusive? Did it change their economic security and create more secure livelihoods? Did it increase social relations and create stronger bonds amongst the community members and between the community members and other entities? Did it decrease the sense of physical vulnerability or threats experienced by the people of the community? Did it give the community members confidence and empower them? These questions and many others have been addressed in this research.

Further details about the methodology, which was heavily influenced by the literature presented above, are forthcoming in the next chapter.

59 A noted exception that focuses on the National Solidarity Programme (NSP) is the series of case studies conducted by AREU that look at the impact of the NSP on women in NSP communities (Echavez 2010, 2012; Echavez and Bagaporo 2012).

2 Situating the Research into a Framework of Analysis

The fieldwork for this dissertation employed a strongly qualitative approach to paint a picture of participatory development in action. Told from the perspective of the community members of a National Solidarity Programme (NSP) Community, an exploratory case study approach was adopted. The research was bounded geographically, but explored temporally. Looking at the NSP community, its members, and the participatory development program implemented from an open perspective helped to create a fuller picture of the research location and how that location and the people within it were impacted by a participatory development project.

This more holistic vision is intended to allow the reader to understand something about the NSP case study community, its history, and its social structures, while exploring how that community feels they were impacted by the participatory development program, the National Solidarity Programme.

Understanding different foundational factors and characteristics of a community can help bring forward a new perspective on what might be influencing the successfulness of a development program that uses a participatory approach. This required spending quality time in the community to become familiar with the community's understanding of their history and their present situation. This viewpoint is rare in the development world, as few researchers have the financial support to conduct such detailed fieldwork, or the time to do so. Therefore, the World Bank, UN, and other development funders are rarely given an inside view into what their programs mean to a community and its people, and not just how those changes are evaluated from an external, quantitative perspective. This research gives a very personal view of how a complex program was integrated into the lives of one Afghan Community.

2.1 Research Stemming from a Historical Base and Specific Levels of Analysis

Understanding what has contributed to the establishment of a community and the events and moments that have shaped it is essential to comprehend how recent events have impacted on it. This research is, therefore, grounded in a historical approach. Only by firmly grounding this research in the historical, can the present be understood, as each of the four areas of investigation that are looked at has been impacted strongly by events of the past.

Additionally, the very idea of participatory development is rooted in the concept that a local perspective is important for planning, implementing, and monitoring development projects, this is due to the people's embeddedness in the community and connection with its history. Understanding the history and norms of a community is one of the fundamental differentials between participatory approaches and more classical approaches, where development experts utilize knowledge from a more generic and generalizing perspective. Therefore, it seems only natural that research that is focused on participatory development should also be embedded in the historical background in which the participatory development program was implemented.

I believe that history is constructed by those who tell it, and therefore would make a distinction between the multiple levels or layers of history that were utilized throughout this dissertation, as diagramed in Figure 8. The first level is the largest in scope. It is the country level perspective: Afghanistan. In order to delve into the other levels of analysis, it is indispensible to be familiar with the broad country-level picture, including political and societal changes, development efforts, and other actions and activities whose impact can be felt throughout the different levels. Therefore, a review of development efforts in Afghanistan and the influencing factors on those efforts will be detailed from 1919 until the present day.

Figure 8: Four Levels of Research Analysis: Country, District, Community, and Household (HH)

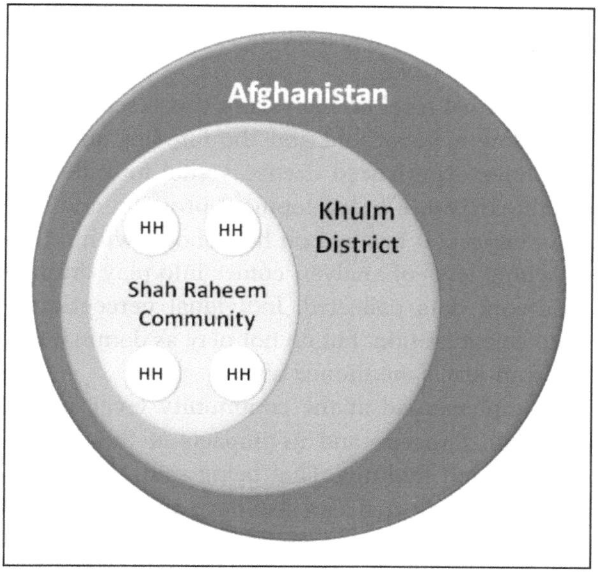

Source: own design

The second level of analysis utilized for this research is the district. Although the next governmental administrative level in Afghanistan is the province, the people of the NSP case study community indicated a stronger connection to the district of Khulm, than to the province of Balkh, both of which they belong to administratively. This could be associated with the transfer of the district from one province to another in recent history, which will be explored more below. Therefore, the second level of analysis stems from Khulm District and the variances of experiences specific to this district. A review of the historical events that shaped Khulm District was also conducted and is presented below.

The NSP community of Shah Raheem is the third level of analysis. Shah Raheem, due to many factors, including its population and its geographic setting, has unique historical experiences that have shaped the community. The NSP community of Shah Raheem actually consists of three individual villages – Chahar Sukh, Shah Raheem, and Yaqubai – which came together for the purposes of the participatory program.

The villages decided to use the name of the largest of the three villages as its identifier: Shah Raheem. The historical forces that impacted Shah Raheem are looked at more closely when the empirical data is analyzed to provide evidence of participatory development programming impacts at the community level.

The fourth and final level of analysis of this research is the individual household. How a household and the families and individuals in that household have experienced events is vital to understanding how they interpret the arrival of a development program and how they feel and explain the impacts it has had on households within their community. The household level of analysis comes into play during the analysis of the fieldwork data collected. Individual perceptions of history were taken into consideration, but do not play as dominant a role in the analysis as the four levels mentioned above.

The NSP is implemented at the community level, in this case the community of Shah Raheem, and its impacts at this level of analysis dominates the research findings. That being said, the spillover effects, both experienced and expected, can also have an impact at the different levels of analysis. Households can be impacted by community level infrastructure projects, which can affect their economic security and their personal security. It can also affect their social security within the community and their social connections outside of the community.

A program such as the National Solidarity Programme is expected to induce stronger relationships between communities and different levels of government, as explored in the NSP Operational Manual (NSP 2009a: 8) and to contribute to the solidification of democracy in Afghanistan by integrating people into a democratic process (Boesen 2004: 2). While the findings of this research concentrate most intensely on the individual household and community levels, the analysis at the district and country levels helps to situate the research into time and place, grounding it, and also allows for room to connect my findings to a wider framework.

2.2 Inserting the Independent Variable: The National Solidarity Programme

In this research, the National Solidarity Programme (NSP) is the participatory development program that was the participatory development focus. The program is looked at as an example of a relatively standard community-driven development approach that has been utilized in other countries, such as Indonesia. It, like many other participatory development projects, establishes a community group to inform the implementation of the NSP in the community, and has been formulated to include as large a part of the community population as is possible. In general, the NSP tries to impart the decision-making responsibility to the people of the community in which the program is being implemented. More details about the program are provided below.

The National Solidarity Programme (NSP) was implemented in Shah Raheem Community, though its affects were anticipated to reach all four levels of analysis, from the household to the country level, over time as the program spread throughout the country, as shown in Figure 9.

Figure 9: Adding the National Solidarity Programme (NSP) to the Four Levels of Analysis

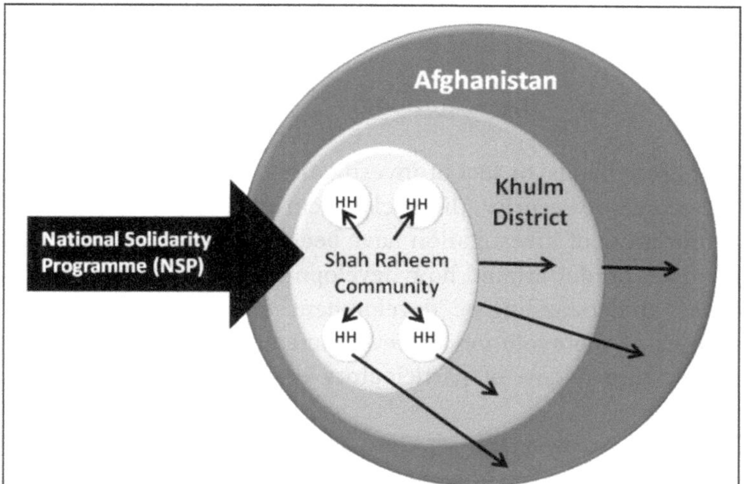

Source: own design

The NSP was the external input that was introduced to the community of Shah Raheem. In order to understand what this meant, I had to combine my understanding of the historical setting of the country, the district and the community, and then integrate that knowledge with the introduction of the NSP into the community. This research, therefore, explores how the program was presented and adopted into the community, who participated in different aspects of the program, how people contributed to the different aspects, and the level of influence individuals felt they had. Participation of the local population is vital to a participatory approach and is a critical differentiating factor that separates it from the more classical approaches. In alignment with Oakley (1991: 240), I determined that, "The evaluation of participation will be concerned with the analysis of a dynamic, qualitative process and not merely the measuring of a static, physical outcome."

Integrating the historical knowledge and the participatory development program implementation process created a platform from which I could begin to unravel the complicated ways in which the program changed or shifted different indicators in the community of Shah Raheem. Therefore, I felt that only from this basis could I begin to analyze the impact of the external variable of the NSP on Shah Raheem Community in terms of the four foci or areas of investigation of this research.

2.3 Four Areas of Investigation

The four areas, fields, or foci of investigation include: (i) personal security, (ii) economic security, (iii) social security, and (iv) empowerment. These four areas of investigation have been established in order to create a defined border around how development is envisioned in this research. My understanding of "development" stems from both a historical analysis of the evolution of development theories and approaches and a literature review of participatory development approaches and experiences.

These four areas, while separate and definable, are also interconnected, with each supporting the existence of the others. The foci and their definitions were explored with the NSP community members individually. They have been embedded in the events that have shaped the landscape and affected the lives of those who have lived there. The areas were also identified as important by the community members and flowed naturally from the literature on development in general, and participatory development in particular.

The first two of these areas – economic security and personal security – originate directly from the human security paradigm of Mahbub ul Haq (UNDP 1994).[60] The third focus of social security also stems from the human security paradigm, where it is referred to as community security. In order to expand its definition, I have combined it with the concept of social capital and relabeled it social security. These three foci of personal security, economic security, and social security, were selected because they address some of the main concerns conveyed to me in the research area and in the country as a whole. They also strike a chord with the expanded definitions of development that emerged in the 1990s. Additionally, I evaluated them to be the most relevant to the research study community and country based on conducting over three years of research in Afghanistan prior to starting this research. The other four human securities presented by Mahbub ul Haq – food, health, environment, and political – are included in the first three foci of this research in different ways.

Once I had established these three areas as important for evaluating the impact of a participatory development project, I had to address the empirical literature. As has been shown above, participatory development has been researched both theoretically and empirically, and while the results have been mixed, there appears to have been little absorption of the less positive findings of some of the research cited above. Recognizing this led me to reevaluate the three areas of focus I had developed to determine if they were expansive enough to cover the impacts of a participatory development program.

60 Human security is rooted in the concept of people-centered development (UNDP 1994: 23), as is participatory development. "It is concerned with how people live and breathe in society, how freely they exercise their many choices, how much access they have to market and social opportunities– and whether they live in conflict or in peace" (UNDP 1994: 23). The UNDP posits that there are seven securities that must be fulfilled in order for one to have human security, and in turn, to be developed. The human securities include: economic security, food security, health security, environmental security, personal security, community security, and political security (UNDP 1994: 24-5).

While I found them to be relevant, it became apparent that through the inclusion of the people, participatory programs were expected to not only impact positively on the three areas of focus I had identified, but they were also expected to have a larger impact on the community itself. Participatory development programs are expected to bring about the empowerment of the people. This differentiates participatory development approaches from top-down approaches stemming from the classical development theories. This was clearly shown in the theoretical literature regarding participatory development and in the empirical literature. I, therefore, added a fourth dimension of investigation to the three areas stemming from human security: empowerment.

2.3.1 Personal Security

The first focus or area of investigation in this research is personal security, or what is sometimes recognized as physical security. This is most often what people think of when one asks them about their own security. It raises questions such as: Is someone being threatened physically? Are they experiencing bodily harm? Are their lives being threatened? Being safe from physical violence is the main determinant of personal security (UNDP 1994: 30-31). Violence can come from the State in the form of physical torture, from another State in the form of war or conflict, from groups of people, from individuals, or even from ones' self (UNDP 1994: 30-31). Violence can be directed at an individual for their personal characteristics or for characteristics they share with an identified group, nationality, religion, etc. The UNDP report (1994: 30) pays special attention to the physical threats directed towards women and provides evidence of women's increased level of personal insecurity, as compared to men. In this research, special attention was paid to women's experiences in the community, as women are often underrepresented in Afghanistan.

In Afghanistan, numerous conflicts have contributed to creating a physically insecure environment. This can be measured by the number of lives lost to fighting, to anti-personnel mines, to droughts made worse by the destruction of irrigation networks, etc. Present day fighting has resulted in threats to individuals and communities via night letters (*šabnāme*), suicide bombers, rocket attacks, night raids, improvised explosive devices (IEDs), and other violent attacks that can tear apart bodies, families, and communities.[61] Violence and threats play a role in the personal security of the population at large and of individuals. Therefore, it is a vital area to be researched in this context. In order to do this, I have used participant identified indicators, such as incidences of violence, threats, *šabnāme*, accidents, and more to capture community members' sense of personal security, historically and in the present day.

2.3.2 Economic Security

Within the confines of human security theory, economic security is defined as an individual having a basic income, be it through paid work or via state support (UNDP 1994: 25). It also entails having access to credit, as even those with jobs often face job insecurity or face real wage losses dues to inflation. The focus of economic security in this research is based on this definition. This means that through the collection of empirical data I worked to extract the economic factors affecting peoples' lives, from learning about their livelihoods to understanding the market structure and market flows. This required the research to also identify different risks to the economic security of individuals and to the community as a whole, both from internal and external forces, and to analyze how such risks and threats were dealt with, both successfully and unsuccessfully. From this understanding, it was then possible to analyze the impact of the NSP on the economic security of the community as a whole and for individual households.

Further elements inherent in the focus of economic security in this research include education and health, based upon the understanding that in Afghanistan access to both education and health services and knowledge require economic security. Additionally, the lack of either of these elements can threaten one's economic security.

Though not mentioned directly in the 1994 UNDP Human Development report, I see education as one of the strongest predictors for economic security in Afghanistan and a vital component of all of the human securities. Knowledge and education be it formal or informal, can greatly impact on one's ability to earn a regular income and to main-

61 *Šabnāme* are letters that are delivered at night that provide warnings or threats.

tain the health of oneself and one's family. It is also an important factor in one's social status in Afghanistan.

In order to be healthy, one must first have access to basic foods, yet this often requires economic security to obtain the food. As noted in the UNDP Human Development report of 1994 (1994: 27), "Access to food comes from access to assets, work and an assured income." Secondly, one must have access to healthcare and health knowledge. Having health security is about avoiding diseases and other health problems that can be caused by a lack of access to safe drinking water, basic foods, environmental problems, etc. It is also about having access to health services, which includes having the means to transport someone to a local clinic or the ability to procure services once at the clinic.

2.3.3 Social Security

The third area of investigation concentrates on social security and social capital, or the social networks and connections between individuals, families, communities, organizations, ethnic groups, etc. According to the UNDP Human Development Report (1994: 31), where this is called community security, these connections create a group that is seen as providing a cultural identify and a set of values, which aligns with Putnam's (1993: 293) focus on social norms and Bourdieu's "*habitus*" (1989: 18).

In agreement with Hayami (2009: 109), who discusses social capital as providing a social safety net for communities against crisis, in this research social security is defined as the support one is able to call upon utilizing one's different social connections. It is a networks oriented view (Woolcock and Narayan 2000: 230-234) that recognizes both bonding and bridging social capital as different aspects that need to be investigated. It aligns with Portes (1998: 6), who believes that a consensus had grown towards the end of the 1990s in which social capital "... stands for the ability of actors to secure benefits by virtue of membership in social networks or other social structures." This concept of social capital as an important asset in one's life, or in the life of a community, has also been highlighted by Woolcock and Narayan (2000:

225) and goes all of the way back to Hanifan's (1916) discussion of social capital in the early 1900s.[62]

This research took into consideration both bridging and bonding social capital, in alignment with Putnam's definitions (Baron, Field et al. 2000: 10). Bonding social capital is related to relationships and norms shared by a group with close ties and high levels of trust, and bridging capital relates to connections between a group and other groups, organizations, institutions, and government agencies (Putnam 2000: 320). Bridging social capital is often between un-equals, such as a patron-client relationship (PREM and World Bank 2002: 12). Another way of terming this is horizontal and vertical social capital (Fox and Gershman 2000: 402), with horizontal social capital referring to bonding social capital among people with similar levels of power, and vertical social capital referring to relationships that breach power equalities.[63]

Social capital is a concept that has been embraced by some of the larger international organizations and institutions involved in development and some NGOs. The World Bank, for example, began to discuss the concept of social capital as an important aspect of development starting in the 1990s, and included it in its *World Development Report 2000/2001* (World Bank 2002). The redistribution of social structures and social connections, including social power, are common expectations connected to participatory development programs and are often cited as one of the main benefits of such an approach. Social Capital is seen as bringing improvements in well-being and economic standing (Willis 2010: 111).

This research looked closely at how the NSP has impacted on the social relations and connections, or the social capital of the community, taking into consideration both the positive aspects of strong social norms, and the negative possibilities. In order to accomplish this, this research investigated instances when an individual or family faced risks

---

62 Hanifan wrote about the importance of community participation in improving school performance.

63 Coleman and Bourdieu both think of social capital as coming in two different forms: primordial and constructed in Coleman's words (Baron, Field et al. 2000: 8), or spontaneous social organization (Bourdieu and Coleman 1991: 1). Primordial social capital is that which stems from birth, such as one's family connections, and constructed social capital or spontaneous social organization is that which is formed through social connections.

and where they obtained assistance or support. It also looked at the number of social activities in which people participated, bringing them together to interact, and participant evaluations of their connectedness and the resources they thought they would be able to access when they face problems they cannot handle on their own. Another contributing indicator investigated was people's perceptions of their relationships and the closeness they felt existed within their community.

It is not unusual for only the positive side of social capital to be explored, as has been done by some of its most famous theorists such as Robert Putnam (1993). It is important, however, to recognize that while social capital can help people to better survive the risks they face in life, it can also act to restrict people in different ways and it can be used for negative purposes. "Closely related with this problem is the fact that mutual trust and cooperation among community members are often supported by rivalry and hostility against outsiders" (Hayami 2009: 111), such as gangs, mafia families, prostitution, and gambling rings amongst others, as noted by Portes (1998: 15) and Fukuyama (2001: 14).[64] Some authors, such as Woolcock (1998: 168-178) see social capital as entirely neutral, in that it can be both positive and negative simultaneously. Woolcock and Narayan (2000: 226) highlight both positive and negative possibilities, noting that social capital can be both a "blessing and a blight." This research aimed to learn about both positive and negative aspects of social capital and how they were impacted in the community of Shah Raheem by the implementation of the National Solidarity Programme.

2.3.4 Empowerment

The fourth focus of investigation is empowerment of the community and its individual members. Participatory development is considered to be empowering, especially by those who advocate for it (Chambers 1994: 1253, 1265; Cleaver 2001: 36; Guijt and Shah 1998b: 9; Oakley

64 One example provided to show the negative possibilities of social capital is the robber's cave experiment written about by Sherif et al. (1961), where two groups of homogeneous boys on a island are separated and form group identities that result in conflict between the two groups. Another example provided is that some Asian immigrants anglicize their name in order to remove themselves from communal obligations to others (Woolcock and Narayan 2000: 231).

1991: 17). It is seen as a means of providing a voice to the people and giving them a chance to analyze their own problems (Narayanan 2003: 2485). Mansuri and Rao (2004: 3) see the participatory development practice of community-driven development as aiming to change power relations and give a voice to poorer people.

While empowerment is a common term in a great deal of the participatory literature, it is a term that is rarely clearly defined. The World Bank loosely defines empowerment as, "the process of increasing the capacity of individuals or groups to make choices and to transform those choices into desired actions and outcomes" (Operations Evaluation Department 2005: 14) or as, "the expansion of assets and capabilities of poor people to participate in, negotiate with, influence, control, and hold accountable institutions that affect their lives" (PREM and World Bank 2002: vi).

For the purposes of this research, empowerment is defined as individuals or a community feeling as though they have the power to influence decisions in their area and actively taking advantage of this power. In summary, it should equate to the people feeling as though they have, "increased control over their lives" (PREM and World Bank 2002: 11). This definition aligns with the final step of Arnstein's (1969: 248-249) ladder of participation, citizen participation, and with Pretty's seven types of participation as self-motivation (Pretty 1995: 172; as noted in Teves 2000: 39).

To further break down the term empowerment, I adopted the World Bank's four defined elements of empowerment: access to information, inclusion or participation, accountability, and the capacity to organize at the local level (PREM and World Bank 2002: 14). Each of these four elements is individually essential to empowerment, yet the four also act "in synergy," supporting one another (PREM and World Bank 2002: 14). Access to information is essential for people to be able to take advantage of services, to exercise their rights, and to be able to put pressure on their leaders to serve their needs (PREM and World Bank 2002: 14). It is especially important for economically disadvantaged people to have information about the rules they are to follow and what responsibilities the government has towards them, amongst other pieces of information. Therefore, I looked at what information the community members have in general and about the participatory development program, including their understanding of the program, its goals, and what occurred throughout the planning, implementation, and post-implementation stages. What information was shared with them and how was it shared? What information did they not have access to, and what acted as barriers for them?

Inclusion and participation are bound together and understanding them involves looking at who has participated or was included, and what role they played in the decision-making process (PREM and World Bank 2002: 15). In this research, in order to avoid the pitfalls of looking only at attendance rates, I worked to understand who was able to actively participate in different aspects of the program. Active participation was defined as individuals feeling they were able to express their opinions and that these opinions were respected, considered, and helped to inform decision-making. Attending a meeting is important, but contributing meaningfully to that same meeting is even more important, and shows a greater degree of empowerment. Therefore this research will look at the inclusion of community members and their involvement in the participatory development program in terms of attendance, voice, contribution, and ownership.

The third aspect of empowerment is accountability, which means that the people are able hold those with power and control responsible for their actions, their ideas, or their use of funds, amongst other things (PREM and World Bank 2002: 16). This research, therefore, investigated the means the community had to hold the project and the people who were running the project to account for their actions and whether they were utilizing these means. This meant understanding how irregularities were addressed, if they were addressed at all, and how complaints and problems connected to the project were handled.

The fourth aspect that was used to indicate empowerment was the organizational capacity of the community. I analyzed historical and current trends in the peoples' ability to work together to resolve problems. As PREM and the World Bank (2002: 17) note, this also entails looking at the community's ability to mobilize resources to address problems, which is closely connected to social capital, both bonding and bridging.

2.4 The Additional Value and Importance of Applying a Gender Lens

One of the main purposes of participatory projects is the inclusion of all voices in a community, not just those of the elite, the educated, or the most powerful, but also those who have traditionally had little to no voice in community affairs (Cernea 1985; Chambers 1983; Gran 1983). The general separation of the sexes and the traditional exclusion of women from the public sphere in Afghanistan make it important to specifically look at how participatory development programs have impacted men and women in the community, as separate entities. Therefore, a gender lens will be used in accordance with the idea that both male and female voices need to be heard to understand how one program has affected their experiences.

This required speaking with women and men separately in order to allow them to speak openly, without the influence of the other gender. In Afghanistan, it is most often men whose voices are heard, as they have greater access to social spaces outside their homes, but this can result in only a male perspective being presented. As the life opportunities and situations of men and women in Afghanistan are often very different, it appeared logical that trying to understand how a development program impacted both genders was critical, as their experiences and knowledge of the program could be divergent, without either gender fully recognizing this possibility. Ester Boserup, a gender specialist in development, highlights how development projects and policies can have very different impacts on males and females or how those that are supposed to support women can miss the target (Boserup 1970: 113, 221; Boserup and Liljencrantz 1975: 37).[65]

Personal experience conducting research in Afghanistan has shown me that only by interviewing men and women separately can one start

65 Boserup and Liljencrantz (1975: 37) provide the example of handicrafts and other small business projects that focus on women and train and encourage women to produce certain products, but with little to no thought put into whether there is a real market for such projects. This is something I witnessed while conducting a market sector assessment of handicrafts made by women for export in Afghanistan in 2006. Boserup (1970: 113, 221) discusses how sometimes handicraft projects can actually hurt women's empowerment and how advocating for special benefits for female workers can lead to fewer women being hired.

to gain an understanding of how a development project has impacted on both genders of a community. While a male focus group might say that everyone in a community participated, this can actually just mean that all males in the community participated. It is no guarantee that women participated, as in some communities women are not considered as a part of the community when it comes to holding community level meetings. In the NSP community of Shah Raheem, the men often told us that they represented both the men and women of their household, but when we spoke to the women of their household separately, we would be given a very different set of answers to the same questions. Therefore, it was crucial to obtain the views of both genders separately, to record as many different versions of their experience with a participatory development program as possible.

2.5 Extracting Meaning: The Analytical Framework

Combining these different aspects together, the historical analysis provides the grounding to help draw a picture of the historical situation of the community and how the four main areas of investigation have been impacted. This then provides the opportunity to investigate how the National Solidarity Programme (NSP) entered into those historically defined spaces and the different impacts it had on the community and on individuals in that community. It allowed me to then analyze whether the four areas of investigation – personal security, economic security, social security, and empowerment – have been impacted by this participatory development program and what that impact looks like at the different levels of analysis (see Figure 10).

By combining all of these different pieces together, a more comprehensive understanding of the community and its experience with the NSP emerged that helped provide a personal and micro perspective of a participatory development program from a decidedly local perspective.

By utilizing this framework, insights were gained regarding the main research question: How has a participatory development program impacted at the community level? This overarching question was then broken down into four sub-questions, which were investigated in turn, and informed my determination to obtain as much information as possible about the historical context into which the participatory program was integrated.

Figure 10: Analytical Framework, Connecting the Levels of Analysis, the External Input, and the Four Areas of Investigation

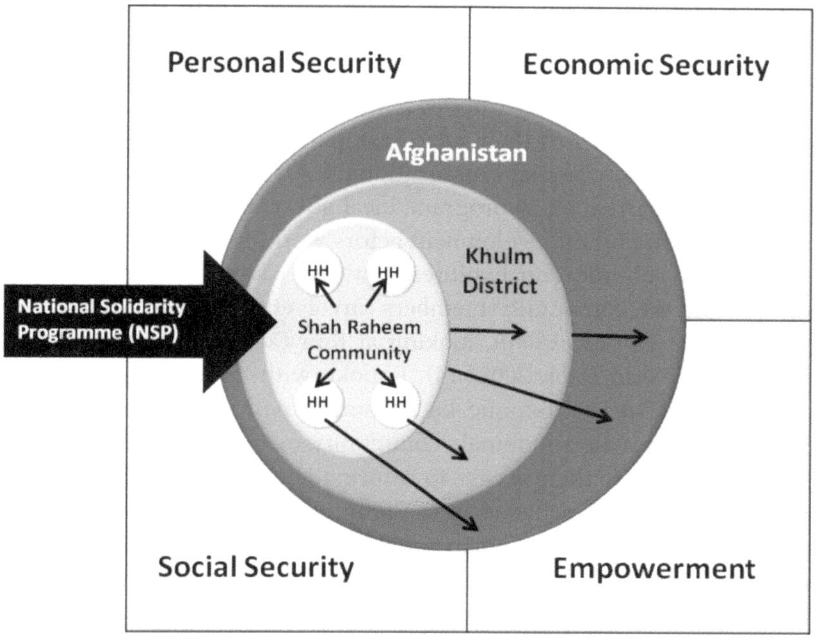

Source: own design

The first set of sub-questions focused on personal security. What impact did the participatory development program have on the community members' understanding of their own personal security? What impact did the NSP have on external views of personal security in the community? Did the process of participating in a development program impact on the number of perceived threats to one's personal safety?

Further sub-questions looked at the second area of interest, economic security. What impact did the participatory development program have on the economic security of the research community? In other words, did the program have a positive or negative impact, helping people feel more secure economically by altering their livelihoods?

Another group of sub-questions focused on the social security in the community, both in the present and in the past. How did the NSP impact the social structure or social security of the research community? Participatory development programs involve the community members and bring them together to discuss the development projects that stem from the program. This creates an environment where social interactions can be increased, which can impact on the connectedness of community members, their trust in one another, and their connection to external entities. Therefore, this research looked at the community members' perceptions of their relationships with one another and also with the non-governmental organization (NGO) that facilitated their inclusion in the participatory program, local government staff, the central government, and other development actors working in the area.

The final sub-question explores how the National Solidarity Programme and the community members' involvement in the NSP shifted their sense of empowerment, looking at four factors of empowerment (PREM and World Bank 2002: 14). I spoke with community members from different socio-economic backgrounds, with different education levels, and both male and female community members of different ages to understand how their access to information had changed, and how they perceived their inclusion in the program. I also obtained information about what was done when problems with a development project or program arose and how people were held accountable if mistakes had been made. Finally, I explored what efforts the community had made in terms of development projects and programs, both prior to the establishment of the participatory program and after its completion.

These main sub-questions drove the empirical collection of data, all with the purpose of being able to extract areas of impact of a participatory development program from a local perspective. The means of gathering this data is explored below.

3 A Predominantly Qualitative Methodological Approach

The research for this dissertation was undertaken from a worldview, or ontology, which is constructivist in nature, meaning that there is the fundamental belief that all research is biased and that knowledge is itself a construction. "Social constructivists hold the assumption that individuals seek understanding of the world in which they live and work. Individuals develop subjective meanings of their experiences – meanings directed toward certain objects or things. These meanings are varied and multiple, leading the researcher to look for the complexity of views rather than narrowing meanings into a few categories or ideas" (Creswell 2009: 8).

This worldview requires open-ended questions, which provide the research participants with the opportunity to construct their own meanings of a situation or an event. "Thus, constructivist researchers often address the processes of interaction among individuals. They also focus on the specific contexts in which people live and work, in order to understand the historical and cultural settings of the participants" (Creswell 2009: 8). In general, this results in research that is more inductive in nature, with the findings leading the research forward, not theories. This research started from a desire to make sense of the research question and the data gathered, and to allow patterns to emerge from the combination of those two.

3.1 Methodological Demands of the Research Question

While a number of research studies have been conducted on the impacts of participatory development, most have focused on measuring only a few indicators and/or utilized a quantitative perspective. This research, in contrast, has utilized a largely qualitative approach, in the belief that in order to understand the impact of the National Solidarity Programme (NSP), the scale of examination could not remain only at the macro level, but must also focus on the micro level, and that evaluating participation requires a qualitative approach. Therefore, this research takes the reader into one NSP community, and the households that make up that community, in order to bring a first-hand account to the forefront. It also explores the four areas of investigation as indicators of development to try to create a more holistic picture of how this approach has impacted on one community and its citizens, with the understanding that this provides greater insights into the experienced strengths and weaknesses of a participatory development approach.

Due to the nature of the research focus, a strongly qualitative methodology was required; as numbers alone could not tell the story I was looking to tell. "Qualitative data consist of detailed descriptions of situations, events, people, interactions, and observed behaviors; direct quotations from people about their experiences, attitudes, beliefs, and thoughts; and excerpts or entire passages from documents, correspondence, records, and case histories" (Patton 1980: 22). Understanding the impacts of participatory development at the household and community level requires information about how different households act, react, and operate. Qualitative data collection techniques allowed me the opportunity to understand how the NSP impacted on a community and its people, and provided me with the opportunity to gain insights into why some of those changes happened and what they meant to the people involved. It was only by using qualitative techniques that I could start to interpret motivations and rationales, which quantitative techniques cannot capture in the same way. The data for this research was obtained using a number of tools: open-ended household and market surveys, unstructured key informant interviews, observations, guided focus group discussions, and participatory mapping. Some quantitative data was gathered in each of the surveys and from secondary sources.

Taking into consideration the main research question, I determined that a case study approach was the best way to obtain the detailed insights the research aimed for, or as Patton (1980: 17) would say, it fit the "nuances of the research question." However, I understood how such an approach would limit the representativeness of my findings. Doing a case study allowed me to spend an extended period of time getting to know the geographically bounded case study location, which resulted in me being able to draw a more comprehensive picture of the situation at hand, looking at multiple and complex variables. This kind of research produces results that are particularistic, descriptive, and heuristic, according to Merriam (1998: 34) and has the benefit of providing more in depth and detailed data according to Patton (1980: 22). It also, normally provides the researcher with the time to not only build trust with the people living in the case study area, but also to gain insights into how descriptions from people contrast or align with lived experiences.

By utilizing a case study methodology, I was able to gain a different perspective of an Afghan community than is most commonly presented. I was able to explore a community in depth, which is often not possible using other methodological research approaches. This approach allowed contradictions in actions, words, and ideas to emerge that would otherwise not have been discovered. These contradictions, while frustrating from the perspective of trying to draft a single, clean picture of village life and the impact of a participatory development program on that life, highlight the incredibly complex situation that exists in rural villages in the country and shows the great level of diversity of views that can exist within one community.

3.2 Limitations of the Methodological Approach

Using a case study approach impacts on the applicability of the research across a broad spectrum, but taking such a localized view provides insights and details that are incredibly difficult to obtain when approached from a larger framework. One of the major drawbacks to the case study approach is a lack of comparability, which prevents one from being able to confidently apply one's findings to similar situations with a specific degree of certainty. This was somewhat mitigated through the review of other research and secondary literature on the National Solidarity Programme (NSP).

The qualitative nature of this case study also means that the research was heavily dependent upon my abilities as a researcher and my understanding of the local environment, culture, etc. Having lived and worked in Afghanistan the three years prior to conducting this research allowed me to immerse myself in the research immediately because I already had a basic understanding of one of the national languages (Dari) and a feeling for some of the traditions and societal influences at play in rural Afghan communities.[66] These understandings allowed me to make connections and create acceptance in the community at a faster rate than otherwise would have been possible, and also meant that things learned were not just taken at face value, but were instead explored to ensure that the meaning was clearly explained.

The integrated nature of the data collection process in a case study, can also influence the data obtained, as those providing information can tell researchers what they think an outsider might want to hear, or they might try to hide some of the less pleasant realities of their lives, or play them up, depending on how they view the researcher and their team members.[67] Expectations and a desire to please can lead to more positive responses to questions about a specific program than the experienced reality. Additionally, some respondents who might not have been heavily involved in a participatory project, but who view that project now as a success, will sometimes indicate that they contributed or participated more than they actually did. This is sometimes referred to as the halo effect, where people would like to be associated with projects that have had a positive benefit in the community or for their household in particular.[68]

Understanding my status as an outsider and that few foreigners enter into Afghan communities without at least the intention of providing them with development projects, distributions, or something of a similar nature, I did my best to preface every interview with an honest discussion of my research and my inability to provide development assistance. Despite this, difficulties were encountered regarding community member's expectations of me in my status as a foreigner (*xārege*). It was important for me to have patience and continue to probe deeper into

66 I conducted multiple research projects over those three years that required me to travel to sixteen different provinces of Afghanistan to work in both provincial centers and in rural villages.

67 During a focus group discussion in Shah Raheem with females, the women explained how different factors were considered before they answered evaluation questions about development projects, and sometimes those other factors resulted in the women covering up failures of projects in an attempt to ensure that the program would continue.

68 The halo effect was demonstrated through empirical research as early as 1920, when Edward Thorndike researched the effects of different attributes or characteristics of soldiers and how they were rated by their commanding officers, finding a strong correlation between commanders' assessments in different categories (Thorndike 1920: 25-29).

situations to determine the thought process behind what was being said or to comprehend the underlying assumptions common in the community. I needed to develop my own ability to uncover false responses, by triangulating my data, and figure out new ways to work both with and around these answers to try to find information grounded in people's lived experiences.

The tools used for this research provided the people of the community with a means to express their impressions of what happened and what was happening. This research is, therefore, heavily based on perceptions, which also maintains room for the exploration of differing opinions, as each opinion is considered and analyzed individually to create a patchwork picture of events, experiences, etc.

3.3 Empirical Research Tools and Sampling

Three separate rounds of fieldwork were conducted in order to obtain empirical data for this research: June through August 2009, January through March 2010, and May through September of 2011. The first two rounds were led by me in the field and the third was led from Germany, due to increased insecurity in the area that prevented me from traveling to Afghanistan personally. In this third round, two research assistants conducted follow-up interviews to fill in holes in the data that had emerged after the first two rounds of research data were analyzed.

In total, 234 people (144 male, 90 female) participated in the research for this dissertation. Household survey participants account for 32 people in total (29 male and 3 female); 94 people (79 male and 15 female) participated in 143 separate key informant interviews, including the shopkeepers; and 98 people (36 male, 62 female) participated in the four male, six female, and one mixed gender focus group discussions.

3.3.1 The Centrality of the Household: The Semi-Structured Household Survey

In order to obtain an understanding of how the community-driven development project impacted on the lives of the community members, a household survey (HHS) was drafted. This survey was created to extract basic demographic information about the research area, and also to provide insights into community members' sense of how different areas of their life have been affected from a personal and historical perspective. Additionally the household survey provided detailed information about a household's participation in the National Solidarity Programme (NSP) and provided ample space and time for the respondents to provide their own perceptions of how the NSP impacted on their household and their community, highlighting the successes, challenges, and setbacks as they understood them. The survey was conducted in Dari and translated back into English for analysis.[69] I designed the household survey based upon having created and implemented multiple surveys prior and reviews of other researchers' surveys, both those conducted in Afghanistan and those conducted in other countries. The household survey template can be found in Annex F: Household Survey Questionnaire.

The household survey was conducted in 32 households, covering 50 families, as multiple families sometimes lived in one household.[70] This means that the survey covered almost half of the 112 families reported to reside within the community.[71] Despite strong efforts to conduct the survey with both men and women, social restrictions on gender interactions resulted in only three of the 32 surveys being conducted with women. In an effort to rebalance this situation, an increased number of key informant interviews and focus group discussions were conducted with women than originally planned. Overall, however, the data is still biased towards the male perspective. This will be discussed in more detail below.

69 All interviews were reviewed again after being translated into English and discussed with the research assistants to ensure that the meanings of answers were clearly documented and explored and that the translation was accurate.

70 Household surveys will be noted in citations as HHS, followed by the ID number of the household survey interview and the date on which that particular interview was conducted.

71 Basic demographic data about the community was obtained through National Solidarity Programme records and forms that were stored in the facilitating partner's district office (NSP Documents).

Photo 1: Household Survey Interview Being Conducted with a Community Member of Shah Raheem Village

Source: photograph MB Wilson, January 20, 2010

Originally, the methodology entailed interviewing the female or male head of household, but this was shifted during the research as it was discovered that some of the younger family members were able to provide more detailed answers, especially in regards to information about the National Solidarity Programme. In order to balance their lack of direct knowledge of some of the historical time periods discussed, I attempted to maintain a mix of interviewees by age and family status. In the end, we spoke with 24 heads of households (including 3 females), with an average age of 49 years, and five additional male household members with an average of 24 years of age (see Annex G: Household Survey Participants for more detailed information on the household survey participants).

A cross analysis of interdependent relationships was made, comparing the answers provided and the interviewees' age, gender, and household status in order to highlight any major differences found. Overall, it can be said that there was little coherent variance that could be attributed to age aside from the increased details provided by some of the younger interviewees regarding the NSP. Those who were older were able to provide more personal stories from previous time periods, but otherwise, the impressions of security and information about their participation in the present-day participatory development project were in alignment with the answers provided by the younger men. Comparing the answers of men and women, however, generated much greater variance, which will be explored when the data analysis is presented. The small sample size of women for this survey would normally require one to leave these kinds of differences to remain unstated, but as this same phenomena was observed during the fieldwork and confirmed in key informant interviews; these differences will be explored further in the analysis.

Figure 11: Sketch of Chahar Sukh Village with Household Survey and Key Informant Interview Locations

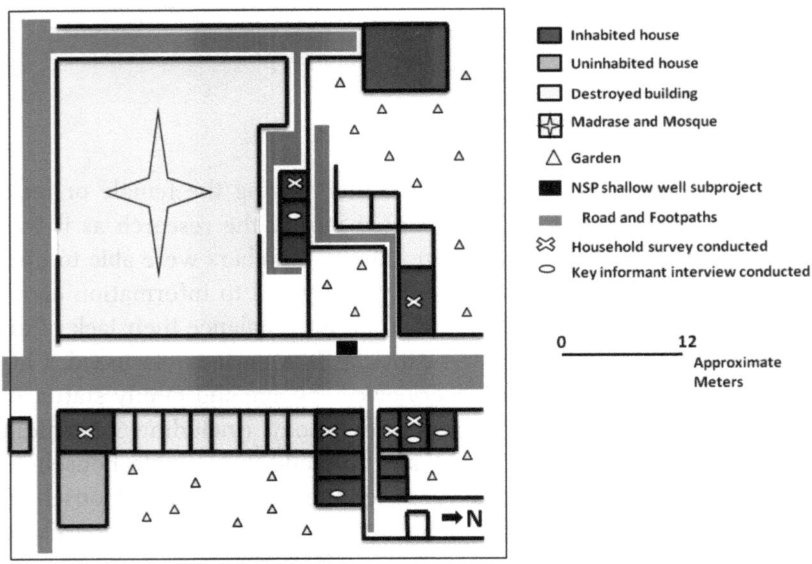

Source: own design based upon participatory community drawn maps

Households that were interviewed for the household survey were selected based upon a geographically stratified, randomized sample. The community consisted of three villages, Chahar Sukh, Shah Raheem, and Yaqubai, which have been sketched out in Figure 11, Figure 12, and Figure 13. The sampling for each village was assigned based upon the number of households in each village, providing a pre-weighted sample. With the help of the community members, the three villages of Shah Raheem were mapped out, including individual houses, gardens, streams, etc. This map was then marked for each household interviewed for the household survey, and on a daily basis the research team would review the map together to ensure that there was a good geographic spread of households covered.

Figure 12: Sketch of Shah Raheem Village with Household Survey and Key Informant Interview Locations

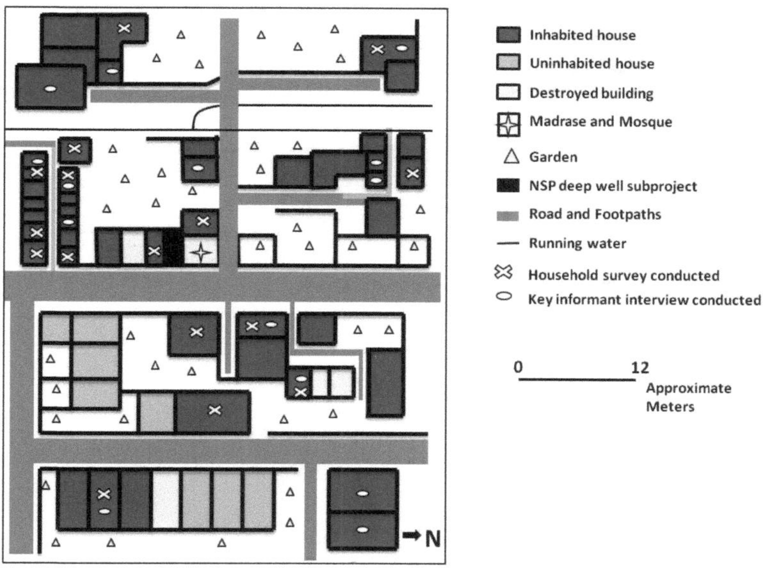

Source: own design based upon participatory community drawn maps

Households were selected in order to ensure that each small street, or *kocha,* was represented in the final sample. Households were selected without prior knowledge of the individuals who lived in each household. This method was used based upon the knowledge that marginalized groups in Afghanistan often live in one geographic area, grouped, and that the same can be said at the community level. This sampling technique helped to ensure that we obtained as many different perspectives as possible and that not all those interviewed came from only one section of the community or from one family or group.

Figure 13: Sketch of Yaqubai Village with Household Survey and Key Informant Interview Locations

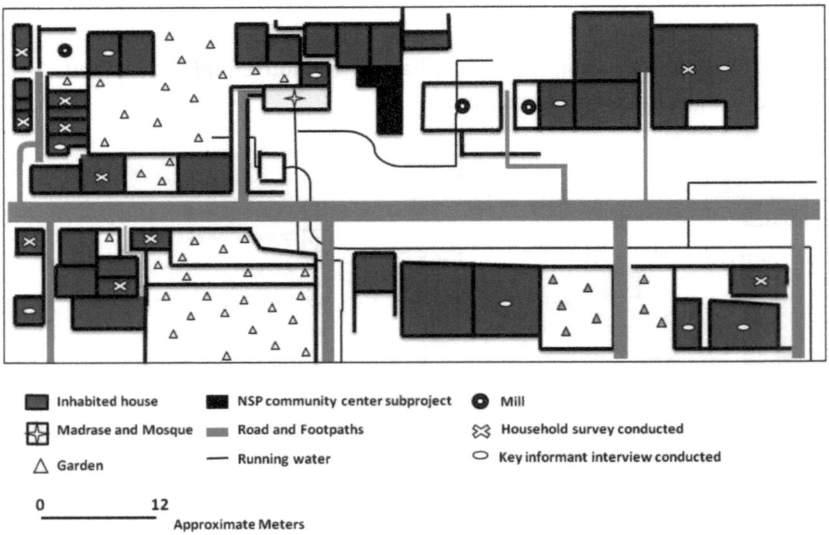

Source: own design based upon participatory community drawn maps

3.3.2 Understanding a Community and Its Environs: The Research Tools

Key Informant Interviews

Key informant interviews (KIIs) were an integral part of this research. Internal and external stakeholders were interviewed to obtain specialized data and information, to triangulate with other sources of data, and to gain a deeper understanding of the community, the district in which it sits, and the province and country as a whole.[72] Many interviews were conducted over the three fieldwork periods with NGO staff, district officials, community members, and additional stakeholders. In total, 143 separate KIIs were conducted with a total of 94 people (79 male, 15 female), including people in the shopkeeper questionnaire. Some individuals were interviewed multiple times.[73] Information on the shopkeeper interviews, which are included in these numbers, is provided in a following section.

NGO staff that was responsible for the implementation of the NSP in Khulm District of Balkh Province was interviewed, as was necessary to obtain the relevant information and for follow-up questions. A total of 38 interviews were conducted with fourteen NGO staff members (thirteen male, one female). They were able to provide detailed information about the NSP and its implementation, not only in the research community, but also in the other districts in which the facilitating partner NGO implements the program. They also shared with me the monitoring and evaluation system of the NSP, which allowed me to see the research community in the district context, and resulted in me creating

72 Key informant interviews will be noted in citations as KII, followed by the ID Number of the key informant interview and the date on which that particular interview was conducted. Some individuals were interviewed multiple times. They will have one ID Number with different dates. Shopkeeper Questionnaires will have the prefix SKQ, facilitating partner NGO staff will have the prefix FP, internal interviews with community members will have the prefix Com, and external key informant interviews will have the prefix EX. Those conducted with men will have no suffix, while those conducted with females will have an f as the suffix. All KIIs are listed in the Annex and can be identified by the ID number provided.

73 Multiple interviews were conducted with an individual when the first interview had not gathered all of the data desired and the participant was willing to be interviewed again or when new issues arose that I wanted to review with someone I had already spoken with.

a database of village information for the district.[74] For a detailed list of NGO staff interviewed by title, see Annex K: Facilitating Partner Staff Key Informant Interview Participants.

Fifty-five separate internal key informant interviews within the community were conducted with 27 males and thirteen females. All of these interviews were open-ended in nature, guided by the participant, their knowledge, and a general outline of themes that was constantly evolving throughout the duration of the fieldwork. I interviewed people who were seen as having a great deal of power and influence, such as the *kalāntar* (village representative), landowners, and elders. I also conducted interviews with those perceived as having less power, including those who were gardeners for landowners, widows, and younger adults. I also attempted to maintain a nuanced view by interviewing those considered highly educated and those who had no formal education. Through these internal key informant interviews, I was able to obtain many different and divergent views that confirmed the complexity of the social, economic, and power relations of the three villages that make up one "community." For a detailed list of internal key informant interviewees, see Annex L: Community Member Key Informant Interview Participants.

External key informant interviews were conducted mainly in order to gain a better understanding of the operating environment of the district and to learn about the historical developments in the district over the preceding forty years. A total of 30 external KIIs were conducted with 21 people (20 male, 1 female). The main areas under discussion were governance, livelihoods (including agriculture and water access), health, education, development efforts, transportation, and the market. I spoke with local government leaders, district ministry representatives, and others who had relevant information to share. The gender imbalance in these interviews stemmed from the positions the people held, and not from selection bias. There were no women holding government positions in the district of research, and most public positions of power or influence were held by men. For a more detailed list of external key

74 This database includes information provided in the NSP records and forms stored in the Khulm District office of the FP for each of the 75 NSP communities active at the time of research. This database allowed for comparisons of the research community with the district as a whole across multiple indicators. Information from these sources will be cited as NSP Documents.

informants see Annex M: External Key Informant Interview Participants.

Focus Group Discussions

In addition to these one-on-one key informant interviews, focus group discussions (FDGs) were also conducted.[75] Focus group discussions are useful because they allow 6-10 people to be interviewed in a room together at the same time, which provides the researcher an opportunity to observe how the different participants interact with one another and allows participants to discuss the questions posed. Four different male FGDs were conducted with 33 people. These male FGDs revolved around attempts to gather the CDC members of the community, or from a single village, together in order to learn more about their experience with the National Solidarity Programme. In order to help balance out the male heavy nature of the research, six separate female FGDs were conducted with 59 people, and an additional mixed gender FGD was conducted with one family (three male and three female). The female FGDs focused on human security issues, development efforts, and more general themes of traditions, social rules, and regulations from a female perspective. The mixed gender FGD was conducted with male and female family members of one extended family who provided a glimpse into the structure of their family and how the events of history had impacted on them. See Annex N: Focus Group Discussions Held for more detailed information on the FGDs conducted.

75 Focus group discussions will be cited as FGD, followed by the ID number of that FGD and the date. Please note that those conducted with men will have no suffix, while those conducted with a mixed gender will have the suffix mix, and those conducted with an all female group will have the suffix f. All FGDs are listed in the Annex and can be identified by the ID number provided.

Photo 2: Male Focus Group Discussion in the NSP Community of Shah Raheem

Source: photograph MB Wilson, February 3, 2010

Observations

Observations were another key aspect of the research methodology. Patton notes that observations provide five added values: they help the researcher better understand the context, allow the researcher to be inductive, provide a different set of eyes that can sometimes see things that participants miss, allow a researcher to learn about things that people are not willing to speak about, and help the researcher move beyond just reporting about the perceptions of people (Patton 1980: 125). These five added values were very important to this research.

In the almost five months I was in Afghanistan conducting this research, I spent time observing the normal workings of the research community, the district market, and the interactions of government and local entities. Internal to the community, I learned more about the social interactions and norms by observing community meetings held in the *masğed* (mosque), daily interactions on the streets and in the homes of the community, and interactions at both community weddings and funerals. Observations of development activities were made by attending microfinance group meetings, agricultural trainings, a three day district development planning session, and the handover over of some National Solidarity Programme communities from their facilitating partner NGO to the Ministry of Rural Rehabilitation and Development (MRRD). Watching the speed with which infrastructure projects in Tashqurghan, the district center, took place over the duration of the fieldwork provided additional insights.[76]

Observations allowed me to contrast what I was told, with what I was able to see while spending time with the community members.

As noted above, observations can help the researcher to obtain information that informants may not have wanted to talk about (Patton 1980: 125). This was true for my own research as I began to observe differences in responses to similar questions based upon who was participating in a conversation, I began to see discrepancies between stories of what had happened in the past, and I was able to learn more about who was actively participating in group activities and who, though in attendance, was not able to contribute to the conversation at hand. I found that the desire to not shame oneself or one's family could lead to community members not being forthcoming about problems they were encountering, some of which I was made aware of through observations and informal interactions with community members.

Participatory Mapping

With the assistance of one community member, the three villages of the NSP community of Shah Raheem were mapped out. Each village was mapped separately, with details of the buildings (those being used, those that were uninhabited, and those that had been destroyed), compounds, houses, gardens, mosques, shrines, waterways, mills, side streets, paths, and other important landmarks located and marked. These maps were shared with others in the community to confirm their accuracy. Though not drawn perfectly to scale, the maps closely approximate differences in the sizes of the land plots. These maps helped

76 I have also seen Tashqurghan written in the following ways: Tashkurgan, Tashkorgan, Tash Qorghan, Taxkorgan, Tash Khoorgaun, Tausk Koorghaun, Tásh Khurghán, and Tashik'ergan.

the research team to visualize the breadth and general physical boundaries of the NSP community and of the three individual villages that make up that community. This ensured that we had the necessary information to verify that our interviews covered the full geographic span of the community. After each day in the field, I would mark the compounds belonging to those we had spoken to, indicating whether a household survey, a key informant interview, or a different form of data collection had been conducted there.

Photo 3: Participatory Community Mapping with Community Elders of the NSP Community of Shah Raheem

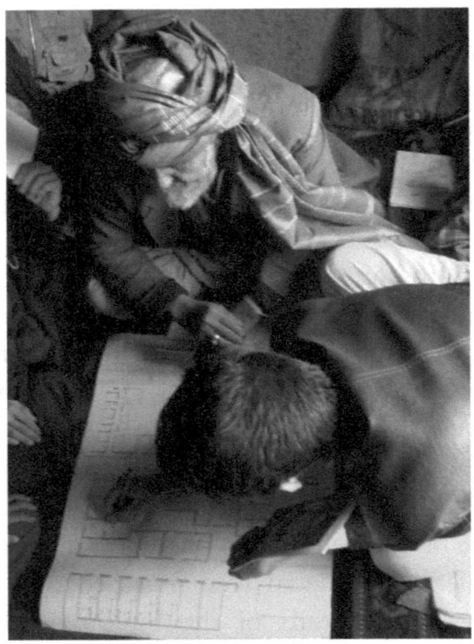

Source: photograph MB Wilson, February 2, 2010

Market Sector Survey and Shopkeeper Questionnaire

A market survey was conducted in the bazaar of Tashqurghan, the center of the research district of Khulm. During the research it became apparent that the people of the community of Shah Raheem were connected to this market for selling and purchasing personal goods and products, with only rare and irregular trips to the markets of the provincial capital of Mazar-i-Sharif.[77] In order to better understand the nature of the market structure in Khulm District, both in the present and from a historical perspective, I conducted a market survey and a number of shopkeeper questionnaires. See Annex H: Market Survey Questionnaire and Annex I: Shopkeeper Questionnaire to view the market survey and shopkeeper questionnaire tools.

The market survey was based on the present day structure and layout of the district center market, which rests on both sides of the main road into town. Starting at the west end of the market, each storefront and the products it sold were documented. This information was used to understand the size and scope of the market itself. It also provided me with insights into the economic status of those living in the area, and provided me with a comparison to previous studies done on the bazaar in the 1970s by Centlivres (1970) and Charpentier (1972).

The purpose of the shopkeeper questionnaire was to better understand trends in supply, sales, and prices over the past forty years for the products that are important to the lives and livelihoods of the people of the community of Shah Raheem. Selection was determined by the types of products they purchased and sold. Based upon information gained through the HHS, KII, and general observations, a list of the most commonly purchased and sold goods in the community of Shah Raheem was established. Shopkeepers were then selected who could speak about those products specifically and provide insights into the current and historical costs of buying and selling each of the products, product sources, and the quantities of goods they bought and sold in an average month.[78]

The necessary commodities included the two major cash crops produced in the community of Shah Raheem and sold to local and foreign markets, pomegranates (*anār*) and almonds (*bādām*), and an additional cash product produced locally and sold to the Tashqurghan market,

77 I have also seen Mazar-i-Sharif written Mazar-e-Sharif, and it is often referred to simply as Mazar or MeS.

78 For each product at least four shopkeepers were interviewed in order to triangulate the information received to obtain local pricing points, and not those sometimes utilized for outside buyers or sellers. When information did not triangulate, additional shopkeepers were interviewed until an evident consensus emerged. Sometimes, up to nine different shopkeepers were asked about one product. These questionnaires also helped me to understand differences in perceived qualities of products and ensuing differences in pricing structures.

eggs (*toxm*). It included the most commonly purchased products mentioned by the residents for regular consumption – wheat, oil, rice, sugar, diesel, and petrol.[79] Understanding the sourcing of the products and the price points helped draw a clearer picture of the costs for local residents and possible opportunities.

Photo 4: Shopkeeper Interviewed in the Main Market of Tashqurghan, the Capital of Khulm District

Source: photograph Waheed Omari, March 10, 2010

In total, 327 shops were surveyed and 19 shopkeepers were interviewed, all male. The shopkeepers had a range of five to 35 years of experience as a shopkeeper in Tashqurghan, with an average of just below 18 years of experience (see Annex J: Shopkeeper Questionnaire Participants for a detailed list of interviewees). There were no female shopkeepers in the market.

79 Most people in the research community do not grow their own wheat and have to buy flour in the market, most of which comes from Pakistan (KII: Com26, March 7, 2010).

3.3.3 Gaining Access and Permission

The research team made all possible efforts to "do no harm" while carrying out this research. This entailed obtaining oral consent from each research participant. The research was originally introduced to the community through a general meeting held with the male elders of the NSP community of Shah Raheem, with representatives from all three villages on January 19, 2010 (FGD01).

In this meeting I talked to the elders about the research and sought their permission to conduct the research in their community. After a two hour-long discussion, where all 14 participants were welcomed to make inquiries that I did my best to answer, the elders decided that the research would not be detrimental to their community or its members and agreed to support it. This "in" into the community meant that almost all members of the community that we encountered were willing to speak with my assistants and me. The permission of the elders removed the largest obstacle to the research at hand.[80]

Prior to each individual interview or focus group discussion, I explained the intent of the research and how I intended to use the information we were gathering. I made sure to clarify that my research was academic in nature and that it would not directly inform any development plans or projects.[81] I also made it clear that each individual was free to participate or not participate in the research; that their participation was entirely voluntary. For all who agreed to answer questions from my assistants or me, we sought and obtained informed consent orally, prior to proceeding.

---

80 Their permission was given more enthusiastically when it was discovered that I was acquainted with one of the most powerful elders of the NSP community, as I had been introduced to him at a dinner in Mazar-i-Sharif only a few nights prior. I believe that it was this relationship that really opened the doors of the community to me and my research.

81 This was made very clear as previous research in Afghanistan had shown me that speaking with a foreigner can raise expectations of receiving humanitarian aid or development projects. I wanted to ensure that the community members were not expecting a direct benefit from me because they shared their insights. The only benefit I provided to community members was copies of photos of them that they allowed me to take. It was a small token of my appreciation for their time and hospitality. Those who participated in my research and those who did not received copies of their photographs, without discrimination.

Photo 5: CDC Members and Elders of the Community of Shah Raheem Following a Meeting to Introduce them to the Research

Source: photograph Waheed Omari, January 19, 2010

Participants were provided with the opportunity to ask questions, and they were promised that their responses would remain anonymous. Though few people in the NSP community of Shah Raheem expressed concerns about anonymity and most were not opposed to having their names connected to what they had expressed during our time with them, I have chosen to eliminate all names to protect the few who wanted their anonymity to be protected.

A few people in the community did opt out of speaking with us, but mostly due to a lack of free time. This happened rarely and could have been either based upon actual time constraints or possibly an unwillingness to speak with the research team. For the most part, the few that said this did appear to be very busy, and I evaluated that this was the main motivating factor for their declining to participate. In general, people of the NSP community of Shah Raheem were eager to speak with me and my assistants and seemed happy to share their experiences. Some thanked us for being the first researchers to spend time in their homes, drinking tea and learning about their lives.[82]

Obtaining permission to work in the community was established at multiple levels, the most important from my perspective being at the community level, as detailed above. However, the most important level, logistically, rested in gaining the permission of the District Governor, which took place during my first trip in the summer of 2009. His permission was granted, as was his support for the research in the form of written letters to the local government administrators requesting their cooperation. While I am aware of the bias this could have introduced into the answers provided by the administrators, this was a necessary step for my security and the security of my assistants, without which, the research would have been impossible to conduct.

3.4 Difficulties and Challenges Encountered During the Fieldwork

3.4.1 Security

Balkh Province, with its location in the North of the country and its strong-handed Provincial Governor, Atta Mohammad Noor, has not faced the great insecurities of other provinces such as Hilmand or Kandahar in the South of Afghanistan. Taliban infiltration is considered to be minimal and as of 2009 was limited to the districts of Balkh, Chimtal, Chahar Bolak, Dawlatabad, and Sholgara (RRD representative July 29, 2009 and confirmed by FP staff in June 2009). As of February 2012, the number was down to four, with Dawlatabad once again considered safe (personal communication, GTZ Staff, February 21, 2012 and ADB staff, February 22, 2012)(ANSO 2011b: 7). Despite this relative provincial security, threats of kidnapping, improvised explosive devices (IEDs), and armed militias still exist in the province, which complicated the logistics of the fieldwork.

82 One man said that we were the first people from outside the community who had taken the time to try to understand his life and his problems and enter his home (KII: Com10, February 7, 2010). He had participated in some previous research efforts, but he said that these efforts only asked about a specific topic, such as the market, which made him feel as though he was not understood.

Security rules, advice, and concern for the community led me to base myself in Mazar-i-Sharif, which is a 45 minute-drive from the NSP community of Shah Raheem. There are no foreigners living in the district and it was determined that living in the community could possibly put me at risk of kidnapping and I worried about the impact my presence could have on the safety of the community members. Despite my lack of affiliation with the government, my existence as a foreigner could have placed the community in conflict with active armed opposition groups who have threatened villagers in other communities for cooperating or collaborating with the government, NGOs, and foreigners, who are oftentimes viewed as all connected.

This restriction, therefore, resulted in my assistants and me having to commute to the research location. As recommended by the Afghanistan NGO Security Office (ANSO) and other security analysts and experts, we attempted to alternate the regularity of the travel, to make it more difficult for anyone to plan anything such as an ambush or an attack. However, being restricted to only driving during daylight hours, also as a security precaution, meant that travel times could not be greatly altered without sacrificing time in the community. In order to attract as little attention as possible, a regular white Toyota Corolla, standard fare in Afghanistan, was used for the commute to the community. On average, the team spent four days in the community and two days in Mazar-i-Sharif each week, with one day off.

Regular security advice was provided to the research team by the facilitating partner of the NSP operating in Khulm District, ANSO staff, other security officers for international NGOs and international organizations based in the province, and local informants in the district itself. Only two times was fieldwork delayed or postponed due to security threats.[83] Once these threats had been evaluated or had passed, field-

83 Towards the end of the second research trip, all vehicular travel within the city limits was restricted and many checkpoints between Mazar-i-Sharif and Khulm were put into place in preparation for the Afghan New Year, Nou-ruz. This caused increased tensions within the city and on the roads into the city, preventing us from going to the field for one day. A second day of fieldwork had to be shifted to the weekend as threats of suicide bombers having entered Mazar-i-Sharif were reported and all international organizations and NGOs required all staff to forgo travel within or around the city for the day.

work was continued. These interruptions never lasted for more than a twenty-four hour period.

Following the second research trip, insecurity in the country and my research district increased greatly, preventing me from being able to return to conduct follow-up research. In December of 2010, the assassination of a German expat in the district capital resulted in the research community being inaccessible even to the Afghan research assistants I had worked with.[84] Additionally, an attack on the UN compound in Mazar-i-Sharif occurred in April 2011 (BBC News 2011), resulting in twelve people being killed. Only in May 2011, did the research assistants feel that the situation had calmed to the point where they would not be endangering their lives by traveling to the research community. During this third round of research, however, their travel to the field was minimized and a few interviews were conducted via telephone to reduce the number of times they had to commute.

3.4.2 Gender Segregation

Current societal norms throughout Afghanistan, and especially in rural locations such as that of the research community, require a distinct spatial separation of the sexes. Though men and women of the same family are more or less free to interact, it is considered improper for men and women who are not related with one another to interact and the community mostly practices *purdah (parde)*, or separation of the sexes. I will not try to determine the roots of this division here, but instead will report it and its implications for the research and the data collection process.

Due to the separation of the sexes mentioned above, it is not often possible for male interpreters or research assistants to speak with or interview adult women. While we found that there was some room to work around this restriction, as some of the older women in the community said that they would like to talk to me and provide their stories even if they knew that most days of the week I only had a male research assistant available to help me with translations. This was only possible after we had spent many weeks within the community and

84 The aid worker was fatally injured when visiting Khulm in order to check on the road project being implemented by the German Government and the German development bank KfW (The Local 2010).

they had become accustomed to us and knew our behavior. The women told me that they could speak with the male research assistant because of the good behavior he had shown, saying they felt he was like a son to them (personal communication, March 2010). In general, however, this was the exception to the rule.

Prior to being able to identify a female research assistant, I conducted a few female focus group discussions using my own limited Dari and a great deal of pantomime (FGD02f, January 20, 2010). I was eventually able to hire a female research assistant who accompanied me to the community once per week. She was fluent in both Dari and English and was given permission by her family to travel outside of the provincial capital with me, two difficult to find characteristics when searching for a female research assistant in Mazar-i-Sharif.[85]

After a great deal of searching through local NGOs, Balkh University, and other means, no female research assistant was found. Therefore, I had to adapt my fieldwork plans and reduce the prevalence of female respondents. The female assistant who worked full-time during the week for an international NGO was hired to work with me on the weekends. This worked well, and provided further insights into the impacts of the programming and of historical shifts on the women of the community, but was less satisfying than being able to conduct research with females on a more regular basis.

Luckily for me, as a foreign woman, foreign females are not held to the same social norms as local females, which meant that I was able to interact freely with male community members. They considered me to be a "third gender" of sorts and I was able to eat meals with them, ask them questions, and be present at public functions, all of which women from the area are not allowed to do. This access to both genders was a benefit for me as a researcher that external male researchers are not commonly able to obtain in Afghanistan.

---

85 The literacy rate amongst women is low throughout the country, estimated to be only 12% nationwide (ICON-Institute 2009: 65). Though the literacy rate is somewhat better in Mazar-i-Sharif, at 32% for women (UNAMA 2007: 10), it was still difficult to find a woman who was well-educated and fluent in both English and Dari. Most women with these skills have full-time positions working for international organizations and NGOs based in the provincial capital. Even those who are allowed to work outside of their home are often not given permission to travel to outlying districts.

3.4.3 Research Assistants and Working in Translation

Working in translation is always a challenge, as it was for this research. While I am familiar with some basic Dari, I do not possess the strength of skills to conduct such intensive research on my own without translation assistance. Therefore, I had to hire two research assistants who provided on the spot translations and written translations as the research required, both of whom were male. A third research assistant, a female, also joined the research team at a later date.

While both of the male research assistants had previous experience as translators and researchers, working with others still complicated the data collection process. A great deal of time was spent training the research assistants in the groundings of the research and the objectives, which was sometimes too theoretical or esoteric for them to fully understand. This caused some misunderstandings in their interpretations of the purpose and meaning of specific questions. These kinds of issues were addressed through regular debriefing sessions and reviews of the data as it came in. Questions were adjusted and clarified to ensure that the data desired was being collected.

Additionally, using research assistants meant that they accepted many answers at face value, without probing deeper, due to their integration into the social, political, and economic structure. This sometimes left me with less dense answers than I would have liked, though this was addressed through follow-up interviews and continued, habitual feedback to the research assistants.

The research assistants had a strong command of English; however, there were multiple moments of confusion and miscommunication that arose due to language barriers of understanding and usage. It required a great deal of patience on both sides of the equation to ensure that mutual understanding had occurred. This was also true in following up on written translations. After reading through each translation I would sit with my research assistants to ensure that I had understood their meanings and words correctly.

Working in translation adds an additional layer between the researcher and the subject matter at hand, which is not ideal. It can prevent the lead researcher from asking relevant follow-up questions and can result in less detailed research. Despite these challenges, my research assistants and I worked hard to establish a means of checking and confirming meanings that helped ease the progress of the research and ensure its depth.

3.4.4 Community Suspicions and Expectations

As was noted earlier, it is not unusual in Afghanistan for community members to harbor expectations of foreigners who enter their domain. One focus group discussion participant told me, "we pray whenever we see a foreigner that they will bring us things" (FGD06, February 3, 2010). These expectations vary from community to community, but the common expectations I have experienced include that foreigners are there to provide development projects and that they have the means to resolve problems. In some areas of Afghanistan, the expectations of foreigners are that they are there for sinister reasons. Luckily, this last expectation was not one that was held in the community in which this research was conducted.

The former expectations, however, were held and did cause some minor conflicts. After having worked in the community for a couple of weeks, I was made aware that there were rumors that I was paying people to speak with me. This could have been very detrimental if not addressed quickly and clearly, as some people who had not been interviewed were beginning to feel discriminated against. If their anger had been allowed to continue to build, it could easily have resulted in me and my research team being unable to work in the community. Luckily, by quickly addressing this rumor and speaking directly to some of the individuals who had started the rumor, we were able to ensure that an accurate portrait of what we were doing in the community was prevalent.

In another incident that could have threatened the research, I was invited to "observe" an all female gathering in one of the three villages. They were going to discuss the fact that a quilt sewing project they had hoped would come to their village was going to be implemented in the neighboring village instead. Unbeknownst to me, I had been invited not to observe, but rather to resolve the problem and force the NGO implementing the project to give it to the village in which I was sitting. The women threatened not to participate in the research anymore if they were not given the quilt project, despite my protestations that, as carefully stated from the beginning, I had no influence with any of the NGOs. The wife of the local *kalāntar* presented the situation to me and I explained that this was not something that I could help them with and that if that meant they would no longer speak with me, that I was willing to accept that as I only wanted to speak to people who were talking to me voluntarily, without expectations of personal gain, for ethical reasons. With some further discussions the situation was resolved and despite the fact that I had made no difference in their receiving the project, the women continued to welcome me into their homes and offered to share their knowledge and understandings with me.

3.5 Identifying the Fieldwork Research Location

This research was based in Northern Afghanistan in a district called Khulm. The NSP community selected as the case study community is called Shah Raheem. It consists of the three villages named Chahar Sukh, Shah Raheem, and Yaqubai.

3.5.1 Selection of Northern Afghanistan

Northern Afghanistan, stretching in the west from Hirat province to Badakhshan in the east, nestles against its northern neighbors in the Turkestan Plain, shown in Figure 14.[86] After the fall of the Taliban, this stretch of provinces was seen overall as more secure than other regions of the country, despite regional conflicts over control of Mazar-i-Sharif between General Dostam and Atta Mohammad Noor.[87] Its current relative security is belied by a history that has been punctuated by intense insecurity and violence as will be detailed when we turn to the historical analysis of the district of Khulm.

---

86 Turkestan is also alternatively written Turkistan.

87 This conflict lasted for multiple years, starting in 2001. Atta Mohammad Noor was named Governor of Mazar-i-Sharif in 2004, after which he has maintained power in the region. He is seen as ruling through both formal and informal means of power (Mukhopadhyay 2009: 536, 540-541, 550).

Figure 14: Research Location: Northern Afghanistan in Relation to the Country as a Whole: Narrowing in on the Research Area

WFP (2001) Afghanistan District Vulnerability Map - Combined (updated 7th May 2002), adapted by Andrei Dörre, October 2012

Historically, not a great deal of data is available about this section of the country, especially in terms of academic writing. Local descriptions of life in Northern Afghanistan are not to be readily found, as continues to be the case in modern times. This lack of local perspective left me dependent upon foreign writings, such as those of British explorers and soldiers from around the turn of the century, who provide biased glimpses into the lives of the Northern Plateau and the political environment (Ferrier, J. P. and Jesse 1858; Hamilton 1906a; Maitland and Talbot 1887; Malleson 1878; Moorcraft and Trebeck 1838; Wood, J. 1872; Yate 1888; Yule 1872). Soviet writings have also come to the foreground, though few have been translated, which have also contributed to knowledge about Northern Afghanistan in the 20th century (e.g., Knyazev 2000; Sharipov 1984; Solov'eva and Svetlov 1984; Teplinskii 1958, 1982).[88] In the 1960s and 1970s, European and Northern American academics conducted research in Northern Afghanistan (Barfield 1981;

88 These sources are discussed by Dörre and Kraudzum (2012).

Grötzbach 1969, 1972, 1990; Humlum 1962), but this came to an end with the invasion of the Soviet Army in 1979. From the Soviet invasion until the fall of the Taliban in 2001, the Northern Region, along with most of Afghanistan, was incredibly difficult to enter and conduct research in. This has left a hole in the academic litany, which this research, because of its historical approach, attempts to help fill.

While there have been a number of programmatic studies and some waterways reports written, the province of Balkh has raised little interest in the academic world beyond the anthropological focus on the bazaar of Tashqurghan (Centlivres 1970; Charpentier 1972) and the ancient city of Balkh (Lee, J. L. 1996; MacChesney 1991).[89]

The lack of data about the Northern Region, combined with the relatively less conservative nature of the people of the North, and the region's relative security, were factors in deciding to conduct this research in the North of Afghanistan. These factors offered me the opportunity to conduct the research in a less constricted manner, providing me and my research assistants with access to a multitude of informants from different backgrounds and different sectors of society, including the marginalized, who can be restricted from providing information to researchers, or unknown to researchers, in other parts of the country. This included women, who are seen as a key marginalized group in the country as a whole, and a subject of specific focus for this research. It was also assumed, inaccurately as it turned out, that the less conservative reputation of the people of the North would have resulted in greater levels of female participation in the National Solidarity Programme.

The greater level of security in the Northern Provinces, as compared to other provinces, also meant that my presence in the research community was less likely to endanger the community members. In Khulm District in general, interactions with foreigners and outsiders were not

89 A French archaeological expedition has been ongoing in Balkh since 2004. It is a project that built upon earlier research of the French Archaeological Delegation in Afghanistan (DAFA) to study the early Islamic period in the Balkh Oasis and the Balkh Ab River Valley (Étrangéres 2009). Additionally Oxford University commenced the "Balkh Art and Cultural heritage Project" in September 2011. A three year project to conduct new research on "Afghanistan's early Islamic history," while also building capacity of researchers in Afghanistan (Faculty of Oriental Studies 2012).

as highly contested or likely to cause problems for the community participants or for me as an outsider.

The infiltration of the Taliban and other anti-government elements (AGE) is lower in the North, and while this infiltration has been increasing since 2008 (personal communication, ANSO Staff, July 2009), it is still limited to specific districts in specific provinces. This means that the Northern Provinces provided me with relative freedom of movement, allowing me to spend a good deal of time in the field working directly with community members and other stakeholders.

3.5.2 Selection of Khulm District, Shah Raheem Community

Khulm District was selected for this research after thorough consideration. Two other districts were discussed in detail regarding their appropriateness, but Khulm was determined to be the most suitable amongst the districts because of its historical importance, its accessibility, and the relative security in the district since the fall of the Taliban, please see Figure 15.[90]

Prior to traveling to the district, its geography and its demographics were discussed with the local NSP facilitating partner (FP) NGO, CARE International, and district level data on demographics were reviewed to assist in the selection of the community. This entailed a full review of all communities involved in the National Solidarity Programme (NSP) in Khulm District and discussions with the FP regarding those communities where the program had not yet been implemented.

90 It was also chosen because of the time-sequenced implementation of NSP in the district, with some communities starting the program in 2006 and some already completed, some communities still in process, and others that were expecting to start implementation in 2010. Unfortunately, lack of security prevented me from being able to fully explore the differences between communities at different stages of the process.

Figure 15: Map of the Districts of Balkh Province, with the Research District of Khulm Highlighted

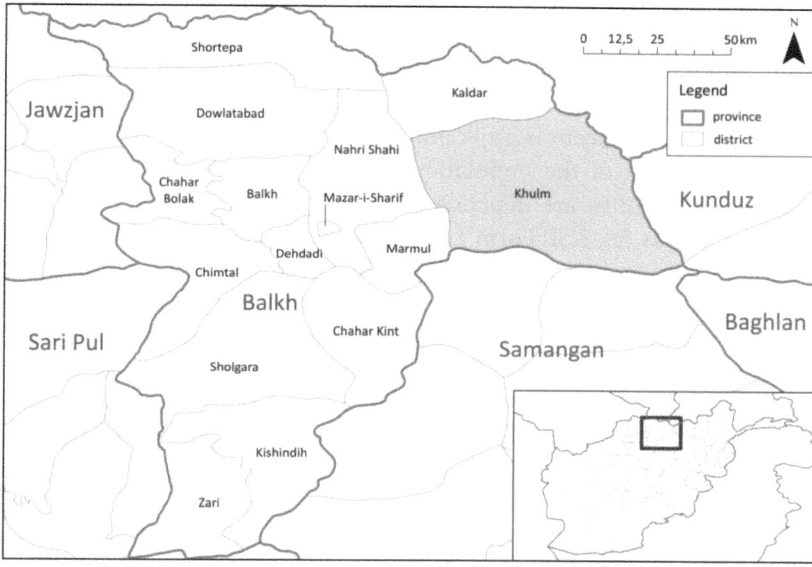

WFP (2001) Afghanistan District Vulnerability Map - Combined (updated 7th May 2002), adapted by Andrei Dörre, October 2012

The district was also chosen as it was open to foreigners and their presence was not a cause for alarm or for the raising of arms. This allowed me, with appropriate support and approval, to interact at ease with the community and also provided some interesting background on how development and development projects have been perceived from a historical perspective. A number of development projects have been implemented in Khulm District over the years, though it has by no means been inundated. This was important to me too, that the community of focus had some experience with development programs, in the hope that the community members might be able to tell me about the differences they perceived between a participatory approach to development and a more standardized approach.

A full database of the communities within the district where implementation had started was compiled and general demographic characteristics and trends were noted.[91] The community of Shah Raheem does not diverge greatly from the other 74 communities in the district in terms of demographics, development goals, and a number of other criteria, such as distance from a main paved road. Almost all of the community of Shah Raheem is Tajik and speaks Dari as their first language. Fifty-two percent of the population is below the age of 19 years, and most of the families are dependent on agriculture to some extent for their livelihoods (cf. NSP Form 1: Community Profile).

Communities were also discussed with the facilitating partner in the hopes of locating a community that was viewed as average in terms of its success with the National Solidarity Programme. While it is sometimes imperative in research to study extreme cases, I wanted this research to focus on what can be considered a normal level of success of a participatory development program in the Afghan context. By avoiding the extreme examples, we can see more clearly how such programming works under non-ideal circumstances, but circumstances that still allowed for the program to precede.

Therefore the NSP community of Shah Raheem was selected. It is situated only a ten minutes drive from the district center, known locally as Tashqurghan or the city of Khulm, and a 45 minute drive from the provincial capital of Mazar-i-Sharif. In total, the community implemented three separate subprojects with their NSP block grant, one in each of the villages of the community.

3.6 Bringing Together the Different Means of Data Collection

The methodology employed for the empirical data collection for this dissertation was accompanied by the use of secondary sources for both the historical perspective and the present day analysis, pulling together information from the different levels – country, district, community, and household. While the fieldwork has provided new empirical evi-

91 There were 91 communities identified in Khulm District, 75 of which had implemented the NSP as of March 2012 (NSP Documents).

dence that captures a micro perspective of how a participatory program has impacted on a single community in Afghanistan, the secondary research has given the empirical research a foundation on which to rest. Used together, they establish a strong picture of how the community of Shah Raheem was impacted by the implementation of the National Solidarity Programme in a specific geographical, societal, and historical context. What emerges is a rich picture of participatory development in a concrete setting, and a new perspective from which to think about the NSP, participatory development approaches, and development in general.

4 A Case Study Situated in Afghanistan

4.1 The Dominance of Modernization Theory in Historical Development Efforts

The development history of Afghanistan is an important backdrop to the story of how a participatory development program impacted on an Afghan community. Through understanding the different development attempts made in the country as a whole, we can properly see just how big a change a participatory approach is to what had come before it. We can also better understand the community, its people, and the social and political systems that influence them.

As will be outlined below, the participatory development approach is relatively new to Afghanistan. It stands in stark contrast to the majority of development efforts made in the country and could be guiding the way for further expansion of such approaches into the future. Steeped in almost a century's worth of foreign assistance, development in Afghanistan has most often been aligned with modernization theory, as espoused by Walt Rostow (1960), with Afghanistan looking to follow in the footsteps of more developed, industrialized countries. The main approach to development has focused on the Afghan Government working with external entities that funded, planned, and implemented development projects that were large in scope, concentrated on the adoption of foreign technology, and addressed the state as the main agent of development. The models they looked towards included countries such as the Soviet Union, the US, Germany, and other European countries. The results were large infrastructure projects that did little to improve the lives of the general population.

Following the end of the Taliban regime in 2001, development organizations, NGOs, development contractors, and funders began to establish themselves in Afghanistan in large numbers.[92] Though many of the projects that followed in the post-Taliban period would remain firmly entrenched in the fundamentals of modernization theory, numerous de-

92 A limited number of development stakeholders remained in the country during the Taliban Regime, but this group was quite small as compared to the large number of NGOs operating in the post-Taliban time period.

velopment actors presented, and advocated for, a more participatory approach to development. They brought with them a newly revitalized appreciation for participatory development approaches, which had been growing in the previous two decades within the general development discourse, as already shown. This approach has spread and has been adopted at multiple levels in Afghanistan today, with donors seeking to support programs that involve the participation of the people.

In order to put into perspective the current trend in Afghanistan towards increased development programming that uses participatory approaches, understanding the political and development history of Afghanistan is important. While the political and military history of Afghanistan has been written about by many authors, from academic scholars to eyewitness accounts (Dorronsoro 2005; Dupree, L. 1980; Ewans 2002; Giustozzi 2008; Goodson 2001; Johnson and Leslie 2004; Rashid 2001; Rubin, B. R. 2002), very little has focused intensely on historical development and modernization efforts in Afghanistan, with a few exceptions, such as Brant (1974) and Zakhilawal and Thomas (2008: 2). Most information on this topic has to be gleaned from other historical accounts, and then compiled. This has therefore resulted in a narrative that, while comprehensive, still retains some periods where confirmed data is not available. However, one can clearly see how the historical development efforts in Afghanistan aligned with the dominant development theories and approaches of the time.

4.1.1 Early Modernization Efforts (1919-1945)

The main story of Afghanistan's modernization begins in 1919, when Amanullah Khan took control of the land-locked country and became King.[93] After liberating the country from its historical ties with Britain and regaining control of the country's foreign policy, Amanullah began

93 Amanullah took over after his father was assassinated on a hunting trip (Barfield 2010: 179; Baxter 2007: 24). His uncle Nasrullah proclaimed himself the *Amir* but Amanullah challenged him (Barfield 2010: 180) and became King and his title changed to Amanullah Shah. He had been sent to the Military Academy for his education (Poullada 1973: 111) and was in control of both the treasury and the army when his father was killed (Baxter 2007: 24). Upon taking the throne, he threw his uncle into jail, where he died a few years later (Barfield 2010: 181).

to focus on the internal development of Afghanistan.[94] During his ten years as the ruler of Afghanistan (1919-1929) Amanullah maintained a strong focus on modernization and how to bring Afghanistan forward. Leon B. Poullada (1982: 231), a noted historian on Afghanistan under Amanullah and a former US diplomat, described the King as having, "strong anti-British sentiments and a frantic desire to modernize his country." Following Amanullah's tour abroad in 1928, his vision of a modernized country was even more strongly based on Westernized countries, such as Germany, Italy, and Turkey. He believed that by imitating them and gaining their technical and financial support, Afghanistan could follow in their development footsteps, aligning with the modernization development theory.

Amanullah's drive for modernization was almost solely supported by the Soviet Government, as the Afghan Government had only very limited revenues to invest in development. The USSR provided Afghanistan with, "aid in the form of cash, technology, and military equipment" (Baxter 2007: 26). These efforts, combined with the USSR being the first nation to formally recognize Afghanistan's independence (Poullada 1982: 231), created a positive image of the Soviet Union in the Afghan Government.[95] This resulted in the USSR supporting small aid projects throughout the country in the 1920s and 1930s and the two countries signing a friendship treaty in 1921 (Goodson 2001: 47). The Soviet Government introduced the telegraph and provided communications equipment (Gregorian 1969: 246), establishing telegraph lines between Kabul and Mazar-i-Sharif, and Hirat and Kandahar in 1924 and 1925 (Dupree, L. 1980: 451).

At the same time, however, Amanullah tried to remain independent and neutral, hoping to avoid becoming aligned with the Soviet Union or their competitors, the British. Therefore, Afghanistan signed similar treaties with additional countries such as Turkey and Iran in 1921 and

94 Britain had controlled Afghanistan's foreign policy since 1879 (Poullada 1982: 231). Amanullah's *ǧehād* (Jihad) against the British, which was declared in May 1919, ended with the signing of the Treaty of Rawalpindi, in which Afghanistan gained full independence (Barfield 2010: 181; Dorronsoro 2005: 41).When the country gained its independence from Britain, it resulted in the loss of British subsidies, which had been around 120,000 to 180,000 pounds annually (Hager 1983: 105; Rubin, B. R. 1994: 201).

95 L. Dupree (1980: 451) notes that the first foreign aid provided by the Soviet Government consisted of technical experts and airplanes provided in 1919.

Italy, France, Belgium, and China between 1921 and 1923 (Gregorian 1969: 233; Wilber 1962: 179). An Afghan-German Treaty of Friendship was signed in 1926, which resulted in the Germans playing a role in the development plans of Amanullah, though to a lesser degree than the Soviets (Ikbal Ali 1928: 253). The Germans sent town planners, engineers, teachers, and other professionals to help support the development of the country (Kreutzmann and Schütte 2010: 3). They also opened up two hospitals in Kabul, one for males and one for females, in 1923, and helped Amanullah to establish a small munitions factory with equipment from the German Knupp Company (Gregorian 1969: 252). According to one historian, Germany was seen in Afghanistan as being the best source for technical aid (Wilber 1962: 180).[96]

In December 1927, Amanullah began his "grand tour." This tour, which would last through July 1928, took him through Europe, Turkey, and the USSR (Gregorian 1969: 256).[97] He was impressed by the modernization efforts he saw and was convinced that, "Afghanistan must imitate the West" (Wilber 1962: 154). "By the Grace of God our sublime Government will employ measures of reform as may prove suitable and useful to the country and the nation so that the Government and nation of Afghanistan may make and gain great renown in the civilized world and take its proper place among the civilized powers of the world" (Amanullah as quoted in Gregorian 1969: 239). Poullada (1973: 168) believes that this trip, "convinced Amanullah more profoundly than before that Afghanistan was actually losing ground in the race for modernization."

Amanullah's policies clearly aligned with the growth theories that became dominant in the development discourse in the 1950s: modernization or growth theory. He focused on building factories with imported technology and large infrastructure projects, whose need was dictated from the top-down. Amanullah also supported small, government industries to produce matches, soap, textiles, etc. with tools and technology imported from Japan and Germany (Gregorian 1969: 252;

96 Barfield (2010: 206) notes that Germany was popular in Afghanistan because it fought against the British and the Russians during WWI alongside the Muslim nation of Turkey.

97 Barfield (2010: 188) dated his travels as having started in November 1927 instead of December, ending in June 1928 instead of July.

Poullada 1973: 79, 139).[98] These industries were protected from outside competition, in the tradition of import substitution industrialization (ISI) (Gregorian 1969: 253; Poullada 1973: 79). By eliminating outside competition and supporting internal industries, the expectation was that these industries could grow and become competitive. According to Poullada (1973: 151), many of these small industrialization efforts were not fully implemented, as some machinery purchased in Europe did not arrive until Amanullah had already been forced from power.[99] Some historians, such as Barfield (2010: 169) see Amanullah's modernization efforts as a failure.

Amanullah did not make efforts to involve the population of Afghanistan in the development decisions being made in Afghanistan. He saw development as coming from external sources and the involvement of the population in such decisions as unnecessary, as would also be thought by future rulers of Afghanistan. During this time, the state was still the main actor in development and local knowledge was not valued in the way it would come to be valued with participatory development approaches.

Unfortunately for Amanullah, his time in command was to be limited, and his rule ended in January of 1929.[100] Though some people associate his fall with the rejection of his social policies – such as the "emancipation" of women, the abolishment of polygamy, and the imposition of compulsory education of both boys and girls by Pashtun and Tajik groups and his legal reforms (Nicosia 1997: 254), Poullada (1973: 158) believes that his downfall was a "tribal revolt in the classical pattern." Poullada (1973: 151) and Dorronsoro (2005: 59) support the view that the rebellion was prompted by the attempt of the central govern-

98 German engineers, architects, and technicians started working in Afghanistan in the spring of 1922 (Nicosia 1997: 247). The number of Germans working in Afghanistan would rise to 100 by 1928, despite attacks on them (Nicosia 1997: 252).

99 Amanullah purchased machinery for tanning, gunpowder, cement, soap, textiles, stone and woodworking, ice-making, vegetable oil, sugar, dairies, foundries, and more during his travels in Europe (Poullada 1973: 139).

100 Amanullah abdicated the throne and gave it to his brother Inayatullah when the Pashtun rebellion was too strong for his forces to hold off (Barfield 2010: 191; Dorronsoro 2005: 59; Gregorian 1969: 265; Magnus and Naby 2002: 42). However, Inayatullah could not hold onto power and Bacha-i-Saqao took over.

ment to put an end to the collection of protection money on the roads of the Shinwari.[101]

Amanullah's overthrow was in part orchestrated by Habibullah Kalakani, who obtained the nickname Bacha-i-Saqao.[102] He was not an elite royal figure, but instead was an illiterate man from Kohistan. His takeover of the government was one of the first rounds of destruction that the country would see. This view is supported by Gregorian (1969: 274) who said that, "The fall of Amanullah and the rise of Bacha-i-Saqao were serious setbacks on the road to modernization." Bacha-i-Saqao promised the Afghan people that he would rule according to the strict principles of the Koran and *Sharia* (*Šari'at*), in contrast to his predecessor who Bacha-i-Saqao believed had led the country away from the true path of Islam with his Western modernization plans (Gregorian 1969: 265). Though his reign only lasted a short time (nine months), he and his men made fast work of spreading destruction throughout the country (Magnus and Naby 2002: 43). They destroyed telegraph lines, roads, schools, administrative offices, and industrial machinery (Guha 1964: 424). Luckily for the country, his rule was shortened by the rise of Nadir Shah (*Šāh*) who helped lead the battle against Bacha-i-Saqao.[103]

Nadir Shah ruled Afghanistan from 1929-1933.[104] A member of the royal family and former commander-in-chief of the army, he was in a strong position to take over Afghanistan (Poullada 1974: 42). In an effort to appease the religious leaders, Nadir Shah retracted most of Amanullah's reforms, with the exception of his educational reforms (Ewans 2002: 139). Though Nadir Shah is noted as having desired modernization for the country, he is seen as having learned lessons from Amanullah's attempt to bring about modernization quickly, without taking into consideration local traditions and ways of thinking (Gregorian 1969: 293). In an attempt to find a different path to modern-

101 The Shinwari are a Pashtun tribe traditionally associated with Eastern Afghanistan and across the border into Pakistan. It is known for traditionally being involved in trade.

102 Bacha-i-Saqao is also referred to as Habibullah Ghazi, the name he chose for himself (Gregorian 1969: 276).

103 Bacha-i-Saqao was hanged as a traitor to Afghanistan November 1, 1929, along with his ten most important supporters (Poullada 1973: 195).

104 Nadir Khan became the King of Afghanistan, which resulted in the change of his name to Nadir Shah, on October 16, 1929 (Magnus and Naby 2002: 44). He was declared king by a national *jirga* (Barfield 2010: 169).

ization, Nadir Shah aligned himself with the religious powers and put forward anti-foreigner policies (Gregorian 1969: 295, 306). The only foreign money that Nadir Shah is said to have accepted during his reign was 170,000 pounds from the British, which he used to build roads, increase education, and to "subjugate" the tribes of the North and South (Dupree, L. 1980: 461). His modernization efforts were less intrusive than Amanullah's, as they focused a great deal on physical infrastructure and less on societal norms. Most of his efforts went into road construction and communications (Baxter 2007: 30). By 1933, 2,000 miles of road had been laid (Gregorian 1969: 366). He also expanded cotton, fruit, and karakul exports along with creating the first bank in Afghanistan (Ewans 2002: 142).

In some sense, Nadir Shah had isolated Afghanistan from other countries, in the spirit of the import substitution industrialization approach, which was also being explored in Latin America at this time, though it would not be advocated for in a wider sense until the emergence of dependency theory in the 1960s. These efforts were supported by the establishment of *Bank-i-Melli* (Afghan National Bank) in 1932 (Barfield 2010: 203). This was seen as "the beginning of modern Afghanistan's economic development" by some (Brant 1974: 92; Dorronsoro 2005: 32). Over its first thirteen years of existence, the bank invested in 50 different companies, some were industrial and some were focused on trade (Brant 1974: 93). The bank helped support the establishment of cotton production, sugar beet production, soap factories, a match factory, a canning plant, and other industrial enterprises known as *shirkats* (*šarākats*) (Brant 1974: 93).

Nadir Shah did not retain power for long. He was assassinated while visiting a school in Kabul in 1933 (Barfield 2010: 197; Magnus and Naby 2002: 44). Upon his death, Afghanistan remained in a less than industrialized state, with only a few factories and very little progress seen in the field of agriculture, except for support of the growth of cotton in the North (Gregorian 1969: 314, 319).

Nadir Shah's son, Zahir Shah, took his father's place and ruled Afghanistan from 1933-1973. He was only 19 years old when he became King, and therefore his uncles held the reigns of control and commanded the government during the first half of his rule (Dorronsoro 2005: xv; Magnus and Naby 2002: 45; Poullada 1974: 42; 1981: 179). Zahir Shah's uncle, Sardar Muhammad Hashim Khan, was Nadir Shah's first Prime Minister from 1933 to 1946. Under Hashim's leadership, transportation and communications remained the main focus of development work, though the government also concentrated a great deal on the military and the economy, with external assistance (Baxter 2007: 31).[105] Small, low-skilled industries began to emerge in the 1930s using the strategy of import substitution industrialization (ISI) (Guha 1964: 438), such as had been seen earlier. Shoes, textiles, matches, and pottery businesses were pursued to fulfill domestic demand, though this never really succeeded.

During this time, the Soviet Government was the main supporter of Afghanistan, providing both aid and advisors (Magnus and Naby 2002: 47). By the middle of the 1930s, Afghanistan had opened up further to external support and was receiving foreign assistance from Germany, Italy, and Japan, with Germany being quite influential (Dupree, L. 1980: 478; Ewans 2002: 143; Poullada 1981: 180). The Afghan Government, seeking a more modernized state, turned to Germans to help it set up factories and hydroelectric projects (Baxter 2007: 31; Kreutzmann and Schütte 2010: 3). In 1939, there were between 100 and 300 German experts located in Afghanistan, working on building roads, hydroelectric plants, and factories (Gregorian 1969: 380). Unfortunately for Afghanistan, most of the work done by Germany and the other two countries proved unsustainable. Many of the small dams and bridges were not strong enough to withstand the spring floods and were washed away, and only one hydroelectric power station in Wardak reportedly survived the floods and remained functional (Dupree, L. 1980: 479).

Moving into the 1940s, development aid was used to implement irrigation works programs and the introduction of cotton and textile production (Lieberman 1980: 281). World War II, however, would have a negative impact on these and other development efforts in Afghanistan.

105 Hashim looked to Germany for assistance in an attempt to avoid turning to the Soviet Union or the British (Baxter 2007: 31).

4.1.2 Cold War Competition, Politics, and Development Aid Intertwine (1946-1978)

World War II caused a slowdown in the economic development of Afghanistan (Dupree, L. 1980: 480; Gregorian 1969: 390; Wilber 1962: 22). Barfield (2010: 208) notes that during WWII, "[Afghanistan's] modernization was stopped in its tracks, and capital fled the country for India." Despite the country's attempt to remain neutral in the war, external pressure from the USSR and Britain resulted in all nationals of the Axis nations being expelled from Afghanistan, taking with them their technical expertise (Gregorian 1969: 390; Guha 1964: 424; Poullada 1981: 181).[106] "The failure of materials to arrive from Germany and other countries halted the construction and equipment of new factories" (Wilber 1962: 23), and the lack of replacement parts and proper maintenance resulted in greater damage down the line (Guha 1964: 424).

The war, which shifted international power structures, resulted in a change of foreign sources of aid in Afghanistan. Starting in 1946, through 1980, the USSR was joined by the US as one of the country's main foreign aid suppliers, though Amstutz (1986: 20) notes that the US Government provided very little economic aid to Afghanistan in the first decade following the end of the war.[107] The Afghan Government repeatedly requested both economic and military assistance and aid from the US during this decade, but the US remained uncommitted (2001: 49; Poullada 1982: 233; Wakil 1989: 361).[108] The US, however, did agree to support the Hilmand Valley irrigation project (Poullada 1982: 233).[109] This limited development aid would expand as Afghanistan be-

106 A *Loya Jirga* was held in Afghanistan in which it was decided that the Afghan Government should abide by the request, resulting in the expulsion of all Axis country foreigners (Gregorian 1969: 388). Barfield (2010: 208) notes that the Afghan's were concerned after seeing Iran invaded and divided up by the British and Russian forces for being too pro-German, and did not want the same fate.

107 The USSR and US, combined, contributed 1.8 billion USD in development aid to Afghanistan between 1950 and 1979 (Williams, Keen et al. 1988: 1).

108 Poullada (1974: 44) asserts that the Afghan request was made with the provision that the US provide Afghanistan with a security guarantee of its borders, which the US was not willing to agree to.

109 Hilmand can also be written as Helmand and is often written so regarding the irrigation project.

came a stage for the Cold War battle between the governments of the US and the USSR.

In 1946, Sardar Shah Mahmud Khan, another uncle of Zahir Shah's, was appointed the Prime Minister of Afghanistan.[110] He served from 1946 to 1953. This, however did not correspond to any dramatic shifts in development policy, and most aid continued to come in from the Soviet Government, which was combining diplomatic efforts with economic, technical, and military aid (Goodson 2001: 49). The two countries of Afghanistan and the USSR signed an important trade agreement in July of 1950 (Baxter 2007: 34). Major development aid stemming from the Soviet Union included projects to build petroleum storage facilities, assistance in exploring for oil and gas reserves, and the free transfer of goods through Soviet territory (Baxter 2007: 34).

US Government aid, at this time, was mostly concentrated on the Hilmand River Valley irrigation plan to build the Kajaki and other dams and provide irrigation for local farming in Hilmand and Kandahar.[111] This was a large scale development project that focused on flood control, reservoirs, irrigation for farming, a power plant, and education (Lieberman 1980: 286). It also involved land reclamation and resettlements, and was expected to positively impact two million people, or one fifth of the population of the country (Williams, Keen et al. 1988: 1). The Kajaki Damn was completed in 1952, with a storage capacity of 1,495,000 acre-feet of water storage (Williams, Keen et al. 1988: 74).

The Hilmand River Valley project was not successful on many fronts, and acted as a major drain on the Afghan Government's finances (Ewans 2002: 149).[112] "It was, in fact, to be one of the first of the

110 Hashim Khan abdicated his position of power to his brother Sardar Shah Mahmud Khan when he became ill (Barfield 2010: 199).

111 The US company Morrison-Knudsen, and eventually USAID, were involved in the Hilmand Valley irrigation project. The project started in the 1930s with support from Japan (Wilber 1962: 236; Williams, Keen et al. 1988: 71). Morrison-Knudsen and USAID became involved in 1946, signing a 17 million USD contract with the GoA (Poullada 1981: 181; Williams, Keen et al. 1988: 1). A few authors assert that USAID only became involved in 1952 (Ruttan 1984: 396; Wilber 1962: 237). The US is estimated to have contributed 54 million USD to the Hilmand project (Williams, Keen et al. 1988: 3).

112 The Afghan Government applied for loans of over 80 million USD from the Export-Import Bank of the US during the life of the project (Cullather 2002: 524). In the 1950s and 1960s, it is estimated that one fifth of the Afghan Government's expenditures went towards the Hilmand Arghandab Valley Authority

economically, socially and environmentally disastrous big dam projects with which the world has subsequently become depressingly familiar" (Ewans 2002: 150). Williams et al.'s (1988: 2) overview of the project indicates that by the late 1980s, it was seen as being too large in scale, with an unrealistic timeframe in an environment where management skills were not up to the task. Very little information was shared with the local population living along the river. Louis Dupree provides evidence of this with interviews he conducted near the Arghandab Dam during its construction. He said, "...the people in the village had never even heard of the project, a common phenomenon up and down the Hilmand and Arghandab rivers" (Dupree, L. 1980: 501). When the water irrigation system delivered water in the off-season the farmers were happy, but they did not know how to deal with the additional water that followed. The extra water flooded their fields, resulting in lower overall production in 1955 (Dupree, L. 1980: 502; Wilber 1962: 241; Williams, Keen et al. 1988: 2).

The Hilmand River Valley project is the best known irrigation project in Afghanistan – it has been the most controversial of the projects and had the most destructive impact on the Afghan economy – but it was not the only irrigation project implemented. Siemens-Schuckert, of Germany, built a damn in Nangarhar to support a hydroelectric plant in Surobi that provided 22,000 kw of power (Wilber 1962: 236). Another damn was constructed upstream at Naghlu with Soviet aid that was expected to produce 60,000 kw (Wilber 1962: 236). The Darunta Dam was also built with Soviet aid and was expected to provide 100,000 kw of power by 1964 (Wilber 1962: 236).

As can be seen, development under Sardar Shah Mahmud Khan shifted back towards a modernization focus, with large infrastructure projects seen as a path towards development. Afghanistan, at this point, also opened itself up to outside assistance, as is shown by the US Government's commitment to provide 1.5 million USD annually for five years starting in 1951, based upon Truman's Four Point program (Williams, Keen et al. 1988: 2). The money was intended to provide agricultural, vocational, and education advisors and trainings to Afghanistan (Williams, Keen et al. 1988: 2).

(HAVA) project and resulted in great internal economic strains for the government (Cullather 2002: 525).

In 1953, Mohammad Daud Khan, a cousin of Zahir Shah, took over from Sardar Shah Mahmud Khan as Prime Minister of the country. During his reign from 1953-1963, development was seen as making serious progress in Afghanistan in terms of industrialization and modernization. Most of the development projects of this time were large scale construction projects, such as hydroelectric power plants, highways, and major civil installations (Newell, R. S. 2007: 137). Daud, successfully funded these projects by obtaining foreign aid from multiple sources, in particular from the Soviet and US Governments (Newell, R. 1974: 80).

The Soviet Union continued to dominate Afghan aid under Daud's rule, as there were tensions between Daud and the US Government (Poullada 1981: 183-4).[113] In 1954, the Soviet Government provided Afghanistan with a loan of 4.5 million USD to construct grain elevators, a flour mill, and a bakery (Amstutz 1986: 24).[114] It also provided Afghanistan with 8.2 million USD to build two cement factories, a fruit canning factory, and more the same year (Amstutz 1986: 24). The focus of these development projects was large in scale, removed from the local population, and technical in nature, requiring outside experts. Development in Afghanistan at this time, continued to be grounded in modernization theory.

In 1955, the USSR would continue their support by providing a 100 bed hospital in Kabul, along with 15 buses, and 100 million USD in aid credit to build hydroelectric plants, the Salang Tunnel, and Bagram Airbase (Amstutz 1986: 24; Poullada 1982: 237).[115] The Soviets also took on the construction of the airfield in Mazar-i-Sharif and improved the airfield in Jalalabad (Baxter 2007: 37; Dupree, L. 1980: 523; Wilber 1962: 197). State citrus and dairy farms were also founded in Jalalabad, but these were not seen as very successful (Amstutz 1986: 24). Projects con-

113 According to Poullada (1981: 184), Daud had information that the American ambassador to Kabul, Angus Ward, was actively working against him and his regime and he took this to be official US Government policy.

114 Williams et al. (1988: 24) say that the Soviet Government constructed 2 grain silos, a flower mill, and a bakery in Kabul between 1950 and 1955, costing 3.5 million USD.

115 Brant (1974: 95) records this as having occurred in 1957 instead of 1955, while Williams et al. (1988: 3) indicate that the Soviet Government provided a line of credit for 100 million USD in 1956. Brant (1974: 95) says that the Soviet Government, following an initial 100 million USD, provided an additional 436 million USD in loans and grants over the subsequent 15 years.

tinued to be announced in the following year, including auto maintenance/repair shops, irrigation dams, a materials testing lab in Kabul, a fertilizer factory, and some other physical infrastructure projects (Dupree, L. 1980: 508-9).

In 1956, the Afghan Government presented its first five-year plan for development (1957-1961), which placed focus on the need for roads to be asphalted (Brant 1974: 95; Wilber 1962: 193).[116] The plan included 32 different projects, not all of which were implemented (Brant 1974: 96) and was expected to cost 280 million USD (Williams, Keen et al. 1988: 5). Brandt notes that this first plan was, "an extremely uneven document" (1974: 96).[117] The Soviet and US Governments provided many of the necessary experts for the projects, along with West Germany and the UN (Williams, Keen et al. 1988: 5). The first five-year plan was funded 90% by foreign aid, the second plan was funded 76% by foreign aid, and the third plan was funded 72% by foreign aid (Williams, Keen et al. 1988: 23).

The second five-year plan (1963-1967) also focused heavily on transportation and was seen by Jensch (1973: 80) as an extension of the first five-year plan. The Soviets and the US supported this plan; both provided technical expertise and funds towards the construction of roads in the country. The second five year plan was budgeted at 800 million USD (Williams, Keen et al. 1988: 5). The third five-year plan, would come later (1968-1972), and was seen as an improvement on the two previous plans (Brant 1974: 97), as will be explained below. The period of the three five year development plans (1956-1978) has come to be called the period of "guided development" in Afghanistan (Lieberman 1980: 272), with the guidance coming from external, not internal sources. Dorronsoro (2005: 64) wrote, "The role of foreign experts was decisive in the selection of aid projects, through which the donors sought visibility to demonstrate their technological prowess, but which succeeded only in creating modernized enclaves."

---

116 Other authors, such as Jensch (1973: 33) document the first plan as having covered 1956-1962, which then meant the second plan covered 1962-1967 and the third plan 1967-1972. These differences in dates are likely due to different interpretations of the Afghan calendar.

117 In the first plan, it was noted that Afghanistan was lacking in statistical data and in experience in planning (Brant 1974: 95). These concerns have remained relevant to the present day.

Under the second plan, the United States and the USSR both contributed to the construction of the ring road that circles the country, with the US Government concentrating on the sections south of the Hindu Kush, and the Soviet Government focusing on sections to the north (Dupree, N. H. and Gouttierre 2007: 59; Wilber 1962: 193).[118] In addition, the Soviet Government paved the streets of Kabul at this time (Dupree, L. 1980: 513). Both countries also helped to build up the higher education system in Afghanistan. The US Government built Kabul University in the late 1950s and 1960s, and the Soviet Government completed the Kabul Polytechnic Institute in 1967 (Dupree, L. 1980: 550, 640).[119]

The US Government, not wanting to be viewed as behind the Soviets in the foreign aid game in Afghanistan (Ewans 2002: 157), took on the project of building the Kandahar Airport in 1956, which it completed by 1962, after spending an estimated 15 million USD (Dupree, L. 1980: 513).[120] They also helped to build up the local Ariana Airlines (Dupree, L. 1980: 514; Ewans 2002; Poullada 1981: 185; Wilber 1962: 197), which PanAm bought a 49% share of in 1957 (Williams, Keen et al. 1988: 46). The US supported the airline with technicians and the provision of training for Afghan technicians until 1962 (Wilber 1962: 197). The US Government also invested 30 million USD in the construction and reconstruction of five airports, helped to train pilots and technicians, and provided planes and ground equipment (Williams, Keen et al. 1988: 18).[121] Between 1950 and 1969, the US Government provided 31%

118 West Germany also contributed to the construction of the ring road in Paktya (Dupree, N. H. and Gouttierre 2007: 59), which sits about 100 km (about 62 miles) to the south of Kabul. The Soviet Government built 75 km (about 46 miles) of road from Mazar-i-Sharif to Khalif and 400 km (almost 250 km) from Kushka to Hirat and from Farah to Girishk (1962: 193). The US Government built 145 miles (about 233 km) of road from Kabul to Torkham, 315 miles (just over 500 km) of road from Kabul to Kandahar, and 65 miles (about 104 km) from Kandahar to Spin Boldak in Hilmand (Wilber 1962: 194).

119 The Kabul University campus was completed in 1964 (Dupree, L. 1980: 599). The US is estimated to have spent ten million USD on education in Afghanistan between 1951 and 1960, which made it the largest recipient of AID education assistance in the world (Williams, Keen et al. 1988: 8-9). Half of these funds went towards teachers and participant fellowships and the other half went towards equipment and the construction of facilities (Williams, Keen et al. 1988: 9).

120 This airport was seen by the early 1980s as a "white elephant" that was criticized by both Afghans and the US (Albert 1974: 254; Poullada 1981: 185).

121 The five airports include: Kandahar, Hirat, Jalalabad, Kunduz, and Mazar-i-

of the total aid budget of the country, while the Soviet Government provided 54% (Dorronsoro 2005: 63).

Some view the development projects conducted by the Soviet Government in this time period as self-serving, with the understanding that they were helping to prepare Afghanistan for the USSR invasion that would come in 1979 (Amstutz 1986: 30; Lieberman 1980: 283). Others noted that some of the Soviet projects were less than successful, such as Louis Dupree (1980: 526); the bakeries produced bread that the local population did not like and many of the construction projects ran behind schedule, with some never being completed.

Daud stepped down from his position as Prime Minister in 1963, leaving behind a legacy of completed infrastructure projects.[122] Louis Dupree says of Daud, " Sardar Mohammad Daoud brought Afghanistan quickly, and sometimes brutally, into a program of modernization" (1980: 556).[123] This sentiment was echoed by Albert (1974: 254), who said that Daud had pushed Afghanistan, "not so gently, into a program of modernization." In general, it can be seen that the projects implemented under his rule were large-scale projects that were focused on economic growth and not on human development. The projects were mostly designed by foreign funders and the distribution of benefits does not appear to have been considered in the planning process. Participatory development was not widely considered.

The Government of Afghanistan did support one known participatory development with the establishment of Village Rural Development Councils in Kabul in 1956 (Carter and Connor 1989: 14). Communities had to have, or create, a community council in order to apply for grants from the Rural Development Department. The projects implemented were supposed to be proposed by the community council and supported by them through contributions of time, money, materials, or labor (Carter and Connor 1989: 14).

Sharif. The total cost was estimated to be 20 million USD (Williams, Keen et al. 1988: 46).

122 Some felt Daud's resignation was because of his stand on Pashtunistan (e.g., Goodson 2001: 52), while others felt that it was to make room for democratic liberalization to take place (Poullada 1974: 45).

123 Though written differently, Louis Dupree refers to Prime Minister Daud. I have seen Daud written in multiple ways: Daud, Dawud, Dawood, and Daoud.

Williams et al. (1988: 56) appear to be referring to the same program when they discuss the Rural Development Department, which they say was established in 1950. This department supported community proposed projects such as water and irrigation, the gravelling of roads, and the building of bridges, with three projects selected per province (Williams, Keen et al. 1988: 58-59). The RRD funds covered 75% of the project costs and the people were expected to cover the rest. Unfortunately, there is little available information on these efforts, how many projects were implemented, or what kind of impact they had, but as it is so rarely mentioned in most academic work from this period, one can assume that it was not the dominant approach to develop throughout the country, or overwhelmingly successful.

After Daud stepped down as Prime Minister, there were five different prime ministers over ten years (1963-1973), but it was really Zahir Shah, who played the largest role (Baxter 2007: 31). During this time, the Soviet Government continued to dominate foreign aid efforts, though the US Government began to assert itself more than previously. Development in the 1960s continued to focus on transportation and communications infrastructure (Dupree, L. 1980: 626), indicating a continuation of the previous trend of top-down, economically focused projects.

The US Government spent five years building the 500 km road (about 311 miles) from Kandahar to Kabul, which was completed in 1966 (Amstutz 1986: 25; Dupree, L. 1980: 637; Goodson 2001: 50; Williams, Keen et al. 1988: 45).[124] The Hirat-Islam Qala Highway was started in 1966, with US Government assistance, and was completed the following year (Dupree, L. 1980: 637; Williams, Keen et al. 1988: 46). US Government development projects were criticized internally as being low in visibility and too long-term in nature, and therefore not impactful enough when compared to the Soviet projects (Amstutz 1986: 25). Poullada (1982: 241) reported that the US development projects took much longer to implement and were much more expensive. The US Government, however, continued its efforts and managed the administration system at Kabul University and provided improvements in transportation, communications, irrigation, and electricity (Dupree, L.

124 The highway reduced the travel time between Kabul and Kandahar from about twenty-two hours originally, to only six hours (Williams, Keen et al. 1988: 45).

1980: 599; Lieberman 1980: 283). US development aid, however, began to taper off in 1966, according to Lieberman (1980: 282).

By 1967, Soviet Government credit to Afghanistan totaled 570 million USD, reaching upwards of one billion USD by 1978 (Goodson 2001: 49). The Soviets focused most strongly on the construction of roads and the production of goods (Lieberman 1980: 283).They built roads in the North, from Termez to Kabul and completed the Salang Tunnel through the Hindu Kush Mountains in 1964 (Goodson 2001: 20).[125] This tunnel created the first all season overland connection between the Northern and Southern parts of the country. The Soviet Government helped build a natural gas pipeline from Shibirghan to the Soviet Union, which was completed in 1968 (Goodson 2001: 50). Additionally, they built a fertilizer plant in Mazar-i-Sharif in the 1970s (Dupree, L. 1980: 638). They helped to build electricity lines between Baghlan and Kunduz, though the fluctuations in flow to the Ghori Hydroelectric Plant were not adequately understood and some factories' electricity needs could not be met (Dupree, L. 1980: 638). They also built the concrete housing blocks in Kabul, known as Microrayon, during this time period (Dupree, L. 1980: 638).

While the main financial aid resources were the governments of the USSR and the US, with the USSR leading, a number of other countries provided development assistance, albeit on a much smaller scale. The West Germans contributed at this time, managing a silo project and electricity projects (Dupree, L. 1980: 641).[126] The French were also contributing by supporting a local high school (Istiqal), a faculty of medicine, and textile companies. Additionally, Yugoslavia loaned Afghanistan eight million USD for the Hari Rud Irrigation Plan (Dupree, L. 1980: 642). The Chinese began to be involved in providing development assistance to Afghanistan in 1965 (Dupree, L. 1980: 641), with a focus on agriculture projects (Williams, Keen et al. 1988: 28). They worked on an

125 The Salang Tunnel goes through the mountains for 1.7 miles (2.7 km) and stands at 3,363 meters in altitude, about 11,000 feet. Its completion reduced the travel time from Kabul to the North of Afghanistan from days to hours (Dupree, N. H. and Gouttierre 2007: 57) by reducing the trip by over 200 km (124.3 miles) (Dupree, L. 1980: 645). At the time it was built, it was the highest motorable pass in the world (Barfield 2010: 210).

126 Williams (1988: 27) notes that the German Government supported power stations, communications, and the regional development of Paktya.

irrigation project in Parwan Province, and small enterprise development projects, such as a textile mill and a pilot fishery (Dupree, L. 1980: 641). Iran also contemplated helping to develop the country. In the 1970s, talk of building a railway link between Kabul and Bandar Abbas in Iran started, but this plan never moved beyond the research phase (Amstutz 1986: 20).

Between 1950 and 1977, Afghanistan received foreign aid that amounted to an estimated two billion USD (Dorronsoro 2005: 63) or 2.3 billion USD (Williams, Keen et al. 1988: 27). As can be seen in the table below (Figure 16), the Soviet Union dominated Afghanistan's foreign aid, but other sources also existed, including the US, the Federal Republic of Germany, the World Bank, China, the UN, the Asian Development Bank, and others. The Afghan government was dependent upon this outside assistance, with two thirds of its annual revenue from 1973 derived from foreign loans and grants (Barfield 2010: 205).

Figure 16: Sources of Foreign Assistance to Afghanistan in Millions of USD (1950-1976)

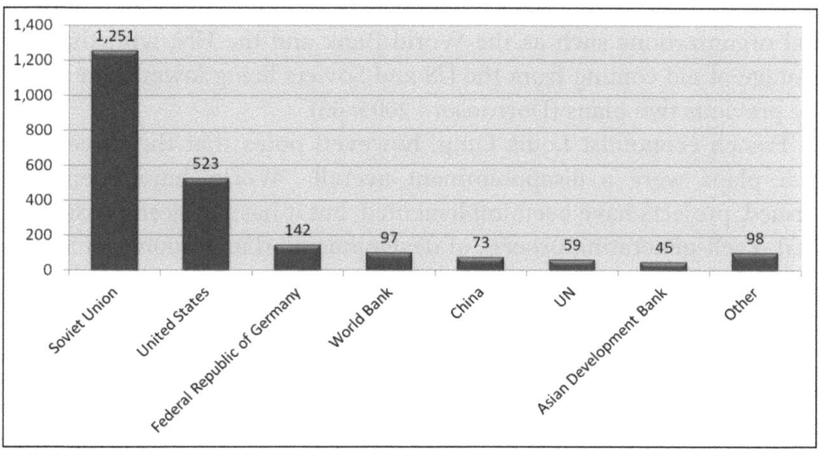

Source: own design based on Williams et al. (1988: 29)

Volunteer organizations had small populations working in Afghanistan at this time, including the Peace Corps, Deutscher Entwicklungdienst, the British Council, and the French National Service, though according to Louis Dupree (1980: 644), none had much more than 200 volunteers. By 1971, there were a few more NGOs and funders operating in Afghanistan beyond those that were already mentioned. The Asia Foundation, the National Organization for Ophthalmic Rehabilitation, and the International Afghan Mission (medical) were all active (Dupree, L. 1980: 630).

The third development plan for Afghanistan (1968-1972) represented a shift from the first two plans and was seen as an improvement. This third plan reviewed the first two plans and also established means for estimating the plan's impact, steps that had not been taken before (Brant 1974: 97). It also had a budget that was one third bigger than the previous plan (Jensch 1973: 167). The development focus shifted from infrastructure and communications to agriculture and irrigation (Dupree, L. 1980: 647), or as Brandt (1974: 97) stated, "emphasis was placed on the need to shift from the predominantly infrastructural composition of the investment programs to more productive projects with more immediate yields." Of secondary importance for this third development plan were industrialization, education, and social services (Dupree, L. 1980: 647). Almost 20% of the aid budget was being funded by multilateral organizations such as the World Bank and the UN, with the percentage of aid coming from the US and Soviets being lower than under the previous two plans (Dorronsoro 2005: 63).

French economist Louis Lung, however, notes that the three five-year plans were a disappointment overall: "Works have been performed, projects have been implemented, but it has not been possible to start a self-generating process of development" (Lung quoted in Sayres 1974: 195). Brandt (1974: 99) agrees, saying, "Afghanistan has received one of the highest levels of technical assistance in relation to its size of any nation in the world. Measured by the results of the development programs, this assistance has not been very productive."[127] Williams et al. (1988: 31-32) found that Afghan leaders and intellectuals were also unhappy with the country's progress and that they believed that the large infrastructure projects were simply not working. A fourth devel-

127 Brant (1974: 103-4) discusses how some projects were poorly planned, "overly costly," and involved outdated technology, and that many projects were not operating at full capacity.

opment plan was drafted for 1973-1977, but a coup prevented it from being implemented (Brant 1974: 98).

In 1973, the former Prime Minister, Mohammad Daud Khan, took over the government in a bloodless military coup while his cousin, Zahir Shah, was in Italy (Barfield 2010: 200).[128] He would stay in power until 1978 and would continue his development efforts from his time as Prime Minister. "President Daoud emphasized in his first proclamation to the Afghan people his intention to increase the pace of economic development" (Brant 1974: 112).

In general, Daud's presidency marked a time of deteriorating relations between the governments of Afghanistan and the USSR, though it still remained the largest source of aid to the country (Baxter 2007: 44). However, due to the strains in the relationship, by 1974, Daud had begun to look for other sources of economic development aid from countries such as Iran, Saudi Arabia, Iraq, and Kuwait (Baxter 2007: 44). Additionally, the government began financing a greater share of its development efforts, with only 66% of development expenditures financed by foreign aid in 1975, as compared to 84.8% previously (Rubin, B. R. 1994: 203-4). It was during this time that the Green Revolution emerged in Afghanistan, with projects to hand out seeds, fertilizer, and pesticides being seen (Magnus and Naby 2002: 48).

Little has been written about the development efforts under Daud's second time in power beyond basic statistics, which simply show Soviet Government aid rising from 470 million USD in 1977, to 1.265 billion USD in 1978 (Christensen 1995: 35). Baxter (2007: 45), however, notes that there was very little economic progress made and that the standard of living for the population did not improve under Daud's regime.

By 1978, the Soviets had finished 71 different development projects in Afghanistan, most of which required Soviet technicians, and 60 more projects had been planned and agreed upon (Goodson 2001: 51; Poullada 1982: 238). The US Government, by 1979, had given 10.6 million USD in grants to Afghanistan (Amstutz 1986: 253). The aid provided from the East and West during the 1970s was lower than it had been in the 1950s and 1960s, the period of peak foreign assistance, as at-

128 Daud took power July 17, 1973 (Dorronsoro 2005: 79). He became the President, Prime Minister, Foreign Minister, and the Minister of Defense (Ewans 2002: 179).

tention and funds shifted elsewhere (Newell, R. 1974: 80; Wakil 1989: 362).[129] This lowering of aid by the two largest aid providers, resulted in increased aid coming into Afghanistan from China, who provided 28 billion USD in credit to develop irrigation, sericulture, and textile manufacturing projects (Newell, R. 1974: 82). However, China's influence did not have a chance to grow very strong.

4.1.3 Communist Rule: Afghanistan Becomes a Battleground (1978-1992)

On April 27, 1978, the People's Democratic Party of Afghanistan (PDPA), a Marxist-Leninist party, inspired a coup that ended in their successfully taking control of the government of Afghanistan.[130] Though they gained power, the PDPA did not hold a great deal of popular support among the population, with only an estimated 10,000 to 18,000 members in 1978 (Barfield 2010: 213; Hager 1983: 110). Following the coup or the Saur Revolution, as it is known, an estimated 3,000 Soviet military advisers from the USSR and an equal number of civilian advisers entered Afghanistan (Hager 1983: 110).[131] The revolution was followed by a great deal of political bloodshed (Barfield 2010: 228).

Internal strife amongst the two factions of the PDPA led to quick shifts in who was in control. The PDPA was made up of two generally antagonistic groups, Khalq led by Mohammed Taraki and Parcham led by Babrak Karmal.[132] Khalq, with Taraki in command, originally gained the upper hand with support from the Government of the Soviet Union, but they did not remain in power for long. Taraki was declared the

129 Brant (1974: 103) noted in 1974, that gross aid had dropped 50% from its 1965 level. Dorronsoro (2005: 65) noted that the US had minimal interest in Afghanistan due to the détante experienced at the time.

130 The coup lasted two days. The PDPA was founded in 1965 (Zaeef 2010: 250).

131 It is called the Saur Revolution because it occurred in the Afghan month of *Saur* (Barfield 2010: 225).

132 The majority of Parcham members were recruited from the Dari speaking Kabul elite, while the Khalq recruits mostly came from Pashtun speaking provincial and rural families (Centlivres-Demont 1994: 364; Dorronsoro 2005: 73). Khalq translates to "the people" and Parcham translates to "banner" or "flag" (Dorronsoro 2005: 70; Magnus and Naby 2002: 101). Dorronsoro (2005: 72) noted that the division of the PDPA into the two parties occurred in 1967.

president of the Revolutionary Council and the Prime Minister on April 1, 1978 (Dorronsoro 2005: 86).[133]

The Khalq Government attempted to implement a number of new policies, including: land reforms in which extra land would be confiscated without compensation, the erasing of most debts, the ending of bride price (*mahr*), and mandatory universal literacy and education courses for both sexes.[134] Many of these policies did not align with the norms of the general Afghan society at the time (Ewans 2002: 191), and therefore were not supported. This resulted in the government increasing its military expenditures to protect itself as opposition groups formed, which in turn resulted in a decline in development expenditures (Rubin, B. R. 1994: 212). These efforts and more, including changing the flag of Afghanistan, united the country against the government and resulted in uprisings (Zaeef 2010: 17), one of the most notable being in Hirat in March 1979.[135] This and other actions forced the government to request the aid of Soviet troops multiple times, though the Soviets did not provide support right away (Barfield 2010: 233; Hager 1983: 110).[136] During the early part of 1979, the abduction of the US Ambassador, Adolph Dubs, and the rescue attempt that ended in his death, resulted in the US Government suspending the majority of its aid projects and pulling its personnel out of Afghanistan (Ewans 2002: 197).

In the months prior to the arrival of Soviet troops in Afghanistan, Taraki was removed from his position as Secretary General, and Hafizullah Amin took over.[137] Amin's government continued to face resis-

133 Nur Mohammad Taraki was a Pashtun born in Ghazni Province. He was a founding member of the PDPA in which he served as the General Secretary in 1965 (Barfield 2010: 213). He became the leader of Khalq after its break from Parcham in 1967 (Dorronsoro 2005: 86-87, footnote 51).

134 Zaeef (2010: 13) said that mandatory education for women and the imposition of a ceiling on wedding dowries was included in edict no.7.

135 The government lost control of the city of Hirat for a few days to a week, depending on the source, following a revolt by a whole garrison in opposition to some of the aforementioned policies (Dorronsoro 2005: 99; Ewans 2002: 197). The siege of the city only ended after being bombed by aircraft. An estimated 20,000 people from Hirat and up to a hundred Soviet citizens were killed (Ewans 2002: 197). Dorronsoro (2005: 99) estimates that 150 to 200 Soviet advisors were killed.

136 The PDPA requested Soviet troops to be sent to the country ten times prior to them entering Afghanistan (Gupta 1992: 281).

137 One month later, Taraki was strangled in his home and was killed as Ewans

tance from the Afghan population, and he additionally was not favored by the Soviet Union (Dorronsoro 2005: 92). The Soviet Government decided that it needed to intervene in order to maintain the gains that had been made in Afghanistan since the Saur Revolution (Ewans 2002: 202). Amin was assassinated on December 2, 1979 and Babrak Karmal took his place.[138] Soviet motorized divisions crossed the Amu Darya River into Afghanistan on December 28th (Ewans 2002: 203). Though the Soviets tried to present their presence in the country as a response to the central government's request for assistance, this does not appear to have been the case, according to Ewans (2002: 204) who debunks each of the stories that were presented by the Soviet Government, concluding that they unilaterally decided to intervene. It is estimated that at the start of 1980, the PDPA controlled only 5,000 of the 35,000 villages in Afghanistan (Barfield 2010: 230).

The Soviet incursion into Afghanistan impacted development projects and funding negatively, and was roundly condemned internationally.[139] Non-USSR or USSR-oriented foreign aid was halted. The US Government stopped all remaining loans and grants in 1980 (Amstutz 1986: 253). The World Bank, the Asian Development Bank (ADB), and the International Monetary Fund (IMF) also cut off their funds following the invasion.[140] The United Nations Development Programme reduced its funding from 14.3 million USD in 1979 to 800,000 USD in 1983 (Amstutz 1986: 253). By 1984, the UNDP had ceased its funding activities entirely (Amstutz 1986: 253). Most UN and international development agencies suspended their work, with the exception of some medi-

---

(2002: 200) tells it, or was strangled a week later in his jail cell, as told by Dorronsoro (2005: 90).

138 Babrak Karmal was born in Kabul in 1929 and died in Moscow in 1996 (Zaeef 2010: 261). He was President of Afghanistan from December 1979 to November 1986.

139 The United Nations Security Council passed a resolution that condemned the invasion and called for Soviet withdrawal starting in 1980, with 104 countries voting for it, 18 voting against, and 18 abstaining (Dorronsoro 2005: 192; Ewans 2002: 209). The actions were also condemned by the Organization of the Islamic Conference (OIC) in 1980 (Dorronsoro 2005: 198). The US, Japan, and West Germany all boycotted the 1980 Olympics in Moscow in protest against Soviet Troops having entered Afghanistan (Ewans 2002: 208).

140 There is conflicting information as to whether this happened immediately in 1979 (Bank Information Center 2011; Carlin 2003: iii) or 1980 as put forward by Amstutz (1986: 253).

cal teams and others, though their work had to be conducted covertly as the Soviets did not allow or approve of their presence (Amstutz 1986: 219, 221).[141]

The main source of financial aid in the 1980s was therefore the Soviet Government and countries closely associated with the USSR. The Soviet Union alone provided 30 million USD in economic aid in 1980 and by 1986 had already provided 1.77 billion USD (Goodson 2001: 63, 143). They even had plans for a 150 million USD project to connect Kabul and a number of other provinces to electricity via the Soviet power grid by 1991 (Goodson 2001: 143). Barnett Rubin (2000: 1793) reports that by 1987, there was greater security, more trade, and greater humanitarian access around the country due to the paved road infrastructure that was built.

Despite the disruption and shifting of funding, there does not appear to have been an impact on the type of development projects being implemented, with some established projects continuing with different sources of financing, such as the Hilmand River Valley project that Czechoslovakia took over in 1980 (Amstutz 1986: 254). Development projects still concentrated on larger scale infrastructure, for example, the communications and power projects that East Germany took over (Amstutz 1986: 254). This, therefore, represented little shift in terms of development approaches.

The PDPA Government and the infiltration of the country by Soviet forces did not go unchallenged. Groups of Afghan men banded together in opposition to the government in Kabul and declared their opposition to the foreign invaders (Ewans 2002: 213).[142] Gathered under different leaders who were mostly based in Pakistan, they concentrated on battling against the Soviet troops in the hopes of freeing their country of foreigners and ridding Afghanistan of the government the Soviets were

141 These medical groups, including Médecins Sans Frontières (MSF) and the Swedish Committee for Afghanistan (SCA), helped establish clinics and hospitals in the 1980s that served the civilians and a minority of resistance fighters (Amstutz 1986: 221).

142 Some of these opposition groups had been active against the central government in Kabul prior to the Saur Revolution, though they gained strength and prominence during the time of the PDPA. The perceived un-Islamic character of the PDPA and its attempt to erase all debt and implement land redistribution helped feed anti-government feelings, leading these groups to action (cf. Magnus and Naby 2002: 146).

supporting.[143] These groups came to be known as the Mujahideen. The battles were fierce and destructive and lasted through the ten years in which Soviet troops were situated in Afghanistan, with many civilians suffering. Their existence made it very difficult for the central government to maintain full control of the country, with an estimated two thirds of the nation under the power of the resistance forces of the Mujahideen as early as 1981 (Magnus and Naby 2002: 130).

During this time, cross-border assistance was conducted via NGOs that covertly entered into Afghanistan from Pakistan, mostly in the form of humanitarian relief and "cash for work" projects (Baitenmann 1990: 71; Christensen 1995: 139).[144] Over 50 NGOs were involved in this kind of assistance, mostly with US Government funding through USAID and the US Information Agency (USIA) (Baitenmann 1990: 71). The US Government provided 120 million USD worth of aid to the Mujahideen in 1984, 250 million USD in 1985, and 630 million USD in 1987 (Wakil 1989: 374).[145]

The US supported limited programs to assist with transportation, medicine, education, food, and agriculture via cross border operations from Pakistan (Girardet 1987; Goodson 2001: 152). This assistance consisted of setting up health, education, agricultural, and logistical programs in resistance areas (Girardet 1987), but was not focused on the larger population of the country.[146] These efforts were needed as, "trade, industry, agriculture and the educational system all began to collapse" in Afghanistan (Ewans 2002: 212). The government was not

143 Afghans began to flee the country after the PDPA came to power, though the numbers were still relatively small. When the Soviet Union invaded in 1979, it produced a large increase in the number of refugees fleeing the country (Margesson 2007: CRS-2). Most refugees then, and now, went to Pakistan and Iran, with a larger percentage going to Pakistan.

144 This cross border aid was illegal in both Afghanistan and Pakistan and, therefore, the requirement that it be covert (Baitenmann 1990: 73). NGOs were not allowed to operate from Iran (Lorentz 1987: 102).

145 Christensen (1995: 129) provides a much lower estimate of only 250 million USD for cross border assistance between 1985 and 1989. Dorronsoro (2005: 208) says that the US provided two to three billion USD to the Mujahideen between 1980 and 1992, which was more or less matched by Arab countries, such as Saudi Arabia.

146 Health programs were the most common focus, as it was easier to receive permission from the Pakistan Government for medical assistance than for other types of assistance (Lorentz 1987: 105).

able to control vast swathes of the country and was very dependent upon the Soviet Union for basic necessities (Ewans 2002: 212).

Prior to 1986, the concentration was on humanitarian oriented projects, as compared to development projects. It was only in 1986 that the delivery of Stinger missiles to Mujahideen groups created enough security on the ground in Afghanistan for the implementation of a few, limited development projects (Baitenmann 1990: 71; Christensen 1995: 128). The focus of such projects included health and education (Christensen 1995: 139). These development efforts were carried out by local and international NGOs who had grown in number in Pakistan, as they worked to provide humanitarian support to the millions of refugees who had fled since the incursion of the Soviets into the country. UN Habitat began to operate in Afghanistan in 1986, focusing on sanitation and urban infrastructure projects (Lister 2007: 4).

The number of NGOs grew so great that a coordinating body for NGOs was established in 1988 – the Agency Coordinating Body for Afghan Relief (ACBAR) – to act as an umbrella organization for NGOs working within Afghanistan and also with Afghan refugees in Pakistan (ACBAR 2010a). When it was first established, there were 40 participating NGOs. The number of participants increased in the coming years. There were 58 NGOs registered with ACBAR by 1989, with 46 of those functioning at least partially inside Afghanistan (Christensen 1995: 136). By 1993 there were 66 NGOs who were members of ACBAR (Christensen 1995: 144).[147]

This rise in NGOs was also accompanied by a rise in participatory development approaches being implemented. The earliest reference to "community development" in the ACBAR Bibliography is in 1978 (Rahimyar 1997: 17). When printed in 1997, this bibliography contained 28 entries that mentioned "community development" (Rahimyar 1997), though it is not clear how participatory in nature these programs and projects were. What is known is that a good amount of the aid entering the country was being directed by commanders that NGOs worked through to reach communities (Carter and Connor 1989: 1). NGOs were able to utilize commanders to gain access to communities so they could work with the local *shura* (*šourā*), or community council, that already existed in seven provinces (Carter and Connor 1989: 25). What Carter

147 ACBAR continues to exist today, with over 106 members (ACBAR 2010b).

and Connor (1989: 25) report happened, is that commanders would act as the point person, but the NGOs would demand that a council be formed if one did not already exist to ease the provision of assistance, which was still mostly humanitarian in nature. The NGOs did not have a great deal of success with getting councils to provide development priorities, as was noted by one researcher, who said that instead of a prioritized list, communities provided shopping lists of needs, ". . . with no understanding of how various aspects might be linked, because shura members [were] not experienced development planners" (Carter and Connor 1989: 33). Despite the frustrated nature of this quote, it does show that local knowledge and information was at least sought, which is a large shift away from more top-down approaches and towards participatory development.

By 1984, the Soviets had completed 170 different economic projects (Amstutz 1986: 253), with more presumably built within the intervening years. By the time they left Afghanistan in 1989, they had helped to build over 2,500 km of road (about 1,553 miles) in Afghanistan and they had built a bridge over the Amu Darya (Amstutz 1986: 256-7). They also left behind a country that had been entrenched in war for ten years, with all of the physical and mental scars that can produce. Over 1.25 million people, or 9% of the population, had been killed, with five million refugees estimated to be living outside the borders and two million internally displaced people (IDP) (Wakil 1989: 359).[148] The Soviet Government estimated that it lost 13,369 soldiers, though unofficial estimates have the number of Soviet deaths at over 50,000 people (Dorronsoro 2005: 192). The land was also littered with anti-tank and anti-personnel mines numbering two per Afghan (Wakil 1989: 359). Wakil (1989: 365) described Afghanistan at that time by saying, "The war has turned Afghanistan into a wasteland." The development situation was dire and the Kabul government was estimated to control only 60 urban centers, with six of the twenty-five provinces already controlled by the Mujahideen (Dorronsoro 2005: 227).

148 Ewans (2002: 235) has a more conservative estimate of 1 million Afghan civilians and 15,000 Soviet soldiers having been killed. Goodson (2001: 93) notes the diversity of estimates, writing that Noor Almed Kahlidid estimated 876,825 had been killed between 1978 and 1987, while Marek Sliwinski had an estimate of 1.25 million for the same period, and Noorzoy had an estimate of 1.71 million.

It was this situation that development stakeholders hoped to address when they reentered Afghanistan following the Soviet departure. Their exit opened the door to international organizations to become directly involved in Afghanistan again. The UN started to take an active role in the country and more NGOs began to operate within Afghanistan, with many of those who had been operating from Pakistan opening offices inside the country.[149] The NGOs and UN worked hand in hand, with many UN projects implemented by NGOs due to the scale of assistance needed (Christensen 1995: 133). Mohammad Najibullah, who the Soviet Government put into power, replacing Babrak Karmal in 1986 (Dorronsoro 2005: 194; Ewans 2002: 229), was not expected to be able to hold onto power after the Soviet troops left Afghanistan.[150] He surprised everyone by remaining in power until 1992. After the PDPA Government of Najibullah fell, it was then followed by a Mujahideen government that was consumed by internal strife and fighting.

The Soviet Union provided both military and financial aid prior to Najibullah's fall, in an attempt to bolster his government after they had pulled out (Arnold 1994: 65). This aid was destabilized, however, by the coup against Gorbachev in August of 1991 (Dorronsoro 2005: 237). Prior to that, the Soviet Government provided an estimated three to four million USD a year (Ewans 2002: 238). As the Mujahideen groups continued to fight for control of the country, the governments of the Soviet Union and the US agreed to end their military support to both sides of the conflict in 1992 (Ewans 2002: 246). In April of that year, Najibullah disappeared after trying to flee the country for India, and he was "stripped of his power" (Dorronsoro 2005: 238). Zaeef reports that Najibullah fled to the UN compound April 16, 1992, where he lived until eventually he was killed in 1996, with his body being publically dis-

149 A number of NGOs were cited as being operational in Afghanistan as of 1989, including: the International Rescue Committee (IRC), Afghanaid, the Australian Relief Committee (ARC), Swedish Committee for Afghanistan (SCA), Save the Children, and Enterprise Works/VITA (Carter and Connor 1989: 25-26).

150 Najibullah was personally selected by the Soviet Government, according to Dorronsoro (2005: 178), who also places him as head of the KHAD from 1980 to 1986. He was a Pashtun speaker from Paktya, and he had married into the royal clan, the Muhammadzai family (Dorronsoro 2005: 193). He was President of Afghanistan from November 1986 through April 1992.

played (Zaeef 2010: 50). This was the end of the Communist government in Afghanistan.

4.1.4 Mujahideen "Governance" and the Internal Destruction of Afghanistan's Infrastructural Assets (1992-1996)

After the fall of Najibullah, the Mujahideen groups attempted to cobble together a government in Kabul. The different groups agreed on a power sharing arrangement in which there was a rotating presidency. Sibghatullah Mujaddidi held the presidency first, taking the position in April 1992, and as agreed, handed over the position to Burhanuddin Rabbani (Magnus and Naby 2002: 150).[151] Rabbani was only supposed to serve a four month term, however, he extended it via an assembly that did not include all Mujahideen factions (Dorronsoro 2005: 239; Magnus and Naby 2002: 163), leading to increased tensions amongst the different Mujahideen groups. The battle for political control did not remain solely political for long, as the already armed factions began to square off. The competition for power amongst the different factions' leaders and their general lack of unity (Magnus and Naby 2002: 163) resulted in a civil war that further destroyed the infrastructure of the county, and of Kabul in particular. By 1996, 40,000 people had been killed and the capital of Kabul was heavily damaged (Dorronsoro 2005: 239). The fighting resulted in little to no development work taking place between 1992 and 1996.[152] The little aid that did exist was focused on humanitarian assistance, and not development.

In order to try to coordinate political conversations and bring about a political solution to the fighting, the United Nations Special Mission to Afghanistan (UNSMA) was set up in 1993 (UN News Centre n.d.) and later became the United Nations Assistance Mission in Afghanistan (UNAMA). The UN put forward requests for funding pledges. The first such request for aid was made in 1994, with a concentration on human

151 Sibghatullah Mujaddidi was born in Kabul in 1925 and led one of the Mujahideen parties from Pakistan in the 1980s. His name can also be written, Sebghatullah. Burhanuddin Rabbani was born in Fayzabad of Badakhshan Province in 1940 and headed *Jamiat-e-Islami* (cf. Zaeef 2010: 262).

152 Mangus and Naby (2002: 159) note that the Mujahideen government received little assistance from abroad, less than the previous regime had received, making the government vulnerable.

development and poverty alleviation in rural communities (UN News Centre n.d.). Only 75% of the aid requested was forthcoming (UN News Centre n.d.). This would remain a trend in the following years.

USAID funding for Afghanistan existed in the first few years of Mujahideen rule, but was terminated in 1994, which resulted in an overall funding decrease (Christensen 1995: 137; Giustozzi and Ibrahimi 2012: 18). Soviet aid decreased after 1989 (Rubin, B. R. 2000: 1791).

Different levels of participation began to be incorporated into the development approaches being utilized by NGOs and other international organizations around this time. These organizations began to involve the community directly in aid projects in order to make projects more sustainable, out of concern that the lack of a functioning government meant that government run projects would not be sustainable (Christensen 1995: 142). The main form of participation at this time appears to have been community contributions.

As the fighting amongst Mujahideen groups continued, a new entity arose from the *madrasa* schools and orphanages of Pakistan: the Taliban.[153] Outraged by the atrocities that were taking place in the civil war, the Taliban emerged and fought the Mujahideen groups for control of the country, bringing more conflict and destruction to the landscape and the people. Ewans (2002: 259) estimates that 100,000 individuals were killed during the time when the Mujahideen groups were "in control" of the country. Especially in the South, the Taliban were seen as bringing order to the chaos that the conflicting Mujahideen groups had visited upon the country (Dorronsoro 2005: 246). They were even able to take over some provinces without firing a shot (Magnus and Naby 2002: 182).[154] There was little development assistance at this time because of the insecurity caused by the fighting, but a few efforts did move forward. In 1995, the UN Centre for Human Settlements helped to rebuild 10,000 homes and the World Food Programme (WFP) continued to provide relief aid with the distribution of over 53,000 tons of food (UN News Centre n.d.).

---

153 *Madrasa* refers to a religious school.

154 The Taliban captured Kandahar in September of 1994, Ghazni at the end of 1994, Hirat in 1995, Jalalabad and Kabul in 1996, and Mazar-i-Sharif in 1997 and then again in 1998 (Dorronsoro 2005: 236, 254, 255, 303; Gordon 2011: 79; Magnus and Naby 2002: 182; Zaeef 2010: 293). Barfield (2010: 258) dates the fall of Ghazni to the Taliban as occurring in 1995.

As of 1995 there were three UN agencies active in Afghanistan, along with 26 NGOs, both national and international in nature, this is evidenced by their participation in creating a development work plan (Dupree, N. H. and Gouttierre 2007: 122). As the fighting intensified around Kabul in 1996, the distribution of humanitarian aid became more difficult (UN News Centre n.d.). This was the year in which the Taliban gained control of the country, having taken over the main urban centers. The Mujahideen were significantly weakened, but they did not give up their fight and continued to battle the Taliban government during the Taliban Reign from 1996-2001.[155]

4.1.5 Taliban Control and Further Destruction of the State's Assets (1996-2001)

The Taliban, as a group, started off as a few hundred theology students, but then expanded to include former Mujahideen members and others.[156] The original intention of the theology students was to "institute a moral order such as it imagined had existed before the modernist reforms of the 1950s" (Dorronsoro 2005: 284).

The Taliban Government was very conservative both socially and religiously. Their social policies banned women from working, attending school, and from having freedom of movement.[157] This had a negative impact on the economic situation and had particularly negative re-

155 Magnus and Naby (2002: 202) attested to the weakness of the Mujahideen at the time, "Holding only some ten to 15 percent of the country and no major city, Rabbani-Mas'ud forces were too weak to demand a coalition government."

156 The Taliban is most closely associated with the Deoband school of Islam from India (Ewans 2002: 265). A distinctive characteristic of the Taliban is that the majority of its supporters are Pashtun. While many argue that it became the name of an established group in 1994, others argue that the Taliban has a history that goes back to the 1980s (Strick van Linschoten and Kuehn 2010: xxvi). From this perspective the Taliban had been a part of the ranks of the Mujahideen in Southern Afghanistan much prior to the meeting held in the fall of 1994, which is considered to be the founding meeting of the group by some (Zaeef 2010: 65).

157 Under the Taliban, there were many restrictions for women that go beyond those mentioned in the text. Women were beaten for showing their arms, there were stoning deaths based on accusations of adultery, and many more restrictions which went down to the level of women being banned from wearing white socks (Ewans 2002: 262).

sults for the health system. NGOs were often pressured by their funders to suspend their work in areas where females were not allowed to participate (Dupree, N. H. and Gouttierre 2007: 123). For example, OXFAM refused to give money to schools that did not allow girls to attend, while SCA decided to give aid to girls' home schools (Magnus and Naby 2002: 207). The Taliban, however countered, saying that they were not against equality for women, but that it had to happen in an appropriate way, and that this would take time and could only be addressed once the infrastructure of the country, the legal system, and the administration had been rebuilt (Magnus and Naby 2002: 208).

Restrictions were also imposed upon men regarding their dress, facial hair, and religious activities.[158] The government made attempts to impose its very stern vision of an Islamic society on the population, with fierce punishments for those who did not obey. They additionally imposed very tight restrictions on aid delivery from 1996-1998 (Afghanistan News Center 1998b). The Taliban's conservative and restrictive nature, and ongoing fighting with the Mujahideen groups, resulted in very little development taking place while the Taliban ruled the country.

It was during this time of extreme restrictions and an embattled government that the UN started a number of participatory development projects in Afghanistan. In the 1990s the UN established five different community-based programs to address urban development, agricultural development, and more, including what came to known as the Community Fora of UN Habitat (Strand, Toje et al. 2003: 26). These five programs were brought under one banner in 1997, the PEACE Initiative (PI) (Bernander 2002: 4).[159] The UN felt that community-based programs were the most appropriate to address the needs of the population in a country that lacked a true central authority (Bernander 2002: 3). The PI was implemented in 30 districts of six provinces, reaching an estimated 10% of Afghanistan (Bernander 2002: 27).[160] While participatory ap-

158 Additional examples of the rules imposed by the Taliban included: men who did not attend daily prayers were punished with up to ten days in jail and those with beards that were less than a fist's width below the chin were imprisoned until their beards had achieved such length (Dorronsoro 2005: 285).

159 FAO, UN Habitat, and UNOPS' Afghan Rural Rehabilitation Programme (ARRP) were all integrated in the UNDP PEACE Initiative (Coleridge 2002: 40).

160 The PEACE Initiative was estimated to have cost 20 million USD between 1997

proaches were gaining in popularity around the world at this time, in Afghanistan it appears the approach was tried in response to the fragile political environment, where sustainability could not be guaranteed by the government, a different motivating factor from that explored in other parts of the world.

The UN was not immune to the restrictions and social policies of the Taliban, and also eventually halted all new programming and anything that was not "life-saving" (Afghanistan News Center 1998a). As of 1997, large international organizations such as the World Bank and the Asian Development Bank were no longer engaging in Afghanistan due to the ongoing fighting (Dupree, N. H. and Gouttierre 2007: 60; Lister 2007: 8). In 1997 the UN evacuated its international staff from the country, though some programs such as the Community Fora continued to function (Lister 2007: 8).

The development environment deteriorated further in the summer of 1998 when the Taliban ordered the remaining foreign NGOs to relocate to the Polytechnic campus, which had been abandoned in 1992 (Goodson 2001: 79; Salahuddin 1998).[161] Out of concern for their safety and an inability to pay for the necessary renovations to the Polytechnic campus, most NGOs evacuated their foreign staff from the country, with very few exceptions.[162] UN and ICRC staff were allowed to remain in country (Salahuddin 1998). This exodus took the expat population from 200, to an estimated 45 in only two weeks (AFP 1998a).[163]

This expulsion, as it was termed by the expats interviewed for newspaper articles, resulted in the European Commission, the largest single donor to Afghanistan at the time, suspending all of its funding to Kabul in protest (AFP 1998b; Salahuddin 1998). It also meant the loss of what

---

and 2001 (Strand, Toje et al. 2003: 21).

161 This was the same Polytechnic that the Soviet Government had built in the 1960s.

162 Only one international NGO decided to concede to the Taliban demands and said they would move to the Polytechnic campus, the International Assistance Mission (IAM) (AFP 1998a). Halo Trust was given permission to remain without moving to the Polytechnic (Afghanistan News Center 1998b). Some of the NGOs that were forced to leave included: Action Against Hunger, Médecins Sans Frontières, Save the Children, Solidarité, and CARE International.

163 While the exodus of expat workers meant the end of services for many NGOs, a few, such as CARE International continued their efforts utilizing local staff and existing resources (Winfield 1998).

was termed the "vital" assistance of NGOs, which employed 25,000 people and brought 113 million USD into the country in 1998 (Dorronsoro 2005: 282). The Taliban Government was *de facto* dependent on the humanitarian aid that had been coming into the country until this point (Dorronsoro 2005: 282).

These organizations had been involved in providing humanitarian relief, such as food, medicine, and clothing to thousands of people living in Kabul. Some were involved in smaller development projects, to rebuild roads, canals, schools, and hospitals, along with the removal of mines (Abdullah 1998; Leopold 1998). NGOs were also responsible for running the city's water supply, subsidized bakeries, and health services (1998a; Salahuddin 1998). While none of these projects were able to satisfy the development needs of Kabul, much less the whole of Afghanistan, without them, there was little to no help coming from anyone else. However, this time period in Afghanistan has not been well studied or documented (Goodson 2001: xii). In his book about the Taliban movement, one of the few from an insider's perspective, the former Taliban Ambassador to Pakistan, Abdul Salam Zaeef (2010: 94), states that the Taliban was actively involved in rebuilding factories that had been destroyed during the previous years of fighting in the country. He reports an increased production of cement during the Taliban reign, along with the construction of industrial parks in Kabul, Mazar-i-Sharif, Hirat, and Kandahar; and the development of coal, salt, and marble mines (Zaeef 2010: 96, 97).

According to Goodson (2001: 95), most of the development aid that entered Afghanistan in 1999 and 2000 was aimed at humanitarian relief, with very few longer-term reconstruction activities taking place. Barfield (2010: 264) notes that the Taliban Government was dependent upon such humanitarian aid from Pakistan and the UN. Most UN funds were earmarked for emergency relief projects and very little was left for projects such as sustainable livelihoods (UN News Centre n.d.). Some small development projects did take place, such as literacy programs to address the low literacy rate among Afghan adults (Dupree, N. H. and Gouttierre 2007: 122).

In addition to these limited development efforts, the UN applied sanctions in 1999 and 2000. The first round of sanctions froze the financial assets of the country abroad and denied Ariana Airlines' planes the right to land in Europe (Ewans 2002: 282). The second round of sanctions placed an embargo on sales of arms to the Taliban, but not the Mujahideen (Ewans 2002: 282).[164]

In March of 1999, international NGOs resumed their humanitarian and development activities in Afghanistan, but only after security guarantees were provided by the Taliban (Goodson 2001: 121-2). Development activities were in decline, however, as security incidents continued to threaten international organizations and NGO workers.[165]

The Taliban government, on its own, appears to have conducted no documented development projects beyond the efforts mentioned by Zaeef. It was only following the removal of the Taliban from power that development efforts really resumed in Afghanistan. Dorronsoro noted that the Taliban government did not focus on bringing development to the country, instead he says that, "The modernizing role was thus transferred from the state to the NGOs" (2005: 289). As the international development community set up their offices in Kabul and throughout the country, a large shift in development was made, as resources, money, ideas, and more flooded into the country.

4.1.6 Post-Taliban: A Tidal Wave of Development Efforts (2001 to present)

The attacks of September 11, 2001 in the US led to the downfall of the Taliban government. The US Government believed that the person responsible for the attacks was Osama bin Laden, who had been residing in Afghanistan. When the Taliban were unwilling to hand Osama bin Laden over to the US, the US Military, along with an international coalition, entered Afghanistan and removed the Taliban from power.[166]

---

164 The second round of sanctions also included a clause that all Taliban offices abroad had to be shut (Zaeef 2010: 272).

165 In 1999, Red Cross workers were beaten and robbed by Taliban soldiers on the way to Bamyan Province and in 2000 there were instances of aid workers being expelled, harassed, and killed (Goodson 2001: 122).

166 The decision of whether to hand over Osama bin Laden to the US was discussed when Mullah Omar called for an assembly of clerics to gather, where it was decided that he would be asked to voluntarily leave, but that they would not hand him over (Barfield 2010: 269).

The Taliban government fell very quickly, and a tidal wave of development aid and attention rapidly followed.[167]

In the post-Taliban time period, development efforts, plans, and funding expanded dramatically. The Bonn Agreement was the first agreement signed after the fall of the Taliban.[168] It helped establish an interim government and was intended to serve as a roadmap for how to transform Afghanistan into a stable, democratic nation (Gannon 2004: 37). Some viewed the Bonn Agreement as a peace agreement, yet it left out one of the essential parties necessary for this to be true: the Taliban. It did, however, pave the way for the establishment of an Afghan Interim Authority (AIA) for six months, and then a Transitional Authority appointed by a *Loya Jirga* for eighteen months until democratic elections could be held (Dorronsoro 2005: 329-330; Magnus and Naby 2002: 209). Presidential elections were held in October 2004, and Hamid Karzai, who had led the interim and transitional governments, was elected president.[169] Presidential elections were again held in 2009, and Karzai was again elected, though the validity of this election was highly contested.

After their defeat, the Taliban disappeared at first, but they regained some of their strength in only a matter of years and began to actively oppose the government and the international forces with violence and force. They continue to battle against the government and its supporters, and have created a security environment that makes development in Afghanistan even more complicated and difficult than it would already be under peaceful conditions. Over the last ten years, the impact the anti-government elements (AGE) have had, and the number of attacks they have successfully executed, have been increasing.

Since the fall of the Taliban, many of the former development stakeholders have reengaged and organizations with no previous experience in Afghanistan have also begun operations in Afghanistan. These organizations, institutes, and NGOs have descended upon the country to ad-

---

167 The military effort was called "Operation Enduring Freedom." It started October 7, 2001 with bombing raids (Dorronsoro 2005: 322). The US waged an air battle and collaborated with the Northern Alliance, which was "a loose grouping of opponents of the Taliban" (Ewans 2002: 286) formed from the Mujahideen groups that continued to exist in the North of the country.

168 The meeting in Bonn was held in November 2001 (Barfield 2010: 283).

169 Parliamentary elections were held in September 2005.

dress the multi-faceted task of developing the country, or to profit off of these efforts, depending on how cynical one chooses to be. The UN alone now has 19 different agencies functioning in the country, with a wide range of focuses: justice reform, economic reform, clean water, sanitation, education, roads, etc. International organizations with deep pockets, such as the World Bank and the Asian Development Bank (ADB) have returned to Afghanistan after having been inactive in the country for twenty-three years (Bank Information Center 2011). The number of NGOs, both national and international has risen dramatically over the past ten years, with over 100 international, and an unknown number of national NGOs, operating throughout the country (ACBAR 2010b).[170]

The establishment of national development policies started in 2006, with the creation of the Afghanistan Compact (IRoA and the International Community 2006), which aimed to create a shared vision of how to make Afghanistan a stable and prosperous country (Goodhand and Sedra 2009: S86). The Compact was signed by 60 different countries and established specific goals, benchmarks, and time-frames for implementation (Suhrke 2007: 1299). The Afghanistan National Development Strategy (ANDS) was built off of the Afghanistan Compact. These two separate processes and documents are not seen as having been entirely Afghan-driven, but instead as "top-heavy, donor-led" in which most of the actual writing was done by external consultants (Carlin 2003: iii; Goodhand and Sedra 2009:S86, S91). Additionally, the documents are seen as having too large a number of goals, and having targets that are not feasible (Goodhand and Sedra 2009: S86).

Billions of dollars have been committed to the development of Afghanistan, with 8.3 billion USD pledged for economic infrastructure reconstruction in 2004 and 10.4 billion USD pledged for the ANDS at the London Conference in 2006, which was finalized as a strategy in

170 The exact number of NGOs operating in Afghanistan is unknown as some organizations that are registered as NGOs are actually contractors, some NGOs are not registered, and government statistics are not terribly reliable. That being said, 1,225 national and 301 international NGOs, or those who have taken on the mantel of being an NGO, were registered with the Ministry of Finance as of May 2010 (Massoud 2010). I have chosen to include the more conservative numbers in the text, out of caution. There were 159 NGOs, 106 of which were international, listed on the Afghan Analysts Network website as of February 24, 2011 (Afghan Analysts Network 2011).

2008 (U.S. Government Federal Research Division 2007: 204). The Afghan Government estimated in 2011 that 57 billion USD has been spent in the country since the start of the US-led invasion (Provost and Cummins 2011), however this does not add up to a great deal when looked at per capita.[171] In 2003, international aid to Afghanistan stood at 50 USD per capita, which rose to 66 USD per capita the following year (Barfield 2010: 316). This compares to 111 USD per capita in aid to Mozambique and 237 USD per capita in Serbia and Montenegro (Barfield 2010: 316).

The ANDS stands as Afghanistan's Poverty Reduction Strategy Paper (PRSP), and addresses various areas for development (Goodhand and Sedra 2009: S85). Many of the projects supported are reminiscent of the 1950s and 1960s modernization model of development in Afghanistan, with a focus on rebuilding roads and larger infrastructure (Suhrke 2007: 1293). The US has focused on, "providing big, highly visible reconstruction projects," such as the Kabul to Kandahar highway, which cost over 250 million USD to complete (Gannon 2004: 42-3), mirroring the previous development efforts funded by the Soviet Government.

Foreign governments have provided 26.7 billion USD in official development assistance to Afghanistan since 2002 (Poole January 2011: 2).[172] Afghanistan was the leading recipient of official development assistance (ODA) worldwide in 2008 and 2009, and has been number one or two in this category since 2004 (Poole January 2011: 2, 6). The development dollars have been spread amongst the different development actors in the country from NGOs (both local and international), to for-profit development companies and contractors, to international organizations, and even the military. Each of these actors has utilized different development approaches. However, in alignment with what has been seen worldwide, development in Afghanistan has seen a shift towards a less top-down approach towards a participatory approach, though this does not mean that all of, or even the majority of, projects are now

171 The OECD estimates that 58 billion USD was spent between 2002 and 2010 (Zyck 2012: 2).

172 All numbers provided should be considered as estimates. Tracking of development aid has been weak, at best, with no accurate record of the total amount having entered the country since 2001 available. Goodhand and Sedra (2009: S99) list the top six contributing donors as including the US, Canada, the European Commission, Germany, Japan, the UK, and the World Bank.

based on participatory approaches. It does, however, mean that there is increasingly room in the development environment of Afghanistan for participatory development projects.

Large physical infrastructure development projects have historically been implemented using a top-down approach, but even some of this work has been done utilizing efforts that include participatory development elements recently. For example, a UNOPS secondary and district center roads project worth 400 million USD relied heavily on consultation and cooperation with community groups (personal communication, UNDP Staff, June 2011). However, many road projects continue to be top-down in nature.[173] Additionally, the Kabul International Airport was rebuilt, as were the airports of Hirat, Mazar-i-Sharif, and Jalalabad (U.S. Government Federal Research Division 2007: 213). As was common historically, development in Afghanistan is currently almost entirely foreign financed, as domestic revenues remain very low (Suhrke 2007: 1301), similar to the 1960s and 1970s (Dorronsoro 2005: 335).[174]

As compared to previous times when participatory approaches were rare, many projects now being implemented are participatory in nature. There are many different community-based education programs, community-based agriculture programs, and more. Zakhilawal and Thomas (2008: 10) explain that, with only a few exceptions, such as the PEACE Initiative, community-driven development, where the community is really involved, is a relatively new phenomena that started in 2001.

As the Taliban and other anti-government elements stand against the government, continued insecurity has played its role in development efforts. In 2004, Doctors Without Borders, a medical organization, withdrew its staff after five of its aid workers were killed (U.S. Government Federal Research Division 2007: 225). Other development organizations have reduced their staff or had to resort to temporary suspensions of activities due to insecurity. For example, IRC suspended most

173 Germany has focused on rebuilding the road from Jalalabad to Pakistan, and India, Iran, and Pakistan are building roads that connect to their countries. The ADB has focused on the northern section of the Ring Road, and the US has rebuilt the road from Kabul to Kandahar. This work has proceeded with little to no input from the local populations where it is taking place.

174 Dorronsoro (2005: 335) noted that the government in Kabul is unable to "mobilize resources on its own" because they are not able to pull in revenue with taxes on agriculture due to the very poor economic situation of the country.

of its activities following the killing of four of its staff in Logar in August of 2008. The UN pulled out its non-essential staff from Afghanistan after its Kabul guesthouse was attacked in October of 2009, though they did not do the same when their office was attacked in Mazar-i-Sharif in April of 2011, which resulted in the death of twelve staff.

While these attacks have negatively impacted on development in general, it has also raised concerns about the strength of the government and the role it is seen as playing in the development of the country. As Afghan history shows, this can be vital to the survival of a government and to the sustainability of development efforts. "The state is virtually absent as a problem solver and service provider. Most Afghans think the state has not contributed to the provision of basic services and does not look after the needs of the rural population," states a report focused on the contribution of international actors in the country (Koehler and Zürcher 2007: 4). This is a sentiment that was heard often during the multiple years I spent living, working, and conducting research in Afghanistan.

4.2 Relics of Past Development Efforts and Conflict: The Current Development Situation

After decades of development efforts that have been mainly grounded in modernization theory, Afghanistan remains mired in poverty and insecurity. While a great deal of this can be attributed to the many years of warfare and the destruction that has been levied on the people and the land of the country, the failure of the development efforts to help alleviate much of the economic suffering of the country or the basic human securities in a sustainable way, attests to the difficulties of development.

Afghanistan's history and present have been characterized by low development indicators. Afghanistan's Gross Domestic Product (GDP) per capita is 216 out of 228 countries, at 1,000 USD (Central Intelligence Agency 2011); the Human Development Index (HDI) ranks it 172 out of 187 nations in terms of human development (UNDP 2011: 129-130); and it has a Gender Inequality Index of 0.797, which results in it being ranked 134 out of 138 countries in terms of gender equality (UNDP 2010a: 6).

Afghanistan stands as one of the least developed countries (LDC) in the world, along with 48 other countries (UNDP 2010b: 227).[175] It had the lowest HDI of all Asian countries in 2011, and was the only non-African country that ranked amongst the bottom 30 countries for HDI (UNDP 2011: 127-130). Additionally, it is evaluated to be the second least peaceful country in the world according to the Vision of Humanity World Peace Index, ranking 157th out of 158, with only Somalia being ranked as less peaceful (Vision of Humanity 2012). Based on most standard indicators of development, Afghanistan is consistently towards the very bottom. Added to this, is the label of "highly corrupt" that has been attached to the country by Transparency International (2011), who ranked it as the world's second most corrupt nation, along with Myanmar, in its 2011 Corruptions Perceptions Index Report. But what do these indicators and numbers really mean for Afghans living inside the country?

The following sections will explore each of the four areas of investigation outlined in the analytical framework from the country level perspective. The personal, economic, and social securities will be looked at in turn with the objective of establishing a background against which the research question can be investigated. The fourth focus of investigation – empowerment – will also be dissected, though with a strong focus on the political situation.

4.2.1 Personal Security: A Country at War

Since the fall of the Taliban, international organizations, non-governmental organizations (both international and national), the interim government, and now the elected central government of Afghanistan have been actively supporting development efforts in the country. Unfortunately, they have not had a peaceful country in which to operate. They

175 The World Bank (2011d: 2) estimates the average per capita income to be 528 USD. Least Developed Country (LDC) is an official status that is given by the Committee for Development Policy of the UN Economic and Social Council to countries with the lowest indicators of socio-economic development. This status provides the countries with special concessions in terms of multilateral trading; special consideration for development funds and efforts; and prioritization for all cooperation programs of the UN, bilateral, and regional development partners (cf. UNCTAD 2012).

function in a very volatile environment where insurgent groups are actively working against the government and in turn against development programs and projects supported or associated with the government. Some anti-government elements (AGE) target development projects run by foreigners indiscriminately, whether they are connected to the government or independent of it, as they see such projects as opposed to their goals of delegitimizing the current government of Afghanistan in order to return to or gain power for themselves.

Although the Taliban did not put up a strong fight when international forces entered Afghanistan in 2001, anti-government elements have been active since the fall of the Taliban, some of whom are believed to be a part of the former Taliban government but many who were not a part of the Taliban but still stand in opposition to the Karzai government. The violent activities of these anti-government elements have continued to increase since 2001.

The Taliban and other insurgent groups reportedly caused 2,332 civilian deaths out of a total 3,021 civilian deaths in 2011, with international coalition forces responsible for 410 (UNAMA and UNHCR 2012: 1-2). This was the highest number of civilian deaths recorded in a single year since the fall of the Taliban.[176] Overall, civilian deaths have risen five years in a row, indicating increased insurgent activities and/or increased "success" rates for the AGE (UNAMA and UNHCR 2012: 1). As the Afghanistan NGO Safety Office (ANSO) notes, the number of civilians impacted by the current fighting increased annually starting in 2006 (ANSO 2008: 12; 2010: 14), and only saw their first decrease in 2011 (ANSO 2011a: 7).UNAMA said that, "IEDs and suicide attacks – tactics used by AGEs – accounted for nearly half of all civilian deaths and injuries in the first six months of [the] year" (UN News Service 2011). These two threats to personal security can be found in city centers, on paved highways, on dirt roads in rural areas, etc.[177] These are threats that are not easily avoided, as bad luck and timing play a large role in who is affected and who is not.

176 The number of civilians killed in 2011 was an increase of eight percent compared to 2010 and 25% compared to 2009 (UNAMA and UNHCR 2012: 1).

177 IED's were the largest killer of Afghan civilians in 2011, killing 967 Afghans (UNAMA and UNHCR 2012: 3).

Figure 17: Sources of Civilian Deaths in Afghanistan (2006-2010)

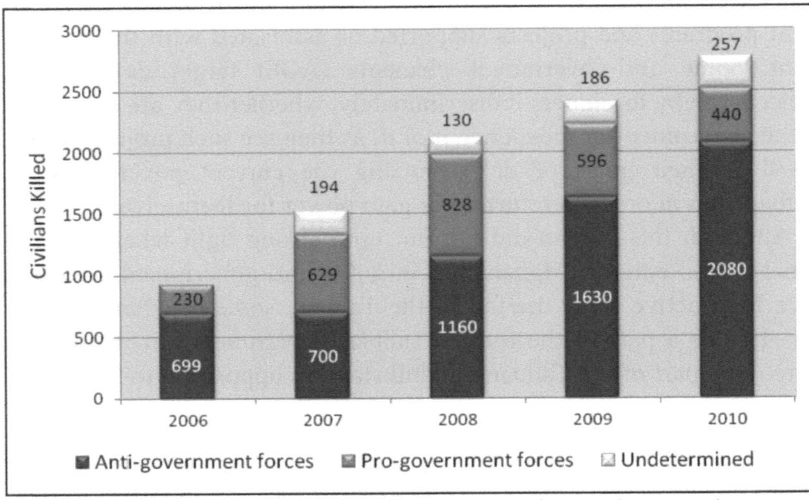

Source: own design based on statistics in The Guardian (September 6, 2011)

UNAMA and the Afghanistan Independent Human Rights Commission (AIHRC) also report that the number of assassinations conducted by AGE increased dramatically in 2010, rising 105% compared to 2009 (UNAMA and AIHRC 2011: i), a trend that continued into 2011 (UNAMA and UNHCR 2012: 18) (see Figure 17). These assassinations are often directed towards individuals who work with, or for, the centralized government based in Kabul.

The fighting between anti-government forces and international troops has spread from a limited number of provinces, mostly in the South and the Southeast, to a larger number of provinces, including some in the North of the country that had previously been considered safe, such as Kunduz. NATO forces increased by 40,000 soldiers as a part of a surge effort to eliminate active anti-government elements in the country in 2010 (Garamone 2010), culminating in a total of almost 132,000 international troops, according to the NATO web portal (ISAF 2011). Plans have been made to hand over different areas of the country from international coalition control to Afghan Government control by 2014 (World Bank 2011d: 1). The first such hand-over occurred with a transfer of responsibility occurring in seven cities or provinces in July

2011 (King 2011).[178] The second handover of 18 areas was announced in November 2011 (AP 2011). It included the complete handover of five provinces, seven provincial capitals, and additional individual districts in different provinces (AP 2011).[179] This is the second of five steps of transition to national control.[180] The third handover is expected to include 230 districts and all remaining provincial capitals (Harooni 2012).

This continuous fighting has had a negative impact on the development of the country, and continues to threaten the lives and livelihoods of the civilian population. Improvised explosive devices (IEDs), unexploded ordinance, active fighting, and threats impact the civilians of Afghanistan negatively. This has resulted in confusion as to whether the country should be categorized as a post-conflict country or a conflict country (Zakhilawal and Thomas 2008: 2).[181] Ongoing violence makes the distinction between post-conflict and conflict hard to make, with many provinces of the country experiencing continuous conflict and others experiencing a more post-conflict environment. What is not up for debate is the concern that Afghans have for their own personal

178 Areas handed over included Lashkar Gah of Hilmand Province, Mihtarlam of Laghman Province, Bamyan Province, Mazar-i-Sharif, the city of Hirat, all districts of Kabul Province except for one, and Panjsher Province (King 2011).

179 No comprehensive list has been released regarding the areas handed over in the second round of transition, but the following can be pieced together: the provinces of DayKundi, Ghazni, Hirat (minus Shindand, Obe, and Chishti Sharif Districts), Nimroz, Parwan, Samangan, Sari Pul (minus Kohistanat and Sayyad Districts), and Takhar; the remaining areas of Balkh and Hirat Province; a number of districts in Badghis (Ab Kamari, Qala-i-Naw), in Badakhshan (Argo, Arghanj Khwa, Baharak, Kishim , Shahri Buzurg,Tishkan and Yaftali Sofla), in Ghazni (Dih Yak and Zana Khan), in Hilmand (Marja, Nad Ali, and Nawa-i-Barak Zayi), in Kabul (Surobi), in Laghman (Qarghayi), in Nangarhar (Bihsud, Kama, Shinwar, and Surkh Rud), and in Wardak (Hisa-i-Awali Bihsud and Jalrez); and the cities of Chaghcharan (Ghor), Fayzabad (Badakhshan), Jalalabad (Nangarhar), Maidan Shahr (Wardak), and Shibirghan City (JawzJan) (Foschini 2012).

180 Originally the plan included six steps of transition, but due to an acceleration of the process it has now been cut to five steps (AP 2011; Riechmann 2012).

181 While DFID officially had placed Afghanistan in the post-conflict category, a recent DFID report noted that there continues to be conflict (Chapman and Vaillant 2010: 4). The World Bank identifies Afghanistan as a post-conflict location (World Bank 2001: i). The UN, however appears to categorize the country as one with continuing conflict (UNAMA and AIHRC 2011: i) in one of its agencies, but a country of post-conflict in other agencies such as UNEP (2003), which entitled its report *Afghanistan: Post-Conflict Environmental Assessment*.

security: 38% of respondents in The Asia Foundation Survey of 2011 identified insecurity in the form of attacks, violence, and terrorism as the biggest problem being faced in Afghanistan (Tariq, Ayoubi et al. 2011: 23). The same poll found that 56% of those surveyed said that they often or sometimes fear for their own personal safety in their local area (Tariq, Ayoubi et al. 2011: 27). This percentage has been increasing over the five years that this poll has been conducted, and is not un-founded, with 22% of those polled reporting that they or someone in their family had been victims of violence or crime in the previous year, a notable increase from 2010 (Tariq, Ayoubi et al. 2011: 28).[182] This violence could take the form of physical attacks such as beatings, or refer to extortion and theft or even to death. These kinds of violent events are reported more often in rural areas, as compared to urban areas (Islamic Emirate of Afghanistan 2011: 30).

Though the state of development of the country cannot be attributed solely to its status as a post-conflict or conflict country, the continuous fighting and the destruction this has caused, and continues to cause families, infrastructure, and the sense of personal security, impacts negatively on development efforts. Conducting development work in an ever changing security environment has proven daunting for implementing organizations. It has meant that development projects have been unable to reach parts of the population in some areas and also that the population has been unwilling to participate in development projects for fear of retribution from anti-government elements in other areas.

There are many threats to security, in the sense of personal security, in Afghanistan today. However, this insecurity is not evenly spread throughout the country and while it paints a backdrop for all development efforts in Afghanistan, it is vital to differentiate the insecurity of the country as a whole and the personal insecurity that Afghans experience in their own provinces, districts, and communities. That being said, physical insecurity in one region can spread to neighboring areas of relative security, which means that insecurity in the country is important to note and to take into consideration.

182 Fifty-nine percent of the survey respondents said that they had reported the crime, and 33% said that they did not (Tariq, Ayoubi et al. 2011: 32).

The personal insecurity that remains in the different parts of Afghanistan has negative impacts on the people, their social structures, and also on the ability of development efforts to reach them and to be able to implement projects in the way they were intended. This is a factor that plays heavily in the minds of the people of communities both situated in insecure areas and more secure areas, and played a vital role in how community members felt regarding their human security overall in the research community.

4.2.2 Economic Security: A Volatile Situation

The economic situation in Afghanistan, for all but a small minority, is not positive. An estimated 36% of the population cannot meet their own basic needs and live below the poverty line (ICON-Institute 2009: 54). Most Afghan livelihoods are connected to agriculture, with almost 80% of jobs being agriculturally related (Central Intelligence Agency 2011). However, agriculture accounts for only 31% of GDP (Central Intelligence Agency 2011), as most agriculture in the country is subsistence in nature. Volatility in the agricultural sector has resulted in great fluctuations in GDP since 2002/2003, though the average growth rate of GDP has been in the double digits, measuring 22.5% for 2009/10 (World Bank 2011a). However, the World Bank estimates that GDP growth will be only 5-6% between 2011 and 2018 (2011d: 1), a significant dip. An estimated 47% (or 5.5 billion USD) of the nation's GDP stems from foreign aid, indicating an economy that is not standing sustainably on its own (World Bank 2010a: 1; 2011a).[183] Funding gaps are expected in the coming years as development aid lessens, which could result in the government facing a gap of 25% by 2021/2022 (World Bank 2011d: 2).[184]

Afghanistan is therefore seen by the World Bank (2010a: 1) as, "one of the poorest and most aid dependent countries in the world," and there are estimates that Afghanistan will need seven billion USD from external sources annually between 2014 and 2021 to fund its security costs and other bills (Graham-Harrison 2011). Adding to these concerns is the opium trade, whose gross revenue is estimated to be as much as a

183 Foreign aid in 2010 was 15.7 billion USD, according to the World Bank (2011d: 1).

184 Aid is expected to drop drastically after foreign troops pull out of active duty in 2014, handing control over to the Afghan forces.

third of the country's GDP (World Bank 2011a). This black economy is dangerous to the stability of the nation as a whole because it undermines the legitimate market and creates incentives and opportunities for corruption.

A large percentage of the Afghan population reports being employed, in some capacity, with an unemployment rate of only 7%, according to the National Risk and Vulnerability Assessment (NRVA) (ICON-Institute 2009: 25).[185] However, other sources estimate an unemployment rate of up to 35% (Central Intelligence Agency 2011).[186] An estimated 91% of all workers' jobs, including daily laborers, are classified as vulnerable, meaning that the employment is not stable nor secure and that it does not result in a sufficient income (ICON-Institute 2009: 31). This leads one to assume that underemployment is quite common, which is confirmed by the World Bank's (2011d: 2) estimation that it is around 48%. The Asia Foundation's annual Survey of the Afghan People indicates that lack of employment was seen as the second largest problem in Afghanistan in both 2010 and 2011 (Tariq, Ayoubi et al. 2010: 3, 23; 2011: 23).[187] It is cited as the largest problem at the local level by 30% of respondents (Tariq, Ayoubi et al. 2011: 25). Almost three quarters said that the availability of employment opportunities in their locality was very bad, with 42% saying it was quite bad and 28% saying it was very bad (Tariq, Ayoubi et al. 2011: 61). The problems of unemployment and underemployment are expected to increase as development aid decreases (World Bank 2011d: 2).

---

185 The NRVA is an expansion of the WFP nation-wide annual assessments that took place the three years prior to the first NRVA conducted in 2003. Two follow-up NRVA surveys were conducted in 2005 and 2007/2008 (ICON-Institute 2009; MRRD and CSO 2005; WFP and MRRD 2004). The World Bank (2011d: 2) provides an unemployment rate of 8%.

186 As was noted before, these kinds of statistical discrepancies are common in Afghanistan as the security situation, the dispersion of the population geographically, and other factors result in a lack of readily available data that is reliable.

187 The Asia Foundation's annual opinion poll is one of the more credible polls conducted in Afghanistan. It was first conducted in 2006. The fieldwork for the 2011 poll was conducted in all 34 of Afghanistan's provinces, with 6,348 Afghan respondents (cf. Tariq, Ayoubi et al. 2011: 175).
Insecurity and unemployment have consistently been the top concerns expressed in the survey since 2006 (Tariq, Ayoubi et al. 2011: 24).

The basic infrastructure in the country has been destroyed through warfare and neglect, and while it has improved in some areas since 2001, many roads are still not leveled or paved, irrigation canals and culverts that were destroyed have mostly not been repaired, and factories that once produced bread, cotton, and other commercial products are barely operating, if functioning at all. Since the fall of the Taliban, progress has been made on the main ring road and other highways, with 2,500 miles of highway (4,025 km) paved, including 78% of the ring road (Katzman 2011: 4). There is very little reliable electricity in most of the country, though this has seen improvement in the last few years in some urban areas, such as Kabul.[188] All of these physical infrastructure problems undermine the security of the economy in the country as a whole, as people are not able to irrigate their land and farm it, utilize machinery they might have for production, or easily transport their products to local and regional markets.

The relatively insecure economic situation has also been impacted by a number of underlying factors, the two most important, I would argue, are health and education.

Health care in Afghanistan is limited and for many people is not accessible, be it because of availability, lack of money to pay for services, or the distances that must be traveled to reach health services. Infant and mother mortality rates are amongst the highest in the world, with 111 infants dying per 1,000 live births and 1,600 mothers dying per 100,000 live births (ICON-Institute 2009: 9, 82).[189] Life expectancy for all children born in Afghanistan stands at only 48.7 years according to the 2010 UNDP Human Development Report (2011: 129). This can be accounted for, in part, by the severe shortage of access to basic health care, with only 65% of the population having access to primary health care.[190] This is a drastic improvement from 2000, when only 9% of the

188 Regular electricity throughout the day arrived in Kabul only in 2009, prior to that city residents only had about four to five hours of intermittent diesel generated power daily (Sarwary 2009).

189 Some mother mortality rate estimates for Afghanistan are lower at 1,400 per 100,000 live births (UNDP 2011: 141). The CIA (2011) Factbook has an infant mortality rate of 142, and a life expectancy of 45.02 years.

190 Access is here defined as a location within a two hour walk from one's household (IRoA MoPH 2009: 6).

population had access to primary health care (IRoA MoPH 2009: 2), but is still quite a challenge for the population.

Illiteracy and a lack of access to education are also major problems in Afghanistan. The literacy rate is one of the lowest in the world, with only 26% of the population being able to read – 12% for women and 39% for men (ICON-Institute 2009: 65). While there has been an increase of primary school students from less than one million in 2001 (IRoA and MoE 2011: 1) to 5.11 million in the 2009-2010 school year (CSO 2011: 71), almost half of the population of school-aged children are still not attending school (ICON-Institute 2009: 65). Additionally, many students are considered to be temporarily or permanently absent, accounting for 22% of girls enrolled in primary school (446,682 children) and 11% of boys (799,822 children) (Jackson 2011: 10). The statistics only become more dire in secondary and higher education, with only 1.45 million students enrolled in secondary school (34% girls and 66% boys) and 541,691 students enrolled in high school (30% girls and 70% boys) (CSO 2011: 72-73). Violence and fighting have played a large role in the fact that 94.2% of women and 66% of men over the age of 25 have not obtained a secondary education (UNDP 2011: 141), with most schools closed down during the Taliban Regime and education for girls banned during this time. Additionally, access was made difficult for boys, and often impossible for girls, under the Mujahideen Government that preceded the Taliban Government.

These aspects of the Afghan economy paint a picture of a country that is mired in economic insecurity. The economy, though growing, has done so in a sustainable way that people believe will be able to continue in the coming years when foreign troops pull out in 2014, followed by the expected pull back in funding for development projects. Low levels of health security and education seem to indicate that the shifts towards creating greater economic security in the country are taking place at a slow rate that will take time to develop. In other words, the economic security of Afghanistan, though improving, cannot be expected to become secure in a short period of time. This economic insecurity filters all of the way down to the rural communities of Afghanistan, as was witnessed during the fieldwork for this dissertation.

4.2.3 Social Security: The Loosening of Ties

Fighting within Afghanistan and the imposition of policies by different political regimes has been accompanied by a decrease in social security in the country. The low level of GDP and the lack of authority and strength of the central government means that families and individuals who find themselves facing difficulties that they cannot manage – from an unexpected illness, losing a job, or the loss of a harvest due to inclement weather, etc. – do not have the option of finding social and economic support from their government. They instead turn to other sources, such as family connections, tribal connections, and at times to social connections within the local village.

Yet many of these social connections have been strained, twisted, and broken during the decades of fighting. Over eight million people are estimated to have fled the country as refugees since the Saur Revolution of 1978 (Margesson 2007: n.p.). Nearly five million refugees have returned to Afghanistan since 2002 (UNHCR 2011), though their return to the country does not equate their return to their home villages. Many of those who have returned have settled in new locations, such as Kabul and other major cities where they hope to find jobs to sustain themselves and their families.[191] Over 350,000 individuals remained internally displaced in 2011 (UNHCR 2012), meaning they have remained within the borders of the country but have had to flee their homes. Additionally, an estimated four million refugees are still living outside of Afghanistan (UNHCR 2012), with the majority based in Pakistan and Iran.[192] This has contributed to the separation of many individuals from their families, their land, their village, and the social norms that bound them to these things.

The years of fighting and war have also impacted negatively on the social connections within the country, as neighbors found themselves on opposite sides of the battle. The constant threat to one's personal security and the physical security of one's family forced some families to

191 Kabul had a population of 500,000 people at the end of 2001when the Taliban fell (Ittig n.d.), and was reported to have a population of 3.573 million in 2011 (Central Intelligence Agency 2011).

192 An estimated 1.7 million Afghans are registered as refugees in Pakistan, with over one million more estimated to be living illegally as unregistered refugees there (Shah, S. 2012).

turn inwards and trust only those within their inner circle, loosening village and community level ties that were not based on blood relations.

Trust in institutions in general has become a commodity that is not easily earned, as years of promises from various government and development organizations have mostly failed to provide what was expected. Alissa Rubin (2000: 1794) sums this up by concluding that, "The years of war had destroyed much of Afghanistan's social capital as communities and institutions were dispersed or destroyed." An increased focus on ethnicity has also further polarized the Afghan society as a whole (Rubin, B. R. 2000: 1793).

The lack of strong social security in the country of Afghanistan has had many repercussions. The lack, or low level, of social cohesion makes it difficult for people to work together, even when they are working for a common cause that will benefit everyone. Fears of being taken advantage of or cheated are not uncommon or unfounded. This can be a detriment to development efforts, as has been noted by development stakeholders previously. That is, in part, why emphasis has been placed on participatory development programs such as the National Solidarity Programme, with the expectation that such programs can positively impact on the social security in individual communities. It is then hoped that this will have a trickle up effect, eventually creating a country with stronger social capital. This research looked at the very first rung of this ladder, the community, to determine whether participatory development programming in one community was able to improve its social security.

4.2.4 Empowerment and the Political Situation: A Government at Risk

The fourth focus of investigation of empowerment, at the country level, can be explored utilizing the four factors of empowerment outlined in the framework of analysis with some adjustments. At the country-level of analysis, this requires looking at the political situation which provides the framework in which empowerment can emerge or in which it can be denied. Therefore, we look closely at the political environment in connection with the four aspects of empowerment: access to information, inclusion and participation, accountability, and the capacity to organize.

There are two different factors to be considered when one is discussing access to information in Afghanistan. The first is the availability of accurate and reliable data about the country, both past and present, and the second is whether there are reliable means of communication. Both of these factors cannot be evaluated positively in Afghanistan, resulting in a low level of access to information for the people of the country.

At the countrywide level, access to information is a major challenge. Even basic demographic information that is accurate is not available, as a complete census has not been conducted in over 30 years.[193] As Barfield (2010: 23) notes, statistics in Afghanistan are validated not through data collection, but instead through repetition. The geographic terrain of the country, its insecurity, and other factors have resulted in a dearth of confirmed statistics for Afghanistan such as the number of people living in the country, the percentage of the population living in poverty, etc. Being able to definitively say that a historical event happened on a certain date can also be difficult, as historians have recorded it on different dates. Historians, both internal and external, also use very diverse spellings to transliterate Dari and Pashtu words, locations, etc., which can increase uncertainties. Therefore, even if one has access to reports or historical documents, the veracity of the information in such resources cannot always be relied upon, but these are the sources that the Government of Afghanistan (GoA), development aid funders, development aid implementers, internal and external researchers, and the Afghan people have to work with.

The means of information transference is also complicated in Afghanistan. Media outlets, while growing in number are still not able to reach the whole population of the country, with those in rural areas having fewer resources to obtain information about actions of the gov-

---

193 The first census was conducted between 1972 and 1974 by the State University of New York (SUNY) for USAID, as noted by N.H. Dupree and Gouttierre (2007: 62). Barfield (2010: 32) notes that even this census was not published. The second census was partial in nature and took place with UN assistance in 1979 (Dupree, N. H. and Gouttierre 2007: 62). The United Nations Population Fund (UNFPA) was tasked with assisting the Afghan Government in conducting a census following the fall of the Taliban. Scheduled to begin in 2008, it was postponed to 2010, and then further postponed to 2011 (UNFPA 2011). As of October 2011, they had completed the data collection in only one province (personal communication, UNFPA staff member, October 15, 2011).

ernment, decisions being made in Kabul, and many other things that affect and impact their lives. Televisions are rare in rural environments, though radios are generally prevalent. A common way in which information travels in rural areas is orally, which can result in inaccurate information being transferred and misunderstandings. While the oral transmission of news, data, and information is also common in urban centers, there are more ways in which a person can confirm or triangulate the information they hear or read. In 2010, there were a reported 175 radio stations and 75 television stations operating in the capital city of Kabul (Altai Consulting 2010: 22). These numbers have been steadily rising since 2002 (Altai Consulting 2010: 51), which has resulted in a greater transfer of information to the whole population of Afghanistan, though perhaps not enough.

Additionally, a media report from 2010, finds that though the media had expanded, there was still a great deal of censorship stemming from the government, insurgents, and other stakeholders, and that self-censorship by the stations is common (Altai Consulting 2010: 27-28). So, while the number of media resources has been growing, the information provided is not always presented in an unbiased manner and some things are not covered at all.

The low literacy rate of rural Afghans and their limited access to messages through oral transmission, the radio, and sometimes television is problematic for the establishment of easy access to information. This limited access to information about what is happening within the country and even externally, impacts negatively on the people's ability to be empowered. The same can be said of the Afghan government and development stakeholders, who are all working off of statistics that are not rooted in solid data collection techniques and able to be documented and confirmed.

In terms of inclusion, the Afghan population has been included in the democratic process through having the right to vote in the elections based upon the Bonn Agreement. However, the legitimacy of these elections has been called into question both internally and externally. Only 54% of those polled for The Asia Foundation Survey of the Afghan People in 2010 said that they felt that the 2009 Presidential elections were free and fair (Tariq, Ayoubi et al. 2010: 6).[194] This percentage was

194 People believe that there was cheating in the vote count and that votes were

even lower for the Parliamentary elections held in 2010, where only 49% reported that they felt the elections were free and fair (Tariq, Ayoubi et al. 2011: 123-4). While 74% of respondents in The Asia Foundation Survey of 2010 said that they would vote in the Parliamentary elections (Tariq, Ayoubi et al. 2010: 12), news organizations found that only about 40% of the population, with 3.6 million votes, had actually voted (BBC News 2010; CNN 2010). The people's main means of participating or being included in the political process has been delegitimized in the eyes of many. Unfortunately, other means of participation are not easily accessible to the population due to the highly centralized nature of the government structure created in the Bonn Agreement, which took place without input from the Taliban, another example of exclusion. This centralized style of government has also historically led to the alienation of the population by the government (Barfield 2010: 223).

The situation of the current government and the political system that has been implemented is precarious, with the population left out. This is not to say that the Karzai Government lacks supporters, because recent polls indicate that his government does have support (ACSOR 2010: 6; Tariq, Ayoubi et al. 2010: 72; 2011: 72), but rather to point out that this appreciation for, support of, and belief in the current government system is not universal, and that not everyone feels that their interests are being represented.[195] Only 62% of The Asia Foundation poll expressed confidence in the administration (Tariq, Ayoubi et al. 2011: 71).

The Afghan Government is not seen by all as legitimate, and its legitimacy is further called into question by the Taliban having estab-

---

bought in some cases, along with some women being restricted from voting, and the intimidation of voters (Tariq, Ayoubi et al. 2010: 13).

195 In one annual ABC, BBC, and ARD poll, Karzai was rated as doing an excellent job by 28% and a good job by 44%, for a 72% positive response, while 19% said he was doing a fair job and 8% said he was doing a bad job (ACSOR 2010: 6). A third of those interviewed, in the same poll, said they were "very confident" that Karzai would work to benefit the Afghan people, and an additional 46% felt "fairly confident" of this (ACSOR 2010: 8). Ninety percent of those who participated in the poll indicated that they would rather have the current government ruling the country today, as compared to the Taliban or any other named entities (ACSOR 2010: 8). The Asia Foundation's annual poll also found a high level of support for, and belief in, the current government, with 73% giving it a positive assessment (Tariq, Ayoubi et al. 2011: 73).

lished shadow government authorities in all 34 provinces, even those in the North that see relatively fewer violent attacks.[196] It appears that this only excludes the population even more, as they are therefore excluded from two governing systems that are in conflict, both of which they have little, to no, control over.

The Taliban call the Karzai Government a "Puppet Regime" and refer to the international coalition partners as "invaders" (Islamic Emirate of Afghanistan 2011). This idea can be supported by estimates that 97% of Afghanistan's gross domestic product (GDP) is related to spending that comes from the international military or the donor community presence (Kerry 2011: 24). This can increase the perception that the government is not in control, which means that the government itself is not empowered on behalf of the population.

The inability of the government to deliver the expected development and general services that the population feels it has been promised and that many desire, has contributed to popular disgruntlement and a greater disconnect between the people and the government. Discussions in the case study community of Shah Raheem have also shown that, at least in some areas, this disconnect is also felt between the people and the local district and provincial governments. This has been raised as a concern by some Afghan historians, such as Dorronsoro (2005: 335) who believe that NGOs in Afghanistan are at times replicating the functions of the state, forming a parallel system. However, NGOs are not highly trusted within Afghanistan. They, and development funders, are seen by the Afghan population as not having Afghans' best interests at heart (Tariq, Ayoubi et al. 2010: 101; 2011: 112).[197] Disappointment in all levels of government and with many development stakeholders was observed within the case study village and throughout Afghanistan.

A common refrain heard in rural villages and urban centers alike is that the people have heard of the billions of dollars being provided to

196 Shadow government actors have been selected by the Taliban to administer to specific areas. They operate in parallel to the government officials who have been appointed by the central government.

197 Forty-seven percent of respondents said that international NGOs were more concerned with their own interests than those of Afghans, and the same was said by 45% of the respondents regarding donor institutions (Tariq, Ayoubi et al. 2011: 112).

Afghanistan for its reconstruction, and they wonder where the money has gone. This indicates that they do not feel they have been included in the development process and also highlights the lack of information they have regarding development. Though Afghans will concede successful development efforts when they are questioned about specific projects, in general the population does not seem to believe that the money promised has gone honestly towards the development of the country, and instead believe that the money has been taken by government officials, NGOs, and others to increase their own bank accounts, however, the people are not able to hold these powerful individuals accountable, the third characteristic of empowerment. The fact that an estimated 75% of all development aid in Afghanistan was delivered outside of government channels (Barfield 2010: 316), means that the government itself is being bypassed by external forces, leading to their exclusion from determining how development aid funding is spent. This weakens the Afghan government's capacity to control the development resources and process (Kerry 2011: 18).

Corruption is often associated with the government, with about half of all Afghan interactions with the government involving some form of it (Tariq, Ayoubi et al. 2011: 101).[198] This corruption affects people's daily lives and their interactions with authorities (Tariq, Ayoubi et al. 2011: 95). Corruption at the local government level is seen as increasingly problematic, with 56% of those surveyed labeling it a major problem in 2011, as compared to only 42% viewing it as such in 2006 (Tariq, Ayoubi et al. 2011: 96). At the provincial level, 64% of those polled said corruption was a problem, and this increased to 76% at the national level (Tariq, Ayoubi et al. 2011: 95). The IMF even suspended its lending program in Afghanistan in 2010 due to corruption at the Kabul Bank, with concerns about bad loans and the mismanagement of bank funds (Reuters 2011).[199] The IMF has since reengaged in the country.

198 Corruption was encountered most often in accessing public health care services and in interactions with the Afghan National Police (ANP), followed closely by dealing with the courts or the judiciary, applying for jobs, and obtaining official documents (Tariq, Ayoubi et al. 2011: 100).

199 The IMF is one of the funders of the Afghan Reconstruction Trust Fund (ARTF), one of the sources of funding for the National Solidarity Programme (NSP).

At the same time, the foreign development and military sectors threaten the validity of the central government in a number of ways. First, anti-government elements are able to utilize this foreign assistance against the government itself, by claiming that the Western support is really dominance and that it is the West, not the Afghan Government that is in control. This propaganda is supported by disputes between the Karzai Government and the international coalition nations. In attempts to mitigate this view, the current government has defiantly stood in opposition to its foreign supporters at times, for example, demanding that they discontinue night raids (Sengupta 2012). The seeming indifference of foreign commanders and troops to such demands, has undermined the Karzai Government, disempowering it, and shows that the government also has trouble holding others accountable.

The final characteristic of empowerment is the capacity to organize, which is also not readily apparent in the country of Afghanistan, though recent demonstrations, both violent and non-violent, seem to indicate that this might be starting to emerge, at least in urban settings such as Kabul, Mazar-i-Sharif, Hirat, and Jalalabad. Women took to the streets of Kabul in 2009 to protest Karzai's signing of the Shia Family Law, organized protests against the burning of the Koran by foreign forces evolved throughout the country in 2012, and other protests all indicate a new sense of empowerment of the people. Some communities have even started standing up to the Taliban elements in their area, fighting back physically or protesting (Habib 2012). They feel they have the right to voice their opinions in a new way.

However, the political system is set up in such a way as to not support the capacity of the population to organize. There are only a few ways in which Afghans are able to organize themselves and give voice or action to their concerns. Even having political parties, where people could join together based upon similar political beliefs or goals is not possible on a grand scale because political parties are not provided legal recognition in Afghanistan (Barfield 2010: 301). Without political parties, those who run for different government offices must make themselves known as individuals, and cannot utilize the recognition of a political party and its policies to gain supporters as is common in many other countries.

Overall, the ability of the Afghan population to organize itself and be heard is not very strong, but it does appear as though this has seen shifts in the last decade, and more changes could be coming in the near future. The Afghan government also faces challenges at the international level, in terms of its own empowerment and ability to organize itself. The plans that the country makes for its own development have often been written in part, or in whole, by foreign advisors and do not necessarily reflect the vision of the Government of Afghanistan, much less the varied wills of the Afghan people, as was noted with the ANDS.

These different factors contribute to the disempowerment of Afghans, as they see their government as weak and unable to provide for them. This has created anger and frustration, which was evident in the research community. When operating in a political environment where information is not easily accessible and inclusion is rare, it is difficult to hold people, the government, and organizations accountable. It also inhibits one's capacity to organize. All of these factors of empowerment are weak in the Afghan context, though protests and demonstrations in the last few years show that perhaps some of these factors are shifting. It will be interesting to see how this develops in the coming years and decades.

The country setting has a large impact on the situation of the people in Afghanistan and their ability to become empowered. This background is important in terms of being able to decipher different steps towards empowerment in the Afghan setting at the different levels of analysis, including at the community level. These steps can be limited or nuanced and can be overlooked if the environment in which they occurred in is not understood and not taken into consideration.

4.3 Is a Participatory Development Approach an Answer to Afghanistan's Development Problems? Introduction to the National Solidarity Programme

One program that has been receiving positive reviews and is seen as a beacon of hope in Afghanistan is the National Solidarity Programme (NSP).[200]

The NSP is a participatory development program that started in Afghanistan in 2003 in only ten districts, and since then has expanded throughout the country (Social Development Department 2006: 17). A community-driven development program (CDD) at heart, NSP aims to put the decision-making power into the hands of the communities, so that they are in charge of determining what subproject they feel is most vital to implement in their community, how it will be implemented, and how to monitor and maintain it. This is drastically different from the more traditional top-down development approaches, where the development intervention and all aspects of the program are determined by development practitioners who are seen as experts in development.

Administered by the Ministry of Rural Rehabilitation and Development (MRRD), NSP is the largest development project that uses a participatory approach ever to be implemented in Afghanistan, in terms of geographic coverage, number of people reached, and budget (Beath, Fotini et al. 2011: 1; NSP 2009a: 8; Zakhilawal and Thomas 2008: 11). It covers districts in all 34 provinces and has provided over 840 million USD to rural communities for development subprojects with contributions of 113 million USD coming from the communities (NSP 2011b). It is considered to be a flagship program of the Afghan Government (Beath, Christia et al.: iv; World Bank 2011c) and is one of the six National Priority Programmes of the National Development Framework (Brick 2008: 1; Torabi 2007: 12). The the program supports subprojects focused on local infrastructure and capacity building, are seen as vital to the development of the country as a whole.

The NSP is sometimes presented or talked about as a "silver bullet" that will resolve the development problems that Afghanistan faces (Benen 2008). Development stakeholders, journalists, researchers, and

---

200 NSP is known locally as *Ham-bastegi melli* in Dari and *Melli Pejwastan* in Pashtu (NSP 2007: 1), the two official languages of Afghanistan.

others praise the program for its successes, calling it the most successful program currently in Afghanistan (Kerry 2011: 27; Nagl, Exum et al. 2009: 2; SIGAR 2011: 2).[201] Alissa Rubin (2010), a reporter for the New York Times, calls it, "one of the government's most successful efforts to spur grass-roots development projects." Representatives of the World Bank have said, "there is no other program which has NSP's reach and scale," and Karl Eikenberry, the commander of US forces in Afghanistan from 2006 to 2007, praised the NSP in testimony in 2009 as having, "proven effective in giving Afghans a greater stake in their government" (Lopez 2011). One news agency stated succinctly that, "In many ways, the NSP is a silver-bullet program for Afghanistan. It gets needed resources to Afghan villages; it promotes stability by connecting local governance and development with the national government in Kabul; and it fosters democracy and accountability throughout the country" (CBS News 2009).

The Feinstein International Center of Tufts University conducted scientific research on the connection between development aid and winning the support of the population, and they found the NSP to be one of the only programs that was consistently seen in a positive light by the Afghans they interviewed in five different provinces (Fishstein and Wilder 2011: 3).[202] World Bank scientists have also reported positively on the program (Beath, Christia et al. 2008, 2011; Beath, Christia et al. 2010; Beath, Fotini et al. 2011; Enikolopov 2011; Enikolopov, Christia et al. 2008, 2009), though as this was a part of their work for the World Bank, it is conceivable that their perspective and presentation contains some bias. Other social scientists who have looked at the NSP, however, have had more conflicting views of the implementation of the program (Azarbiajani-Moghaddam 2010; Brick 2008; Goodhand and Sedra 2009: S89; Kakar 2005; Nixon 2008a; Zakhilawal and Thomas 2008). These authors tend to present a more nuanced picture of the program and its implementation, not ignoring, but also not usually emphasizing, areas where the program is not meeting its goals and objectives

201 A Senior Operations Officer of the World Bank said that the NSP was one of the most successful programs in the world to date when he was interviewed in December of 2004 (Kakar 2005: 4).

202 The Feinstein International Center of Tufts University researchers spoke with 574 Afghans (Fishstein and Wilder 2011: 10).

or where there is a divergence between the program aims and framework and what was found in the field.

The quantitative data regarding the NSP is impressive, with over 29,194 communities out of an estimated 38,000 in the country (NSP 2009a: vi) having been mobilized to partake in the program (IRoA, MRRD et al. 2012b: 6).[203] A total of 48,159 subprojects had been completed and 12,982 more were considered "ongoing" as of March 2012 (IRoA, MRRD et al. 2012b: 6). The program had already reached 358 districts of the 34 provinces as of March 2012 (NSP and MRRD March 2012), as shown below in Figure 18.[204] It was estimated that the program was reaching 70% of Afghan villages (MRRD and NSP 2011), and 91% of the country's districts by SIGAR (2011: 16) in their audit of the program in 2011.[205] The NSP had disbursed 1.01 billion USD in block grants as of August 2012 (IRoA, MRRD et al. 2012a: 1).[206]

203 The total number of villages in Afghanistan varies across MRRD reports. According to the Operational Manual Version 4 (NSP 2007: vi) there are 38,000 villages in the country, but according to the weekly report from March 12-18, 2011 (NSP 2011b) there are 56,000 villages. Discrepancies such as these are common due to the lack of reliable data and the lack of a census mentioned previously.

204 According to the MRRD, Afghanistan has 402 districts, however, this number is not agreed upon universally across ministries, with the Ministry of the Interior (MoI) saying there are 401 districts (Brown, F. 2012: 4; MRRD 2012).

205 The numbers provided by MRRD do not always agree, as can be seen above, with one report indicating that 70% of Afghan communities had been integrated into the NSP (NSP 2011b), and another report saying that 26,856 communities out of a total of 56,000 communities had been integrated into the program, which would mean that 60% of communities had been reached (NSP 2007: vi).

206 NSP participating communities are required to provide 10% of the project costs as the community contribution to the subprojects they select (NSP 2009a: 21).

Figure 18: Map of the National Solidarity Programme's Coverage in Afghanistan by District (March 2012)

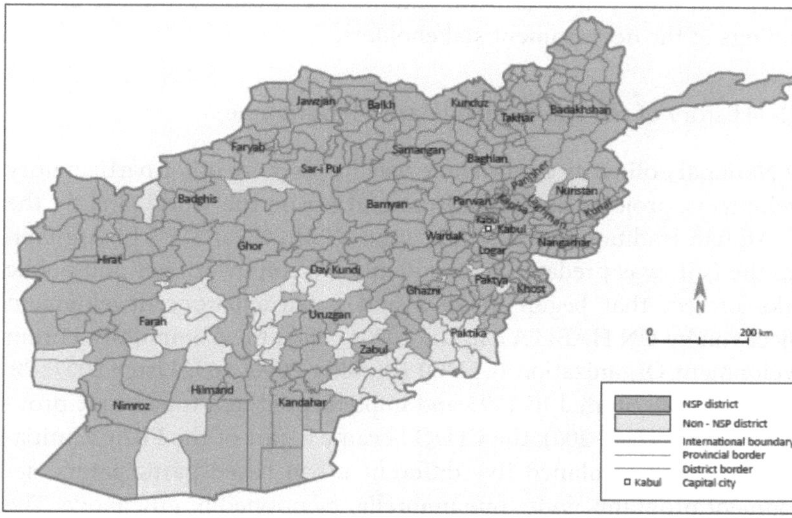

Source: (NSP and MRRD March 2012), cartography Bernd Hilberer

These numbers are often presented as proof of the "success" of the NSP, though how success is defined is rarely clarified. The success of the NSP is often shown through its positive comparison to other development programs running in the country, but is the program really making the impact that it is said to make? Is it really a "silver bullet," or is its reputation larger than its delivery? What kind of impact is this program having in Afghan communities?[207]

This dissertation will explore these questions by utilizing the analytical framework developed and looking at each of the four areas of investigation to see if these quantitative numbers translate into changes in an individual Afghan community. It presents the perspective of how

207 These kinds of questions are rarely broached in literature on the NSP, however, similar questions have also been raised by Monustti (2012: 587) who said, "While there is no doubt that bridges have been built and wells dug with resources the NSP has channeled, we must ask if the program has truly created better living conditions for rural people. Has it brought more social justice, or more economic and political equity? Has it enhanced transparency and popular participation?"

the community members believe the National Solidarity Programme impacted on their personal, economic, and social securities in the community, and their empowerment, and provides insights into the understandings of the development stakeholders.

4.3.1 History of the National Solidarity Programme

The National Solidarity Programme was inspired by other participatory development projects in Afghanistan and across the world, and by the two Afghan traditions of *hašr* and *jirga* (NSP 2009a: 8).[208] In Afghanistan, the NSP was predated by a community empowerment and public works project that began in 2002 and a participatory development project run by UN HABITAT in the 1990s called the Community Forum Development Organization (CFDO) (Barakat 2006: 163; Lister 2007: 22; SIGAR 2011: 2). Started in 1995 and implemented in a total of six provinces (Barakat 2006: 200), the CFDO became a part of the PEACE Initiative (PI), which combined five different urban based participatory development programs under one umbrella, mentioned briefly above. The PI focused on education, protection (including jobs), women's empowerment, and social services, such as health, sanitation, and kindergartens (Barakat 2006: 200). Each community in the PI was responsible for drafting proposals for projects, and was connected to the government through different ministries, as has been replicated in the framework of the National Solidarity Programme. One major difference between the two programs is the NSP's concentration on rural development, versus the PI's focus on urban development.

The NSP was also inspired by participatory development efforts in other countries. Similar World Bank community-driven development projects in Indonesia and other Asian countries stimulated the creation of the NSP in Afghanistan, and informed the establishment of its framework (Brick 2008: 3; Carlin 2003: 9; Kakar 2005: 19; World Bank 2011c; Zakhilawal and Thomas 2008: 15).[209] These programs focused on em-

208 *Hašr* is voluntary communal service, for example, to repair irrigation canals, mosques, and schools and *jirga* refers to the traditional means of dispute resolution, which involves gathering people to discuss things as a group (Boesen 2004: 27).

209 The Kecamatan Development Programme (KDP) was started in 1998, and is often seen as the origin of the term CDD and inspiration for CDD programs in

powering local communities by giving them decision-making authority over the selection, planning, implementation, and maintenance of local small-scale development projects.

The NSP follows in the same vein and is intended to, "develop the ability of Afghan communities to identify, plan, manage and monitor their own development projects" (NSP 2009a: 8). It is hoped that it will help bring development, reconstruction funds, and infrastructure to rural communities and inspire these communities to spearhead local development, by providing them the experience and capacity building necessary to be able to run small infrastructure projects on their own. The goals of the program include alleviating poverty in rural areas of Afghanistan and establishing "a foundation for improved local governance" (NSP 2007: 1).

The NSP was also created by the Government of Afghanistan (GoA) to address the dire poverty affecting many of the rural areas (NSP 2007: 1), and create unity amongst the Afghan people, in part by finding a way to reach rural populations who had seldom been offered development assistance historically. The first phase of the program, NSP I, involved the mobilization of about 17,300 communities and lasted from 2003-2007 (NSP 2011c; World Bank 2010d: 2).[210] The program was extended with a second phase, NSP II, that spanned from 2007 to 2011, and was expected to mobilize an additional 5,900 communities (SIGAR 2011: 2; World Bank 2010d: 2).[211] NSP Phase III is running simultane-

other countries, such as the NSP. Individuals who had helped to come up with the idea for the KDP, also helped to plan the NSP (Kerry 2011: 26; Warner 2007). While the KDP and NSP are similar in nature, there are major differences between the two projects. For example, the KDP proposals come from the hamlet level and need to be approved at the village and the sub-district level prior to money being released for the project (Bebbington, Guggenheim et al. 2004: 53), while in the NSP a proposal comes from a community and only requires approval from those who run the program. Additionally, the KDP only provided 1 USD per capita in assistance, while the NSP provides about 10,000 Afghani (about 200 USD) per family (Brick 2008: 44). Torabi (2007: 7) argues that the NSP is based upon programs such as those in Rwanda and East Timor and McCarthy (2006: 32) says that it was based upon programs in Indonesia and Laos. I was, however, unable to find further independent evidence of this.

210 Other World Bank documents note that 17,223 communities were contacted during NSP I and 16, 914 were mobilized. There were 10,410 subprojects completed (World Bank 2010b: 3).

211 World Bank (2010b: 3) documents note that 5,866 CDCs were established during NSP II as of June 2010, with 28,409 subprojects completed.

ously with Phase II, and is expected to continue through September 2015, having started in September 2010 (SIGAR 2011: 2; World Bank 2010b: iv; 2011b). Phase III will implement the NSP in 15,800 communities that were not yet mobilized as of 2010, and plans to provide block grants to 11,900 previously mobilized communities for a second round of funding (World Bank 2011b).[212]

4.3.2 Practicalities of How the National Solidarity Programme Works

Understanding how this program is supposed to work and the secondary literature and research that exists regarding the program will provide necessary background for this research. The NSP is overseen by the Ministry of Rural Rehabilitation and Development (MRRD) in Kabul and in the provinces, though actual implementation is done by 29 different Facilitating Partners (FPs). These FPs consist of local and international NGOs, UN Habitat, and one for-profit development company (NSP 2011a).[213] It is funded by the Afghanistan Reconstruction Trust Fund (ARTF), the International Development Association (IDA) of the World Bank Group, and other donors (NSP 2009a: 1).[214]

The financial aspects of the program are handled by the Oversight Consultant (OC), which is also responsible for supervising FP contracts, approving sub-project proposals, managing the distribution of block grants, and updating the Operational Manual (OM) (Torabi 2007: 5, 9).

---

212 The original documents for NSP III state that NSP III would cover 10,320 communities not covered by NSP I or II, and provide additional grants for 17,4000 additional communities (World Bank 2010b: v).

213 When the program started, there were only 22 (Carlin 2003: 11) or 25 (Owen 2006: 21) facilitating partners, depending on the source. Almost all of the FPs had previous experience working in Afghanistan (Torabi 2007: 7). NSP facilitating partners as of 2011 included: Afghan Business Capacity Development (ABCD), ACTED, ActionAid, Afghanaid, AKDN, AREP, Afghan National Reconstruction Coordination, BRAC, CARE, CHA, Concern, DACAAR, Flag International, Future Generation, GAA, Ghazni Rural Support Programme, IRC, Interco Operation, MADERA, Norwegian Project Office/Rural Rehabilitation Association for Afghanistan, OXFAM, People in Need, Partners for Social Development, RI, SCA, SDO, Solidarities, UN Habitat, and ZOA.

214 Over 95% of the funding for the first two phases of the NSP, stemmed from the ARTF (Torabi 2007: 6). The ARTF, established in 2002, is a fund supported by international donors to support the reconstruction of Afghanistan (SIGAR 2011: 1).

This position was filled by the German technical cooperative Deutsche Gesellschaft für Technische Zusammenarbeit (GTZ) from 2003 through 2007, and since then has been filled by the London consultancy, Maxwell Stamp (Monsutti 2009: 19), though the position has been re-labeled as the Management Support Consultant (Barakat 2009: 115).[215]

Though the program has been extended, the original timeline for the NSP was four years, in connection to the timeline provided for the ARTF, the program's largest source of funding (Barakat 2009: 119). Barakat (2006: 20) noted that while the NSP was intended to be a development program, it was rolled-out with a relief mindset, which resulted in greater focus being placed on program coverage and the disbursement of block grants, instead of the many components of the program that make it a truly participatory development project guided from the bottom-up. One example of this was that the OC was expected to immediately start disbursing funds after it was hired, before the establishment of an Operational Manual or planning for other aspects of the program (Barakat 2006: 120). The OC told Barakat (2009: 120) that they were overwhelmed by the scale and pace of the program and that this led them to delay and de-prioritize capacity-building, a major aspect of the CDD framework. "As such, a long-term, institution-building project was collapsed into an abbreviated period of time, which often resulted in the need to compress the most significant and time-intensive part of the process, facilitating and building the capacity of CDCs" (Barakat 2009: 119).

While the MRRD and OC have oversight roles within the structure of the program, the twenty-nine facilitating partners (FPs) are actually responsible for project implementation. This means that they are in charge of informing communities about the NSP, mobilizing them, and helping them with the NSP process, including capacity building. The FPs act as a community's guide through the program, assisting them in all aspects of project planning, implementation, and monitoring, including providing engineers and other experts necessary to implement quality infrastructure subprojects, as necessary. "The facilitating partners are key at nearly every stage in the process and some are con-

215 GTZ joined forces with the Deutscher Entwicklungsdienst (DED) to form Deutsche Gesellschaft für Internationale Zusammenarbeit (GIZ) in January 2011 (GIZ 2011).

cerned that they will be pressured to move communities along too quickly, especially if more time is needed to increase the participation of women or encourage more creative approaches to development" (Carlin 2003: 12). This concern was shown to be legitimate by Kakar (2005: 28), who said that a number of FPs had complained that the short timeframes for mobilization meant that they were unable to include women.[216] The FPs of CARE International, GOAL, ActionAid, and ACTED all found that it took longer to mobilize communities than was originally planned for under the NSP guidelines (Boesen 2004: 42).

This concern about meeting implementation deadlines negatively impacted the actual mobilization process of getting communities involved and preparing them to take on their responsibilities in the program. It has been raised as a concern in a number of different reports, including that of Zakhilawal and Thomas (2008: 22) who found that the speed of implementation was too fast and did not allow for well thought out planning.

Mobilization is generally estimated to take approximately six months (Beath, Christia et al. 2010: v; Nixon 2008a: 21), though Nixon found that it was taking place in as quickly as one month in certain communities, depending on the availability of the FP staff. In the district where the research took place, one facilitating partner staff estimated that it took about two months to facilitate a NSP community (KII: FP06, July 6, 2009).[217] The FPs were found to usually hold between two and five civic education meetings prior to the establishment of a community development council (CDC) (Nixon 2008a: 21).[218] These civic education meetings are when the NSP program is fully explained to a community and they are provided with the opportunity to ask

216 Some FPs decided to focus on the men in a community first in order to gain trust in the community before, starting to include women (Boesen 2004: 14; Liu 2007: 30).

217 The head of the NSP for Balkh Province said that ideally the facilitation process took about four to six months as the communities and the FP staff became more familiar with the program, but that at the beginning it had usually taken longer, about eight months. "Now, People have more knowledge about NSP and are more interested in participating, so it takes less time" (cf. KII: FP06, July 7, 2009).

218 Meetings reportedly lasted anywhere from 15 to 100 minutes each (Enikolopov, Christia et al. 2009: 11).

questions of the FP and to discuss whether they want to participate in the program or not.

Prior to this, however, communities must be identified. The NSP defines a community as a rural settlement that consists of at least 25 families (NSP 2007: 5).[219] It does not provide any guidelines about the geographic or social connections within a community. Individual communities were often identified through government records from the 1970s and sometimes records from the 2002 *Loya Jirga*, or if these documents were not helpful, the facilitating partners helped to define communities geographically (NSP 2007: 5). Each community is required to be registered with the MRRD, who is responsible for confirming the number of families present in a community (Torabi 2007: 6).

This size-based definition of a community has resulted in some villages coming together to form one NSP community, as was the case with the case study community of Shah Raheem.[220] Differences in borders can also result from larger communities sub-dividing themselves. Some communities with more than 300 families have split into multiple communities to optimize their gains from the program, which Monsutti (2009: 21) worries could exacerbate local fault lines and "reignite old tensions."[221] In some areas, therefore, NSP communities do not align or represent the pre-NSP established villages, which can lead to problems.

Once communities are identified, they must be mobilized. This starts with the introduction of the NSP to a community, including its goals, expectations, and the responsibilities of the community in civic education meetings. The purpose of this is to provide information and to gain the support of the community to implement the NSP. It is during this time when the FPs are expected to make it clear that the NSP is a Government of Afghanistan (GoA) program. While Beath et al. (2011: 6)

---

219 A "family" is defined by the NSP as, "consisting of a husband, his wife (or wives), and unmarried children; or a single head-of-household (male or female) and his/her unmarried children" (NSP 2007: 5). This, as noted by Monsutti (2012: 572), does not align with the Afghan conception of a domestic unit, which focuses on the people who eat food cooked from one pot.

220 This was also noted by previous researchers who looked at the NSP, such as Brick (2008: v).

221 Block grants are calculated at 10,000 Afghani (about 200 USD) per family, up to a limit of 3,000,000 Afghani (about 60,000 USD) in total, which means that communities with more than 300 families would gain more if they broke into separate communities.

found that FPs did do this, Barakat (2006: 171) found that they did not, which left people in some communities reporting that the program came only from the FP. Whether all of the FPs actually presented the program as such, is not clear, but what is clear is that in some communities, the case study community of Shah Raheem included, there was not a strong understanding of the connection between the NSP and the GoA. This was also found by Boesen (2004: 22).

The process of mobilizing a community has not always run smoothly, as is evidenced by the different stories of resistance, skepticism, and antagonism that FPs met in select villages (Boesen 2004; Kakar 2005; McCarthy 2006; Nixon 2008a). Several FPs told Boesen (2004: 20) that they had faced "outright hostility" in some of the communities they tried to mobilize when they first introduced the program.

It is likely that misinformation about the program and its purpose could have contributed to its inhospitable reception in particular communities. Kakar (2005: 14) was told that the NSP was being called Communist by those in opposition to it in some communities, and that some people spread the idea that the NSP was intended to convert Afghans to Christianity. In some communities they were told that each family would receive cash in hand (Kakar 2005: 13), which led to anger and frustration, along with mistrust, when it was found that this was not the case. One FP said that they had seen CDC members become upset when they realized that the position had no remuneration (KII: FP06, July 6, 2009).

Other communities were skeptical of the NSP because it was being implemented by NGOs, who they said had made promises before but had never kept them (Nixon 2008a: 19). Kakar (2005: 12-13) also found that there was skepticism about the program and whether it would deliver what it promised. This skepticism was observed to have faded only when the first payment installment of the block grant arrived in the community (Nixon 2008a: 20). Two additional reasons were provided by Kakar for why some communities were not interested in participating in the NSP. These included that they did not want to create a new council as they already had one or they did not want women to participate (Kakar 2005: 12-13).

As can be seen by the large number of communities that have participated in the NSP, the vast majority of the population either worked through their skepticism or clarified the misinformation they had and invited the NSP into their communities. Only once a community's interest in the program was established, including obtaining their commitment to participate in its implementation and contribute 10% of the project costs, did the NSP proceed.[222] The first step that followed obtaining the commitment of the community was a participatory community assessment. This assessment, once completed, is then used to create a Community Profile in which detailed demographic and geographic information is gathered in each NSP community, along with livelihoods data (see Annex A: NSP Form 1: Community Profile). While this form is supposed be filled in by the community members themselves, often the FP helps to fill it in, as was seen in the research community of Shah Raheem.

After these steps have been taken, democratic elections are held for a Community Development Council (CDC) that consists of members of the community (see Annex B: NSP Form 2: CDC Election and Composition). The elections are supposed to be secret ballot, universal suffrage elections (Beath, Fotini et al. 2011: 7).[223] All adults over the age of 18 are eligible to vote. Generally, a community is broken into clusters to ensure that different sectors of the community are all represented.[224] In order to prevent elite capture, the NSP has a "no-candidate" policy (Torabi 2007: 5), which means that no one is allowed to campaign for CDC membership, and instead each voter is supposed to write down their preferred representative. Additionally, in an attempt to ensure the participation of at least a selection of female community members, a vote

---

222 Each community is required to provide a contribution worth 10% of the total block grant (NSP 2009a: 21).

223 An additional means of maintaining transparency is that FPs are expected to monitor the CDC election process. One FP staff told us that he had never heard of there being problems with any of the CDC elections held in Khulm District (KII: FP06, July 7, 2009).

224 A community is separated into clusters either by geographical area, wealth, or a different indicator of difference (Enikolopov, Christia et al. 2008: 2). These clusters generally consist of five to 25 families and the adults in each cluster are supposed to vote for a male and female representative to become CDC members from their cluster (Enikolopov, Christia et al. 2008: 2). The facilitating partner in Khulm District said that there were usually five to 15 clusters per community (KII: FP06, July 6, 2009). Ideally, each of these members serves on the CDC as a group together, representing the community as a whole and acting on its behalf.

is only considered valid if 60% or more of the community has participated (NSP 2009a: 17). However, in some communities, women were not allowed to participate in the program at all (Kakar 2005: 23), including voting for CDC members, while in other communities the women were required to hold separate meetings.[225]

The randomized impact survey being conducted on behalf of the World Bank, a quantitative survey covering 500 villages in total, found that the NSP election regulations were generally followed (Enikolopov, Christia et al. 2008: 1), however, other qualitative research on the NSP has shown a great deal of variance in the elections in different villages, with elections sometimes giving way to selection. Nixon (2008a: 24) identified eight different variations of elections, only one of which followed the NSP regulations. Barakat (2006: 72) indicated that voting rarely occurred as directed, with some community members being told how to vote and others not being able to vote secretly. Kakar (2005: 15) noted that CDCs in Nangarhar were not elected, but were instead appointed due to insecurity. The evidence seems to indicate that implementation is not in perfect alignment with the imagined ideals of the program in terms of introducing democracy into community governance and organization.

A number of reports on the NSP found that despite efforts to change the leadership in communities, many communities still elected customary power holders (Boesen 2004: 39-40; Zakhilawal and Thomas 2008: 19). One FP estimated that 50-75% of its CDC members were local elites, while two other FPs estimated this percentage to only be around 5% (Maynard 2007: 47). The randomized impact evaluation found that around 25% of all male CDC members in the NSP communities had been identified as village leaders prior to the elections (Beath, Christia et al. 2011: 11).

However, two different facilitating partners found that having local elites involved in the NSP did not always result in elite capture in the traditional sense, and that it could even be positive for the community, the program, and for the less empowered people in a community. ACTED found that traditional leaders were not always antagonistic to

225 "Pashtun and Tajik do not allow men and women to sit together in a meeting, but I do not know about Uzbeks," said one FP staff member (KII: FP13, July 6, 2009).

the cause and were not always looking to capture things for themselves, but could also be good leaders, and that their inclusion could be a positive support to the legitimacy of the CDC (Liu 2007: 39). The International Rescue Committee (IRC) found that involvement of commanders, religious leaders, etc. was not always negative and in fact resulted in these people taking on greater ownership of the project (Maynard 2007: 38).[226]

As can be seen, traditional or customary power holders were not always replaced or displaced by the establishment of CDCs (Azarbiajani-Moghaddam 2010: 49; Beath, Christia et al. 2010: 18). Kakar (2005: 15) noted that in one community the customary leaders disputed the elections because they had not been elected. A second round of elections was held, and only then did they accept that they would not be a member of the CDC. This pattern was also reported by the FP staff of Khulm (KII: FP06, July 7, 2009). The Provincial Manager of the NSP in Balkh Province described the different efforts some people would go to in order to be declared a member of their local CDC: Some people turned to the local Khan (*Xān*) or commander to try to get them to use their influence to gain a position and he claimed that some even went as far as speaking with their Parliamentarians who would then try to put pressure on the RRD to select their chosen candidate (KII: FP03, June 29, 2009).

Additionally, there is evidence that despite the fact that traditional leaders often play a role in the NSP, the program has also expanded the pool of people involved in decision-making (Beath, Christia et al. 2010: 23), making decision-making more democratic than previously. Even young people are being included in the NSP meetings and, sometimes, as CDC members themselves (Boesen 2004: 44), which is noteworthy due to elders generally holding decision-making authority and power in Afghanistan, especially in rural areas. Affolter et al. (2009: 323-324) also saw a shift in the understanding of who can be involved in decision-making, which can increase the democratic nature of community interactions.

226 IRC included local commanders, social leaders, and political heads of communities in their advisory board, which was in charge of selecting the ordering of communities to participate in the NSP (Kakar 2005: 17; Maynard 2007: 26).

The voting process for CDC membership is led by the facilitating partner and involves the gathering of the men and women of a community together, or separately, to vote for the individuals they believe will best serve the community and its development needs. Where a mixed-gender gathering is not possible, separate gatherings were to be held for the male and female adults of the community.

The council is supposed to consist of both men and women from the community. However, in some communities there is one male CDC and the women do not participate at all; in others there are two CDCs, one for each gender, or one male CDC and a women's advisory group; and in rare cases both men and women serve on a single CDC.[227] It is difficult to determine how many communities fall into each of these categories, as many communities are documented as having a female CDC, but this ends up being true only on paper, such as the case study community of Shah Raheem. Studies on the NSP have supported this with further testimonies (Azarbiajani-Moghaddam 2010: 18; Brick 2008: 28).

One woman told a researcher that the female CDC was a lie told by men (Azarbiajani-Moghaddam 2010: 18). Azarbiajani-Moghaddam (2010: 35), in her report on gender and the NSP, found that some female CDCs are selected by the men without any consultation or input from the women of the community. Other women, who are a part of a CDC, view their inclusion as tokenistic, testifying that they are not allowed to participate in regular meetings or be involved in decision-making (Azarbiajani-Moghaddam 2010: 35; Brick 2008: 13). These women are only included in meetings when FPs or outsiders come to visit the program (Brick 2008: 40; Kakar 2005: 24; MRRD and AIRD 2006: 28). Kakar (2005: 24) concludes that some women were elected just to comply with the NSP policy, but that they are given no meaningful way to participate. This was supported by Brick (2008: 31) who wrote that, "Women [CDC members] frequently complained that they were selected to satisfy an FP requirement and had no actual role in the community nor in how NSP funds were spent."

227 When there are separate councils for each gender, there is generally a family go-between who is supposed to facilitate communication between the two groups (Nixon 2008a: 30). This was confirmed by the FP staff of the research area, who added that male and female FP staff also share information across genders (KII: FP06, July 7, 2009).

Zakhilawal and Thomas (2008: 21) found that there was a distinct lack of meaningful participation of women in the NSP, except in some villages of Hirat. The World Bank (2010b: 12) found that women's involvement varies greatly throughout the country, noting that 38% of all CDC members were women. Only 10% of communities where Boesen (2004: 53) conducted interviews had any women serving as CDC members. It is interesting to note, however, that Kakar (2005: 14) found that some communities where women were not allowed to participate in CDCs when the NSP was first introduced had started to form women's councils two years into their involvement with the program.

The most common reason for women not being involved in the CDCs is a lack of permission from male family members to participate, though older female family members can also block younger women from participating (Boesen 2004: 51). Additionally, a limited access to information about the NSP contributes to women not being involved, and some women simply have no interest in participating (Boesen 2004: 50, 57-58). Kakar (2005: 27) believes that gender biases in FP staff, cultural norms, a lack of educated women, a lack of access to information for females, limited leadership experience, and illiteracy add additional barriers to women becoming involved in the NSP.

This is not to say that people don't believe women are involved in the NSP. MRRD believes that, "NSP significantly increases involvement of women in local governance to women's needs," as written in the SIGAR report (2011: 40). In communities where women were already established as power holders, they have been elected to CDCs and are actively included in decision-making, according to Kakar (2005: 23), though this was the exception, not the rule. One FP described a community in Malistan District of Ghazni Province, where the women were completely in control of the project, including conducting all of the manual labor, because the males had travelled outside of the community to find work (KII: FP07, July 12, 2009).

After a CDC is established, it must be registered with the Rural Rehabilitation and Development (RRD) provincial line ministry (see Annex C: NSP Form 5: Certification of Registration of CDC).[228] Only then is a community meeting held with both male and female adults to dis-

228 CDC members are elected for a three year period (Enikolopov, Christia et al. 2008: 2).

cuss the development needs of the community. Due to social and cultural norms, it is often not possible for the men and women to attend one meeting, and two separate meetings will be held.[229] This meeting is supposed to allow the people the time and space to discuss the needs of the community and determine what development projects they believe are needed, and then to prioritize those needs into a community development plan (CDP) (see Annex D: NSP Form 6: Community Development Plan).

In this process the community identifies problems that it faces and suggests solutions, including cost estimates. These problems are then divided into those that the community will address itself and those which require outside assistance. Most project selection processes were consensus based, with some communities meeting multiple times over weeks in order to finalize their priorities, according to Nixon (2008a: 32).[230] Though it should be noted that even when there was a high level of attendance at such meetings, Enikolopov (2009: 3) found that this did not automatically translate into a high level of participation, with only one out of eight men and only one out of 20 women in attendance voicing an opinion publically. Participation was not always high either, with poverty, lack of time, illiteracy, not wanting to contribute economically, and a lack of knowledge of the meetings noted as preventatives by Barakat (2006: 83).

Once a community development plan has been established, the community, with assistance from the FP, writes a proposal for a specific subproject that the community wants to implement, preferably one of its top prioritized projects. This requires the community to calculate the cost of each subproject, the amount of contribution required from the community, the logistical details of the project, and how the community plans on maintaining the project post-implementation (see Annex E: NSP Form 7: Subproject Proposal).[231]

229 Little information was provided in any of the NSP literature regarding the actual participation of women in community prioritization meetings, nor how much influence they had on the projects that were selected.

230 It is also common for this prioritization to take place utilizing a voting system, where individuals write their votes on cards or where people place their vote into a box with a picture of the type of project. The later form of voting was established in order to assist those who were illiterate to vote in the research district of Khulm (cf. KII: FP06, July 6, 2009).

231 On average, communities are estimated to have contributed more than the

It is the job of the FP to help the community to come up with concrete subprojects that can be supported within the budget constraints of the program. Two kinds of projects were allowed, those that focus on infrastructure, such as drilling wells, creating irrigation canals, building bridges or schools, and projects that focus on increasing human capital, such as embroidery or tailoring training, literacy courses, carpet weaving, etc.[232] While the community is supposed to take responsibility for drafting a community development plan, due to the unfamiliar nature of the proposal process, an often low level of development program knowledge or experience in the rural communities, and low levels of literacy, this responsibility is often shifted to the FP.

The completed proposal is submitted to the NSP Provincial Management Unit (PMU), where the PMU decides to approve the proposal for funding, reject it because the subproject is not permitted (i.e., is on the "negative list"), or request additional clarifications or revisions of the proposal (NSP 2009a: 23).[233] If the proposal is seen as adequate and within the confines of the program, the community will be granted a block grant, consisting of 10,000 Afghani (approximately 200 USD) per family to implement the subproject or subprojects (NSP 2007: vi).[234]

Sometimes education or health projects are rejected because the appropriate ministry is not able to provide the teachers or doctors that the construction of a school or a clinic would require.[235] Another example stems from a community wanting to implement a solar power project,

minimum 10% required, with an average contribution of 17% of project costs (SIGAR 2011: 4).

232 Please note that human capacity building projects were no longer supported by the NSP under NSP III (World Bank 2010b: 31).

233 The NSP established a list of projects that its funds could not support, known as the "negative list." The list includes: diesel generators, equipment or materials funded by other organizations, political campaign materials or donations, weapons, chainsaws, land purchase or lease, money for government office buildings, payment of salaries, any activity that utilizes child labor, any efforts that support drug crops, religious buildings, vehicles, solar power subprojects past a certain date, and more (NSP 2009a: 19-20).

234 A block grant is, "The grant provided by NSP to eligible communities, used to fund approved subprojects. It is calculated based on Afghani 10,000 (approximately US$ 200) per family (subject to a maximum of Afghani 3,000,000, i.e. approximately US$ 60,000 per community)" (NSP 2009a: 2).

235 All NSP projects involving schools must be approved by the Ministry of Education (MoE) and all health clinics require the approval of the Ministry of Public Health (MoPH) (KII: FP06, July 8, 2009).

which was not on the approved list for NSP projects by the time the proposal was submitted. Despite attempts to explain the restrictions placed upon NSP funding, the community refused to give up and went to different ministries to obtain support. Eventually, the community gave up hope and proceeded to submit a new proposal for a different project, but it took three months of increasing tensions between the FP and the community members (KII: FP06, July 8, 2009). In Marmul District of Balkh, 25 CDCs had joined together to fund a power line that would be 27 km long (about 17 miles), however, the maximum number of CDCs that are allowed to join forces is only three (KII: EX04, January 25, 2010).[236] The head of the NSP in Balkh for the Ministry of Rural Rehabilitation and Development (MRRD) said that they had tried to convince these communities to choose alternative projects, but the communities would not listen and were waiting to negotiate the situation with the World Bank, though he suspected that, "these 25 CDCs will be out of NSP" (KII: EX04 January 25, 2010). When issues such as these arise, most often the community moves on to one of its alternative prioritized projects.

In order to receive the block grant, a community must set up a local bank account where the grant money is sent. Multiple members of the CDC are required to witness the extraction of money from the account to ensure transparency. In the beginning of the program, the distribution of money became an issue, with long delays. Interruptions in the payment of block grant installments cost the program trust in some communities and also impacted negatively on the quality of infrastructure in some areas because construction was postponed or held up (Barakat 2006: 172).[237] The SIGAR report (2011: 19) estimates that almost 10% of projects were delayed between 2003 and 2009 due to a postponement in the disbursement of funds in connection to funding shortages. Delays in payments to the facilitating partners also occurred with an estimated 14% of FP payments, resulting in FPs sometimes having to pay millions of USD from their own pockets in order to maintain the NSP projects they were helping to facilitate (SIGAR 2011: 21)

236 Another facilitating partner staff member said that up to seven communities were allowed to combine their block grants (KII: FP05, January 14, 2010).

237 Sometimes work was only partially completed or the winter weather ruined what had been started.

There is a large variety of subprojects that have been pursued under the NSP, both in infrastructure and capacity building. However, the vast majority of the projects have been infrastructure related, in alignment with the requirement that at least 85% of the block grant be committed to public infrastructure (NSP 2009a: 19-20). While most often the subprojects chosen have been one of a community's top choices, this is not always the case, which has led to some frustration. For some communities the problem stemmed from the fact that they were not able to afford the project they felt they needed the most with their allocated block grant (Barakat 2006: 177). Others sometimes wanted something that was not allowed by the program, such as a mosque, and their proposals were rejected (Barakat 2006: 60). A few communities asked for help with electricity problems, but they were told they would have to wait a few years (Barakat 2006: 176). Though happening rarely, these kinds of negative responses to subproject proposals have led to accusations of corruption and/or a loss of trust in the program itself (Barakat 2006: 176).

More generally, however, communities were successful with their subproject proposals. Many communities chose to implement projects such as irrigation, rural roads, electricity, and drinking water (Owen 2006: 48-9). An MRRD analysis of completed subprojects shows that the top three types of subprojects implemented included: transport projects, such as the reconstruction of roads and bridges; water supply and sanitation projects; and irrigation projects (NSP 2009b: n.p. Annex C). These however were not the only projects completed. Over 23,000 km of roads (about 14,291 miles) have been constructed or rehabilitated through the NSP, along with the production of 39,343 kilowatts of electricity through micro hydro and solar power projects (World Bank 2011c). Over 2,000 hand pumps have been installed and 515 schools have also been built (World Bank 2011c).

One can see that there has been a shift in the types of projects being implemented between the different phases of the program. While the percentage of block grants committed to transportation and water supply and sanitation has not changed dramatically, the percentage committed to irrigation has increased under phase III of the NSP, along with the percentage of "other" projects being implemented. At the same time, the percentage of funding committed to power, livelihoods, and education has decreased, as shown in Figure 19.

Figure 19: Allocation of NSP I and NSP II Block Grant Funding as Compared to NSP III Funding

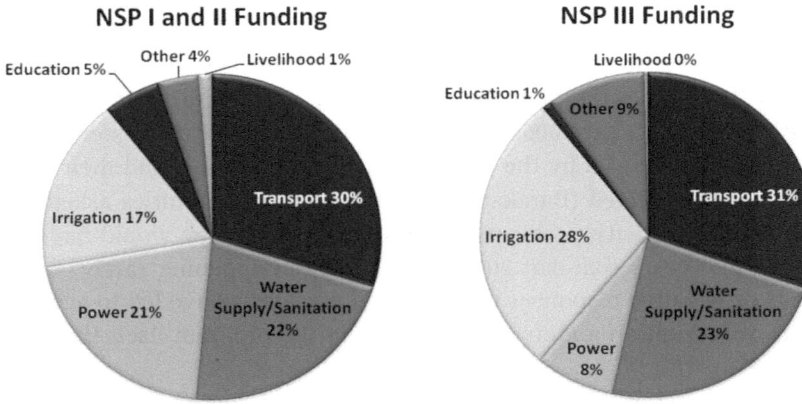

Source: own design based on NSP statistics (IRoA, MRRD et al. 2012c: 11)

Many communities conduct multiple subprojects with their singular block grant, and not just one larger subproject. However, a few communities have pooled their block grants together to construct one larger project that benefits each of the communities that contributed, such as water and electricity projects or the building of a school (Nixon 2008a: 33).

While some researchers have called into question the quality of the subprojects implemented (e.g., Barakat 2006: 173), one facilitating partner evaluation found the subprojects implemented with their assistance to be of good quality, though it is quite possible that a self-assessment of this variety would lead to a positive bias (Liu 2007: 32). Little research has been written on subproject failures, though there is evidence that projects have failed by virtue of certain types of projects no longer being allowed within the program, such as generator projects. One example of a failed NSP project, provided by the FP of the research district, revolved around a human capacity building project where the community had proposed a tailoring project for women (KII: FP06, July

8, 2009). Instead of using the money they received to buy new sewing machines, the all male CDC decided not to purchase sewing machines for the women, saying that they already knew how to sew. They instead used the money for a micro hydro power project, which was not approved and the human capacity building project never took place.

It has become evident that subprojects that require communities to make regular contributions to the upkeep or functioning of the project have often not been maintained (Brick 2008: 22, 23). Some projects that were susceptible to this include generator projects for the production of electricity and deep well projects that require a generator to bring the water up. This was true in the case study community of Shah Raheem, as will be explored later. Barakat (2006: 182) found that communities did not plan for the maintenance and upkeep of their subprojects and that they could not afford to maintain their wells. In contrast to these examples, Ehsan Zia, Minister of the MRRD in 2008, said that 95-98% of all NSP projects were well-maintained (Higashi 2008: 44). Though his estimation is likely a bit more optimistic than realistic, he believes that the people feel a sense of ownership of the subprojects and that they were enthusiastic about them because they selected the subprojects themselves (Higashi 2008: 44). A number of other authors also found that there was a high level of local ownership of NSP-funded projects (Affolter, Richter et al. 2009: 312; SIGAR 2011: 13), though it should be noted that neither appears to have conducted fieldwork in the communities to confirm this.

Once a subproject has been selected, it is up to the community development council, with assistance from the FP, to implement it. This requires capacity building of the CDC members, as few of them have experience in planning and running development projects. The CDC members, at the very least the four Executive officers – Chairperson, Vice-Chairperson, Treasurer, and Secretary – participate in capacity building trainings, which include day-long workshops that inform the participants about the NSP program and its many rules and training efforts focused on project management, accounting, procurement, reporting, monitoring, and other topics (NSP 2009a: 17; Torabi 2007: 19). These training efforts are intended to help prepare the CDC members for the management role they are expected to play during the implementation of each subproject, including maintenance once it is completed. In some provinces, such as Balkh, female community members complained that they were not given as much training as their male counterparts, indicating that this was in part due to women not being allowed to travel to attend the events by their families (Barakat 2006: 190). Azarbiajani-Moghaddam (2010: 19) said that 5% of those interviewed in her research said that there was no mobilization or training for women despite official records that said otherwise. Zakhilawal and Thomas (2008: 21) found that communities had not received much training and had not built the capacity that would allow them to handle future projects. Barakat (2006: 73) came to a similar conclusion, evaluating that less than half of the CDCs interviewed were able to handle all aspects of managing a subproject without external assistance.

While the training events are of great importance to the functioning of the subprojects implemented in a community, it is also important for the long-term building of skills and knowledge regarding development projects and development efforts. It is additionally important for the sustainability of the CDC as a council of action and abilities. If one of the purposes of the program is to empower the people to address their own development needs, this capacity building is essential and it must be strong. Unfortunately, no publically available research has been presented on the content or efforts of such trainings provided through the NSP. In the most recent phase of the National Solidarity Programme, NSP III, 19.3% of the budget of the program has been dedicated to capacity building, while 72.6% of the budget goes towards block grants, and 8.1% goes towards implementation support (World Bank 2010c: 1). It would be useful to know what this money is being spent on and what kind of results the FPs and the communities have observed.

The CDC is in charge of all procurement, hiring, and monitoring that accompanies each subproject, and reporting on these things to the community at large, to maintain transparency. Specific guidelines about how to share program information via small billboards and such is outlined in the Operational Manual, of which there have thus far been six different versions (MRRD and NSP III 2012; NSP 2003, 2004, 2006, 2007, 2009a). It is hoped that this will create greater transparency in project implementation and prevent corruption, thus, reducing the opportunities for elite capture (Owen 2006: 34), though the NSP does not appear to gather or track information on the number of signboards posted in a community (SIGAR 2011: 18). The community is directly responsible for these things, but the facilitating partner is also in charge of providing technical assistance during the planning and implementation phases, as required, and overseeing the utilization of each communities' block grant (NSP 2009a: 24).

Torabi (2007: 21) found that only one FP was able to say that they had not faced any problems with corruption in their implementing communities. He estimates that 1-2% of community block grants involved some kind of corruption, though how this compares to other participatory and non-participatory development programs in Afghanistan is not known (Torabi 2007: 21). Brick (2008: 22) found that projects where things could be expropriated resulted in allegations of corruption aimed at CDC members, concluding that there was not enough monitoring of the CDCs by external authorities.

In general, the project timeframe of the implementation of the NSP in a community is usually between 18 and 24 months (Beath, Christia et al. 2010: v), though it can be shorter or longer depending upon the functioning of the community development council, the facilitating partner, and the funding channels.

When a community has finished its subprojects successfully and used all of their block grant, a certificate of completion is issued by the district office of the MRRD. The community is then "handed over" from the facilitating partner to the provincial line Ministry of the MRRD, who is then responsible for working with the CDC to find funding for future development projects. There is very little evidence that the RRD is ready and able to take on such a large responsibility, but this will only be determined in time.

The National Solidarity Programme, as a whole, is a complex program that has been informed by past participatory development program successes and failures. As can be seen from the review above, its implementation strategy has many steps and processes that are intended to place the power into the hands of the people of the communities in which the program is implemented. As can also be seen from this review, this is not always happening in the neat and defined manner presented in the NSP Operational Manual. This background is necessary to proceed forward in investigating how the NSP has been implemented in the case study community of Shah Raheem and its impacts on the community. Understanding the degree of alignment or misalignment with the prescribed ideals of the program can provide insights into the strengths and weaknesses of the program structure.

4.4 Extracting Meaning: Embedding the Research in Afghanistan

This chapter focused on establishing an understanding of context: The context of the rise of participatory development in Afghanistan historically, the current security and development context of the country, and the context of the National Solidarity Programme. These varied contexts are all vital to the research conducted for this dissertation. Without them, the viability of understanding the impact of a development program at any level of analysis, but especially the community level, would be incredibly limited.

As has been shown, the National Solidarity Programme is a very large development program with a great deal of political support. The positive rhetoric surrounding this project relayed by development stakeholders, journalists, and politicians around the world has overwhelmed the academic and programmatic research that should be tempering it.

This dissertation reveals a micro-level perspective of the program, a perspective that remains lacking in the research about the NSP thus far, and for participatory development efforts in general. It highlights the ways in which the program has positively impacted on the research community of Shah Raheem, where it impacted negatively, and where little impact could be found at all. It is a nuanced description, based upon a historical understanding of the country, the district, and the community.

5 A District at the Crossroads: Exploring Khulm District

The fieldwork for this research was conducted in Khulm District, in the north of Afghanistan. Situated firmly in the Turkestan Plain, Khulm is bordered internally by Samangan Province; Kunduz Province; and Kaldar, Nahri Shahi, and Marmul Districts of Balkh Province.[238] While formerly a part of the province of Samangan, Khulm is now the largest of Balkh Province's 15 districts, geographically. Despite its land coverage of 3,045 square km (about 4900 square miles) (MRRD and NABDP 2006), Khulm's population is not vast. The district has an estimated population of just under 65,800 people, according the Central Statistics Office of the Government of Afghanistan (2004).[239]

Khulm District's geographic location is a consequence, in part, of the natural oasis that is formed by the Tashqurghan River, around which the population is centered. Based upon interviews and reviewing satellite images and maps, it can be seen that despite the large size of the district, most of the population resides in and around the oasis town of Tashqurghan, and large areas of the district remain unpopulated.

The district has played an important role in Northern Afghanistan at different points in history, and its internal history has helped to shape the social structure, infrastructure, and the people who reside in the district today. By understanding their history, a clearer understanding of how the NSP has impacted on the environment can be established.

When the history of the district is examined, three main themes emerge that will be explored: (i) Khulm is a geographical oasis, which has provided the means for the agricultural livelihoods of the population for centuries, (ii) Khulm emerged as a marketplace of importance, situated at one of the crossroads of the Silk Road, and (iii) Khulm has become a battleground multiple times, inundated with fierce fighting

238 The plain is mostly uninhabited land, punctuated by small villages which are situated along the road or nestled into the foot of the mountains which bound the plain to the west.

239 This figure is higher than the National Area Based Development Program's (NABDP) estimate of 49,207 people living in Khulm (MRRD and NABDP 2006), and the NSP population numbers for the district, which state that Khulm had a population 40,693 people in 2006 (NSP Documents).

and the effects of war. These three themes emerged from discussions with key informant interviews at the district and the community level in the household survey and in observations. For many in the district, these three themes help to define the character and the characteristics of a life lived in Khulm.

This section, therefore, will review the history of Khulm District to provide the necessary background for understanding the research community of Shah Raheem. From this, one can begin to look at how the participatory development program of the National Solidarity Programme (NSP) has impacted this community and its population. A special focus is placed upon the four themes of the analytical framework – personal security, economic security, social security, and empowerment – when it has been possible. In addition to this, the major development trends, efforts, or accomplishments at different time periods have also been noted.

5.1 A Historical Perspective of Khulm as an Oasis, a Marketplace, and a Battleground

5.1.1 Exploring the Establishment of Khulm and its Early History

It is presumed by individuals from Khulm District that the natural oasis stemming from the Tashqurghan River is what drew people to establish the town of Old Khulm.[240] Water sources have historically been scarce in the arid and desert-like Turkestan Plain, which is defined as a "mid-latitude steppe and desert climate" (Ashworth 2005: 12).[241]

It was quite common for settlements to emerge and become established around naturally occurring springs and oases in the Turkestan

240 District elders explained that the geographic setting made it an attractive place for a settlement (FGD11, July 10, 2011).

241 The annual mean temperature is 17.4 °C, with an average temperature of 32.1 °C in July, as compared to an average temperature of 3.6 °C in January. There is a reported average of 7.2 hours of sunshine daily, and an average monthly rainfall of only 15.8 mm (cf. Ashworth 2005: 12-13).
This label as a mid-latitude steppe and desert climate means that 70% or more of the annual precipitation falls in the winter; the average annual rainfall is less than 20 times the average annual temperature; and the annual average temperature is less than 18 °C (Ashworth 2005: 12).

Plain, providing access to water for consumption and for irrigated agriculture. This was true for the settlement of the town of Old Khulm, whose existence anchors the district of Khulm. Prior to its inclusion in the present-day nation of Afghanistan, the northern Turkestan Plain was governed by Khanates whose power bases were situated in these oases (Centlivres 1970: 18).

In the 1800s, the Turkestan Plain was run by independent and contentious Uzbek Khanates who were based in the small oases of the northern River Basin, such as Ali Beg, the Khanate of Khulm (Favre and Kamal 2004: 88). The main Khanates, or kingdoms in the north at this time, were Maymana, Gurziwan, Andkhoy, Shibirghan, Sari Pul, Aqcha, Balkh, and Khulm. The Khan of each area was in regular competition for preeminence, causing conflicts to be a regular part of the lives of the people living in the plain, creating personal insecurity in the area (Favre and Kamal 2004: 88).[242]

The foundation of the modern town of Tashqurghan, the current district capital of Khulm, was built upon the site of a city that had seen extreme destruction and battle during the Turko-Mongol invasion, according to Adamec (2002: 308).[243] It was located about seven miles from Old Khulm, which was the first population center built in the seventh century (Charpentier 1972: 23).[244] The town of Old Khulm was trans-lo-

242 For example, Khulm was seized by Beghee Khan in 1751, though only after multiple incursions were conducted (Ferrier, J.P. 1856: 186).

243 The name Tashqurghan is Turkic in origin and translates as "the fort of stone" (Adamec 2002:308; Charpentier 1972: 17; Wood, J. 1872: xl). The town was originally called Khulm, but this was changed under the rule of Zahir Shah who gave the town the name of Tashqurghan in 1946 according to Adamec (1979: 334) or 1941 according to local elders from Shah Raheem Community (FGD11, July 10, 2011). In 1963 its name was again reverted to the original medieval name of Khulm, in what has been considered an attempt to return to the original roots of the town (Charpentier 1972: 17; Dupree, L. 1980: 35).
This continues to cause confusion as both names are regularly in use today, though the name Tashqurghan is more commonly heard on the streets of Khulm District. Henceforth, to increase ease of understanding and adopting Patton's belief in indigenous typology (1980: 169), I will utilize the name Tashqurghan for the district center and Khulm for the district as a whole. While this is not in alignment with the AIMS system, it is in alignment with daily spoken references made within the district and the surrounding provinces.

244 Captain John Wood of the Indian Army reported that Tashqurghan was founded by the ancient Turanian King Afrasiáb. Though he provided no date

cated about seven kilometers to the current location of Tashqurghan because its location was not seen as conducive to securing water access for the population and their needs (Maitland and Talbot 1887: 105), highlighting the importance of water access to the town.[245]

While the exact year in which Tashqurghan, or New Khulm, was established is unknown, it is clear that by 1824 Old Khulm had been more or less vacated (Charpentier 1972: 23).[246] "It is said that Khulm was abandoned, and the town shifted to its present site, because the water-supply is here better commanded. An enemy besieging Khulm could turn off its water into the Angarik branch, but it would apparently be impossible to divert the stream from Tashqurghan, except by very extensive works" (Adamec 1979: 571).

After the death of Ali Beg in 1817, the area that is now Khulm District faced an internal war that erupted between the sons of Ali Beg over control of the district and the town of Tashqurghan. Then in 1840, Dost Mohammad, the Emir of Afghanistan, established temporary headquarters in Tashqurghan in order to support Northern tribes who were battling the British (Charpentier 1972: 23; Ferrier, J.P. 1856: 395; Malleson 1878: 400).[247] Dost Mohammad had an estimated 60,000 troops stationed in Khulm, all of whom were foreigners to the district, in order

---

for its establishment, it can be assumed that this took place prior to the seventh century because the area was already populated and active in agriculture by 645 AD as described by the Chinese traveler Xuangzang referred to in Wood's book (cf. 1872: xlviii).

245 Amir Kalich Ali Beg, the *Khan* of Khulm, laid the first stones of a citadel, which marked the town of Tashqurghan in the early 19th century (Adamec 2002:308). As described by the Gazetteer, Ali Beg moved the town to its current location after the Mir of Aybak diverted the stream that fed the town of Old Khulm, leaving it with no water for eight full days (Adamec 1979: 567). Eventually, Ali Beg broke the dam by force, but recognizing the vulnerability of the location, he determined that it had to be moved to prevent competing Khanates from cutting off the town's water supply. Therefore, he moved the population to its current location, closer to the mountains and to the main source of water that feeds the oasis (Charpentier 1972: 23; Maitland and Talbot 1887: 105).

246 Charpentier is a Swedish social anthropologist and cultural journalist who wrote a book on the Bazaar of Tashqurghan in the 1970s. He and his wife have spent decades conducting research in Afghanistan and have published numerous books on the country.

247 Centlivres (1970: 19) provides a contradictory date of 1850 for the seizure of the town of Khulm by Dost Mohammed.

to attack the British (Ferrier, J.P. 1856: 338). This further decreased the personal security by adding to the conflicts in the area.

Outsmarted by the British and their use of "monetary persuasion," and after nearly being killed by his mutinous troops (Ferrier, J.P. 1856: 338), Dost Mohammad handed over control of Tashqurghan to the youngest son of Ali Beg, Mir Wali. This was followed by the outbreak of war between Tashqurghan and Kabul in 1845, with Kabul eventually withdrawing their forces. Mir Wali battled against Muhammed Akram Khan for power, with Mir Wali hoping to gain control of Kunduz and Badakhshan Provinces. He was stopped by Ghulam Haider Khan, Dost Mohammed's son, who took over Tashqurghan, forcing Mir Wali to flee to Uzbekistan (Adamec 1979: 573).

As can be seen, warfare and strife were common in the district of Khulm in the first half of the 19th century. Personal security, however, began to improve slowly after Khulm became a part of the Afghan State in 1851, along with the rest of Turkestan (Adamec 1979: 20, 35). By the time Maitland entered the plain in 1914, the situation had calmed, with the *Khans (Xāns)* integrated into the Kingdom of Afghanistan.[248] Maitland noted that at this time there was only one gun in the citadel above Khulm and that this was only used during the *Eid* celebrations (Adamec 1979: 570), indicating that a renewed level of peace had returned to the district and its capital.

At this point in time, economic security was ensured largely through agriculture. The Khulm Delta was irrigated by the waters of the Tashqurghan River, also known as the Samangan or Khulm River. The water derives from the rain and snow that falls on the central highland mountains that drain from Ruyi Du Ab and Khuram Wa Sarbagh Districts of Samangan Province, close to Bamyan (Favre and Kamal 2004: 37, 79, 116).[249] Small streams then converge to the west to form the river that flows towards Khulm District through the Tang-i-Tashqurghan Gorge, with the water spreading out to irrigate the delta-shaped Khulm Oasis, shown in Figure 20 (Favre and Kamal 2004: 116). Though incredibly important to the people of Khulm, the river is not

248 The last Khanate capitulated to the King of Afghanistan, Abdur Rahman, in April 1884 (Favre and Kamal 2004: 88).

249 These central highland mountains can be referred to as the Alburz Mountains (UNIDATA 1991: 3). Ruyi Du Ab can also be written Rui Duhab.

very long or very large and therefore not particularly noteworthy in the country as a whole.[250]

Figure 20: Soviet Era Topographic Map of the Khulm Oasis, Tashqurghan, and the Surrounding Settlements

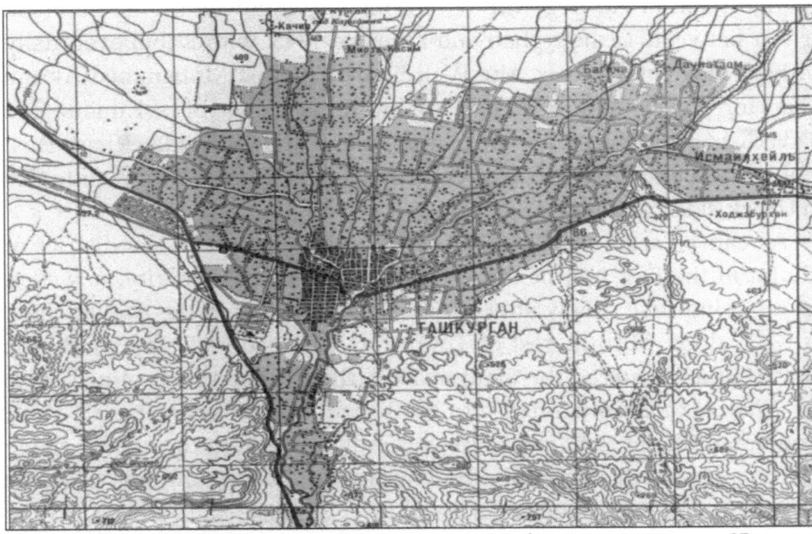

Source: (Generalny Stab. 1984-1986) 1:50 000 Topographic Series 1984-1986, CD-2, sheet J_42_111_3

The oasis of Khulm has been actively involved in agricultural production for many hundreds of years going at least as far back as the seventh century, when the Buddhist monk Xuangzang described Khulm as an area of wheat and pulse production (Wood, J. 1872: xlviii).[251] Captain Wood (1872: lxviii, 265) compared Tashqurghan and its surrounding areas to the city of Balkh, and said that both towns had gardens that were then met by the dry plains: "The fields around it were green as emeralds, while all beyond was desert." Major C.E. Yate (1888: 317) described

250 The river is 190 km in length and forms the smallest river basin in the North of Afghanistan (Favre and Kamal 2004). A map of the Khulm River Basin can be found in Annex O: Map of the Khulm River Basin.

251 Xuangzang was a Buddhist monk who travelled through India and parts of Central Asia in 619-645 AD (Xuangzang 1996: 1). His name can be written Hwen Thsang, Xuán Zàng, or Hsüan-tsang.

the town as a "... succession of lanes, bounded by 8-feet-high mud-walls on either side, with just the tops of the apricot and fig trees beyond peeping over them."[252]

Tashqurghan utilized an irrigation system within the town that flowed outwards from the mountains northwards towards the Amu Darya River, though running dry long before reaching it.[253] A British diplomat wrote in 1928 that, "The whole town is thickly planted with fruit-trees, and through the middle of the streets run irrigating channels" (Ikbal Ali 1928: 15). The irrigation system, which was man-built, enabled the waters from the mountains to spread to the dry, but arable lands. Orchard "suburbs," as Adamec (1979: 11) calls them, surrounded the main oasis town of Tashqurghan at the end of the 19th century. They contributed to the production of goods that were exported in the early 19th century and at later dates.[254] These "suburbs" continue to exist today and play an important role in the local economy.

While the population benefits from the natural delta of the Khulm Oasis, the people of the oasis face economic security risks and vulnerabilities. Droughts and flooding have both contributed negatively to the livelihoods and securities of the population of Khulm. Situated at the foot of the mountains, Khulm has suffered from flooding that washes down the nearby hills, destroying vast tracts of gardens and infrastructure as it passes. This is as true today as it was during Major C.E. Yates' time in the 1880s. He reported that, "the town was overtaken in the spring by a great flood, which washed down from the hills right across this *maidan* and into and through the town " (Yate 1888: 316).[255]

Economic security in Khulm has not always been solely connected to agriculture, however. The town of Tashqurghan was also the largest

252 Old Khulm's agricultural history is also evident in descriptions of the abandoned town and all of the fruit trees that remained when its population moved to Tashqurghan (Hamilton 1906b: 254; Wood, J. 1872: 266). "The town of Khulm stood out on the Oxus Plain, surrounded by a belt of very productive land. The irrigation of this was highly developed; even now there are numerous orchards and cultivated fields around the cite [sic] of the ruined city," said Hamilton (1906b: 254) of the Royal Geographic Society in 1906.

253 The Amu Darya River is also known as the Oxus River, the Amu River, the Amudarya, the Gozan River, and the Vaksu River.

254 The main exports that were produced in Khulm District in the early twentieth century included: grains, pulses, rice, ghee, wood or timber, seeds, fruits, vegetables, and nuts, followed by animals (Hamilton 1906b: 294-5).

255 *Maidan* can also be written ma'dan, and it refers to a geographical plain.

of the Afghan Turkestan towns in its heyday, and the richest, according to Adamec (1979: 11). During the times of the caravans, Tashqurghan was one of the stops along the northern route of the Silk Road.[256] This was in large part due to its geographic location at the crossroads of so many regions, making it an ideal market town. The establishment and growth of Khulm District was also, therefore, connected to its status as a market center both internally and externally.

The Silk Road connected Asia to Europe and many locations in between as early as the second century BCE (Werner 2003: 152), if not prior, and continued to be of great importance until the middle ages, when transportation via sea became more common. The Silk Road was not a single road connecting the two continents, but was instead a number of different routes along which traders traversed, stopping in different towns and cities along the way to their destination, which Christian (2000: 1) denotes through his usage of the plural Silk Roads instead of Silk Road.[257] He describes them as "... a constantly shifting network of pathways for many different types of exchanges" (Christian 2000: 2).

Even after the preeminence of the Silk Road had passed, around the end of the Mongol Empire (Christian 2000: 5-6), Tashqurghan and other Silk Roads outposts remained centers of trade and commerce. This was noted by Yule (1872: 452): "Several roads converge from various parts at Tashkurghan, which is the centre of trade from eastern, northern, and southern Asia." Goods from India, China, Bukhara, and other areas would arrive in Tashqurghan and then be sent to the rest of Afghanistan, making it a vital trading hub in the early 19th century (Charpentier 1972: 156; Ikbal Ali 1928: 15; Yate 1888: 315).[258] "Tash Khoorgaun is the

---

256 Shibirghan, Balkh, and Khulm were principal stages on the route, with secondary stages consisting of water points (Centlivres 1970: 16).
The term "Silk Road," *Seidenstraße*, was coined by the German Geographer Ferdinand von Richthofen in 1877 (Forět and Kaplony 2008: 14).

257 Christian (2000: 1) highlights that the Silk Roads not only facilitated trade between civilizations, but also between the pastoral and agrarian worlds (Christian 2000: 1). He stated, "that the Silk Roads should be seen as a complex network of exchanges that linked different ecological zones of the Afro-Euroasian landmass into a single system" (Christian 2000: 1).

258 Along with specialized bazaar streets, Tashqurghan also hosted a traditional *Tim*-building and about 27 kār(e) wān-sarāj, along with several open market places (Adamec 1979: 335). The *Tim* was a center for goods made both locally and internationally, and was known for its multiple domes and its ornamentation.

grand thoroughfare, and has been so during many years, for karrovans passing and repassing between northern Asia and India. It is the last general mart before entering the range of Hindoo Kush and the natural boundary between the Uzbecks and the Avghauns" (Charpentier 1972: 159).[259]

Tashqurghan was a large center of trade even in the second half of the nineteenth century, much larger than Mazar-i-Sharif, which is now the provincial capital (Maitland and Talbot 1887: 106).[260] It was described as a "a large town . . . possessing a fine bazaar of 450 or 500 shops" (Maitland and Talbot 1887: 105). The international nature of the town, due to the market, was noted by Adamec (1979: 567), who wrote that there were a fair number of Hindu merchants, calling Tashqurghan the "emporium of Central Asian trade," this was also supported be Levi (2002: 160) who noted the dominance of Hindu merchants in Tashqurghan.[261] Another British officer, Yate (1888: 315), described Tashqurghan as: "the great trade-mart of Afghan Turkestan, and about its most important place. Here the caravans from India on the one hand, and Bokhara on the other, all break bulk, and from here the merchandise is distributed all over the country."

In the mid-1800s, Tashqurghan began to see some of its importance as a commercial center diminish as more and more goods began to be transported through Mazar-i-Sharif (Charpentier 1972: 23). As the provincial capital started to become the large and powerful city that it is today, Tashqurghan continued to have a lively market, but its focus shifted inwards and it became a smaller national market with no direct relationship with its former trading partners in other countries, according to Charpentier (1972: 160). This was attributed to the increase in the transportation of goods via air, and the establishment of a customs point in Mazar-i.-Sharif. By the 1970s, Tashqurghan had long lost its primary status as the center of trade in Northern Afghanistan (Charpentier 1972: 159).

---

259 Hindoo Kush can also be written Hindu Kush.

260 In the nineteenth century there were two market days per week, on Sundays and Wednesdays, and the bazaar dues were the largest part of the *Khan's* revenue, or collected taxes (Adamec 1979: 567). Transaction costs were charged on each sale and transit costs were also charged (Adamec 1979: 567).

261 As further evidence of the international nature of this bazaar, Chinese saucers and plates were embedded in its walls (Yate 1888: 317).

5.1.2 The 20th Century, Prior to the Saur Revolution

Finding only limited written descriptions of Tashqurghan from the mid-19th century to the mid-20th century, I supplemented these secondary resources with the memories and stories of people from the district.[262] The elders of the case study community and other individuals living in Khulm District described the area as peaceful and prosperous in the beginning of the 20th century. They painted a picture of a district filled with gardens and plenty of water for irrigation. The active market continued, ensuring employment opportunities and connecting Khulm to other regions and countries. One key informant said, "Before the revolution our life was okay, it was a normal life. We had no security or economic problems" (KII: Com18, February 23, 2010). This sense of peace and personal security would alter notably after the PDPA came into power and the Soviet troops entered the district, as discussed in relation to the development history of Afghanistan previously.

Prior to the Saur Revolution in 1978, and the conflict that arose with those who did not agree with it, there is some evidence that the relative peacefulness described by Shah Raheem male elders existed. Research informants remembered this time period as one of good personal security, with little to no restrictions of their movements. Charpentier (1972: 155), however, reported in 1972 that, "Women seldom visit the bazaar to buy things themselves," indicating a lack of freedom of movement for women.

Based upon planning and development reports from the area, it is evident that agriculture was important in Khulm at this time. A UNIDATA report (1992: 42) notes that Khulm District had an agricultural surplus in 1968, indicating that cereal production was strong.[263] This would change, however, when water access and usage rights shifted in the 1980s, resulting in a production decrease, with only seven percent of Khulm farmers able to sell excess wheat by 1998 (UNIDATA 1991: 15).

---

262 This provides a localized version of history, which is subjective and selective in nature, as is all memory.

263 Khulm District produced 50.8 metric tons of cereals, but only consumed 11.8 metric tons, resulting in 39 metric tons of surplus cereals in 1968 alone (UNIDATA 1992: 57).

At this time Khulm was dependent on water stemming from the spring in the Tang-i-Tashqurghan. Centlivres (1970: 13) notes that it was not only the gardens of Tashqurghan that relied upon irrigation from the stream, but also the surrounding villages, emphasizing that the villages and town would not have survived without the existence of a functioning irrigation system. Prior to the 1980s, the irrigation system and its source were adequate to support agriculture in the district.[264]

Another source of economic security that often worked hand in hand with agriculture was the market. Though it was not as large or active as during the days of the caravans, the bazaar in Tashqurghan was still significant enough to have a reported 800 shops and workshops in the early 1970s according to Charpentier's fieldwork (1972: 167).[265] While the market still contained products from other countries, such as Japan, Czechoslovakia, Pakistan, and India (Charpentier 1972: 150), there had been a shift in importance of Tashqurghan as a crossroads. Most of the foreign products in the Tashqurghan bazaar were brought to Tashqurghan via Mazar-i-Sharif, though some direct importation was still taking place. Some people sold almonds to wholesalers, which then made their way to markets in Kabul and India (Charpentier 1972: 153), a tradition that continues to this day.

This participation in international trade in Khulm began to diminish as more and more trade took place in the provincial capital of Mazar-i-Sharif. At this time many economic opportunities began to shift from Tashqurghan to the provincial capital. However, Khulm was still a major center for handicrafts during the 1970s, attracting tourists who visited not only for the products on sale, but also to see the architecture of the covered bazaar and its domed circular center.[266]

---

264 There was even enough water in the stream to support a fish industry in the district capital (Charpentier 1972: 30), something that is almost unthinkable today.

265 Adamec (1979: 335) has a higher estimate, and writes that there were 1,155 shops and workshops and an additional 27 *caravanserai (kār(e) wān-sarāj)* at the end of the 1970s.

266 Some of the traditional goods made locally, according to the shopkeepers still working in the bazaar today, included homemade ropes, leather shoes, wooden cradles for babies, embroidered wedding hats, woven towels, jewelry, and metal goods such as water ewers. Charpentier (1972: 166) notes that caps, shoes, leather, and almonds were produced in Khulm and were traded to other parts of the country in the 1970s.

In terms of social security, Ikbal Ali (1928: 45-72) wrote in the first half of the 20th century of different social functions, which today were proffered by the research informants as evidence of closeness between community members, or of good social security: the primary being attendance and participation in sadness and happiness ceremonies. These ceremonies include weddings, funerals, and other events where family and community members gather to celebrate or mourn.[267] He notes that some families used to throw such big weddings that they would make themselves poor (Ikbal Ali 1928: 51), which would indicate that social events were so important that people were willing to put their economic security at risk for their social security.

At the same time Ikbal Ali also notes that he witnessed a lack of trust amongst Afghans, even amongst friends (Ikbal Ali 1928: 113-114). He illustrates this by telling the story of a blind man who found a snake. Due to his lack of trust in his friend who told him it was a snake, the blind man was killed. Whether this story was based in reality or not, it illustrates that there was mistrust that called for or produced such moralistic tales.

As with social security, it is not easy to piece together evidence of the status of empowerment of the people of Khulm District prior to the Saur Revolution. The governments that ruled Afghanistan during this time were noted for ruling without consulting the people, which resulted in the people not being able to actively play a role or participate in their own governance. Barfield (2010: 12) notes that Afghan governments have never been truly inclusive with a direct line of responsibility to the people, and that this has been the case since the late 19th century when Abdur Rahman ruled.

267 Generally there are set traditions for honoring the dead in Afghanistan. The funeral is most commonly three days when everyone gathers and prays together, though often with the genders separated. Food is distributed to those who join the family in their mourning. After fourteen or twenty days have passed a smaller group of relatives is gathered to pray and everyone shares soup or rice, depending on the economic status of the family. They gather again on the 40th day to pray and another meal of rice is shared. During the holidays of *Eid* and *Nou-ruz*, the family in mourning returns to the grave to pray. This is also done one year after the death (cf. Ikbal Ali 1928: 72)(cf. KII: Com03f, February 10, 2010 and personal observations). Weddings are celebrated in diverse manners in different regions of Afghanistan (Landinfo 2011: 6).

This time period, from the start of the 1900s to the Saur Revolution in 1978, is one during which different waves of development efforts took place. However, despite the millions of dollars in development aid provided by countries such as the Soviet Union, the US, West Germany, and others as outlined in the review of development efforts in Afghanistan, there is little evidence or testimony of these development efforts having been felt in Khulm District. Only three development projects were discussed in the secondary literature or by district informants: (i) construction of an electricity plant in Khulm District in 1971, (ii) the construction of a hospital in 1972, and (iii) the reconstruction of Saighanchi Village of Khulm District with earthquake-resistant housing after the large earthquake of 1976 (Charpentier 1972: 17-18; Sibtain 1981: 79).[268]

Additional development efforts that impacted Khulm, though not carried out in the district directly, included the completion of the Salang Tunnel in 1964 and the paving of the road between Kabul and Mazar-i-Sharif (Goodson 2001: 20), discussed previously. The road and the tunnel made transportation to the country's capital city of Kabul much faster by establishing a 1.7 mile (about 2.73 km) tunnel through the mountains that was passable in all seasons. The Salang Tunnel reduced the trip from Mazar-i-Sharif to Kabul down to hours, instead of days (Dupree, N. H. and Gouttierre 2007: 57). This reduced the costs of transportation of goods, people, and ideas.

268 The village reconstruction project was an externally-led participatory development program where the villagers did the work. The project included the construction of 25 houses, a drinking water supply system, sanitation, a mosque, a school, and more (cf. Sibtain 1981: 79-89).
The people of Shah Raheem showed me the old electricity poles, and fondly remembered having regular electricity, though they did not have electricity for long, as the power lines and the plant were destroyed following the Soviet invasion in 1979.
The hospital currently utilized is the same one built in 1972, and it is said to have been mostly active in the intervening years (KII: EX08, August 3, 2009).

5.1.3 Khulm District under the PDPA Government: Violence Erupts (1978-1992)

The Soviet Invasion of Afghanistan had a strong impact on the personal security felt by the people living in Khulm District. Local community elders from Shah Raheem said that the Soviets troops entered Khulm in December of 1979, coming through Hayratan (FGD11, July, 10, 2011). Khulm's location on the northern highway between Mazar-i-Sharif and Kabul meant that the district was easily accessible to the Soviets, who stationed themselves on the edge of Tashqurghan.

Personal Security under the PDPA

Though conflict would eventually overwhelm the district, it was not imminent upon the arrival of the Soviet troops (FGD11, July 10, 2011). Khulm residents who resided in the villages surrounding the town of Tashqurghan said that they continued to go about their normal lives in peace, including continuing to sell their agricultural production in the market (KII: EX13, January 31, 2010).[269] On the whole, people in Khulm reported having a good relationship with the Soviets when they first arrived (HHS 2010). One district elder even described how the children of the communities would play and joke with the Soviet troops, making them laugh (KII: FP08, February 7, 2010).

However, these relatively positive feelings would be challenged by local males arming themselves and forming or joining the Mujahideen groups who were opposed to the PDPA Government and the Soviet troops and advisors who supported them. They emerged as a force to be reckoned with in 1982/1983 (Afghan calendar year 1360) (KII: FP08, February 7, 2010), with the worst of the fighting having occurred between 1982 and 1986 (Afghan calendar years 1361-1364) (KII: Com15, February 17, 2010).[270] Once the Mujahideen organized themselves, they

269 One community member described selling figs to the Soviet troops in exchange for bullets, which they would then use on the next Soviet caravan that would come through town (KII: EX13, February 7, 2010). This story was a source of pride and was supported by another key informant (KII: FP08, February 7, 2010).

270 The main Mujahideen groups operating in Khulm District included: Jamiat-i-Islami (Society of Islam), led by Burhanuddin Rabbani; the Shia party of Harikat-i-Islami, led by Sayed Hussein Anwari; Hezb-i-Islami (Islamic Unity Party), led by Gulbuddin Hekmatyar; Najat-i-Melli (National Liberation Front

began to fight against the Soviet troops, bringing violence into the villages and communities in the area. It was at this point that the security situation in the district of Khulm deteriorated drastically. Known locally as the "revolution," this uprising of the populace against the Soviet troops devastated the infrastructure, the social structures, and the livelihoods of many people living in Khulm District, and in the research community.

The violence and regular battles that took place, led the vast majority of the household survey participants to indicate that personal security was not good when the PDPA was in power, with the Soviet troops based in Tashqurghan. The fighting between the Soviets and the Mujahideen was reported as being regular, incredibly violent, and actively present in the NSP community of Shah Raheem (HHS and KIIs 2010). The fear of physical conflict and fighting contributed to a negative perception of personal security that continued into the Mujahideen Regime.

The heavy tactics of both the Soviets troops and the Mujahideen led the people, young men and women in particular, to stay within their own compounds, if they did not flee the area completely. Some households said that they would spend full days in the bunkers of their houses to protect themselves and their families from the battles (HHS: SR002, January 26, 2010, SR005, February 2, 2010, YB007, February 28, 2010, KII: Com20f, February 26, 2010, Com22f, February 27, 2010).[271] Some district residents talked about losing family members and their homes to the fighting.[272]

---

of Afghanistan), led by Sibghatullah Mojaddidi; and Jumbesh, led by General Dostum (Amnesty International 1992)(HHS 2010, KII: FP08, July 14, 2009, EX04, July 29, 2009).
One KII clarified that no single Mujahideen party was ever in control in Khulm District, indicating that they all had a great deal of access to the district (KII: FP08, July 14, 2009).

271 One head of household described an evening when he had invited friends over for food in the garden of his house. The dinner was disturbed by a rocket that landed in the garden. He and his guests all went down into the bunker. When the rocketing stopped they returned to the place they had been sitting and everything was gone, it had been hit by a rocket (KII: Com15, February 17, 2010).

272 One head of household lost his brother (KII: Com18, February 23, 2010) and another head of household's aunt and son were both killed by a bomb dropped on his house (KII: Com40, February 26, 2010). One community member testified that intense battles and bombings could last upwards of 16 days in a row (KII:

During the time of the PDPA Government, many families of the area moved out of their villages to try and find safer places to live, be that in other parts of Khulm District, other provinces, or outside of Afghanistan. The household survey revealed that this was a time of great migration, with multiple respondents saying that only a few families remained (HHS: SR005, February 2, 2010, CS004, February 20, 2010).[273] The UN estimates that anywhere from 30-40% of the population of Khulm District migrated during the years of fighting, with an estimated 5% having gone to Pakistan (UNIDATA 1992: 113).

Almost all of the households I spoke with indicated that women's freedom of movement was limited by their male family members during the time of the PDPA, with the stated objective of keeping them safely away from strangers and from the physical dangers that existed.

Women said that they had some freedom of movement after the Soviets entered Khulm, but that when the Mujahideen started to fight the Soviet troops stationed on the edge of Tashqurghan their movement was restricted (GD05f, January 27, 2010). Freedom of movement was also curtailed quite strongly for young men during this time period. Household survey respondents said that the Soviet troops would take males between 18 and 39 years of age directly off the street and force them to fight on behalf of the Soviets against the Mujahideen groups (HHS and KIIs 2010). This contributed greatly to the fear young males and their families had of them leaving the house, on top of the fighting.[274]

The Mujahideen were also cited as forcing people to be soldiers for them, though this was reported less often, as "Most of our villagers

Com20f, February 26, 2010).

273 More households immigrated from the research community at this time period than any other covered in this research (HHS 2010).
One man explained, "We left to get away from the fighting" (HHS: YB009, February 28, 2010). Most often entire families or households relocated because of personal security concerns, with all who reported leaving at this time, blaming their relocation on the continual fighting. All of the migrants from the NSP research community who left at this time moved to the provincial capital of Mazar-i-Sharif or to a different village within Khulm District (HHS 2010). All of these families returned to Khulm later during the PDPA Reign or when the Mujahideen came to power (HHS 2010).

274 One respondent said that people who did not fulfill their mandatory military service were unable to work for fear of being captured and made to carry out their legal duty to serve as a soldier for the State (KII: Com26, March 3, 2010).

were Mujahid at that time" (HHS: CS002, February 16, 2010).[275] This support was noted by the Soviet troops, creating great tension between them and people in the area, with the people being threatened for supporting the Mujahideen groups.[276] Support for the Mujahideen, however, was not universal in the district, highlighting the diversity of perceptions that exist and existed.[277] Some household survey respondents even described being threatened by the Mujahideen groups, indicating that this happened throughout the district.[278] They said that the Mujahideen soldiers would threaten community members with torture if they associated in any way with the PDPA Government or the Soviet troops, including accepting donations, such as food (FGD04mix, January 26, 2010, HHS: SR001, January 20, 2010).[279] The Mujahideen were

275 One community member said that the Mujahideen were fully supported by the people of the area, and that the people would even sell their harvest in order to give money and food to the Mujahideen forces (KII: Com15, February 17, 2010). Community members told us about their own personal involvement as soldiers for different Mujahideen groups, or how their family members had fought against the Soviets. Within families, there were sometimes divergent alliances to different Mujahideen groups (KII: EX07 and Com39, February 23, 2010, Com26, March 3, 2010), which was rarely seen as causing tensions within families, as was noted by Barfield (2010: 244), and was employed as a security measure in preparation for any side of the conflict to win. One man described how he had served as a government soldier, while his two brothers were Mujahideen, showing how diverse alliances could be within one single household (KII: Com25, March 3, 2010).

276 One man remembered being accused of providing the Mujahideen with information about Soviet positions (HHS: YB006, February 27, 2010). One male head of household was arrested for similar accusations and his house was burnt down (HHS: SR007, February 3, 2010).

277 Some household survey respondents explicitly stated their disdain for the Mujahideen and voiced their belief that the Mujahideen had brought a great deal of suffering into their lives (HHS 2010). They focused, for example, on stories of theft, saying that the Mujahideen would steal half of each household's harvest and steal household goods from them (e.g., HHS: SR003, January 27, 2010). One of the households said people did not want their children to be Mujahideen, but the Mujahideen would beat them and force them to give them as soldiers (HHS: SR002, January 26, 2010).

278 One community member described how he had been forced, upon threat of death, to walk to Pakistan from Khulm, a distance of over 500 km (about 311 miles), in order to procure and deliver weapons to the Mujahideen (FGD04mix, January 26, 2010).

279 A few households reported that the Mujahideen groups confiscated any food in a household which had been distributed by the Soviet troops for their own consumption (e.g., FGD04mix, January 26, 2010).

also cited as punishing those who attended any of the government schools.[280] One man voiced his fears, saying, "There was no law or justice for Mujahideen actions" (HHS: SR003, January 27, 2010).

Though the actual number of troops in Khulm at any point in time has been difficult to track down, anecdotal evidence indicates that Tashqurghan saw a good number of Soviet soldiers. They were mostly stationed on the outskirts of the district center, in the *Bagh-e-Jahan Nama* palace and garden complex.[281] Originally built by Amir Abdur Rahman in the late 1800s (Centre for International Heritage Activities 2011), the complex provided the Soviets troops with a good vantage point from which to face their opposition, who were situated just outside of the villages surrounding the district capital of Tashqurghan. The Directorate of Agriculture testified that while Tashqurghan was under the control of the Soviet troops, the rest of the district was controlled by the Mujahideen (KII: EX13, January 31, 2010), making the district ripe for conflict.

Some of the Mujahideen groups, meanwhile, were located just north of the research area. From their position in the garden, the Soviet troops would launch rockets towards the Mujahideen groups just beyond the research community, and sometimes into it. The community members and their lands became a battleground, as they were stuck in the crossfire between these two armed forces, resulting in a great deal

280 I was told a story about a twenty year old boy from the community who was chosen to serve as a warning to those thinking about attending a government school or being involved with the government. He was captured by the Mujahideen on his way to school. The Mujahideen then hung him from an electricity pole and skinned him alive, ending in his death (FGD04mix, January 26, 2010).
A community member said, "They thought that if a person went to school they had joined the government" (FGD04mix, January 26, 2010).

281 Alternative spellings for the garden are *Bagh-e-Jehan Nama* and *Bagh-i-Johan Ama*.
The palace and gardens were renovated under Daud in the mid-1970s in the hopes of making it into a museum, but this was prevented by serious damages suffered during the 1976 earthquake (Centre for International Heritage Activities 2011).
The earthquake was strong and resulted in a number of deaths: "In 1976 Tashkurghan was hit by an earthquake which killed at least 70 persons. The gorge, Tang-i-Tashqurghan, collapsed as did the *Bagh-e-Jahan Nama* Palace, the government hotel, the White Mosque, the Tim, and parts of the covered bazaar" (Adamec 1979: 335).

of fear, physical damage, and deaths within the NSP community of Shah Raheem and the surrounding villages. Many houses were destroyed and many people lost family members (FGD05f, January 27, 2010).[282] People's sense of personal security was generally not good.

During the PDPA Regime, many people remember feeling that their "lives were at risk."[283] They told stories of losing their family members, of having their gardens and homes destroyed by grenades and bombs from both sides of the battle, and other evidence of the physical threats that existed in the district during this time.[284]

One respondent summed it up, saying, "We were in the middle of a battlefield at that time" (HHS: CS002, February 16, 2010). A district elder said that Khulm was on the frontlines of the Soviet-Mujahideen war (KII: FP08, July 14, 2009). This was especially true in the NSP community of Shah Raheem, with a Soviet checkpoint situated across the road from the *madrasa* at the entrance to the community (KII: Com15, February 17, 2010).

The situation was made even worse for the people of the communities situated geographically around the town of Tashqurghan because the Soviet troops placed antipersonnel mines in the gardens and farmlands of these communities in an attempt to cause collateral and economic damage to the Mujahideen groups sheltered in and around them. This mining, however, did not just impact upon the Mujahideen soldiers, but also affected the civilian population.

---

282 One man reported seeing a Russian plane drop a bomb that killed seven of his family members (HHS: SR003, January 27, 2010), while another reported seeing thirteen people killed in the city in just one day of fighting (HHS: YB008, February 28, 2010).

283 They told stories of losing family members or their homes (HHS 2010). One man in his 60s described how his wife had been seriously injured by a rocket propelled grenade (RPG). She spent 50 days in the hospital in Mazar-i-Sharif and barely survived an operation (FGD04mix, January 26, 2010).

284 Community members lost brothers (HHS: YB004, February 24, 2010), sons-in-law (HHS: SR003, January 27, 2010), uncles (HHS: SR010, February 9, 2010), and many others due to the fighting.
One man described craters of one to two meters in depth being created by the bombs that would come crashing down (FGD04mix, January 26, 2010). When the bombs hit inside a garden, he said that the gardeners would simply keep building the craters back up. One head of household told the story of a rocket having been launched into his home (HHS: SR009, February 7, 2010).

It effectively limited the mobility of the people, as they could no longer go to the fields to work, impacting on social security, and it also put them and their children in harm's way even within their own compounds.[285] The mining of the local gardens and fields impacted on the women strongly, evidenced by their emphasis on the mines in interviews and their descriptions of how their lives were limited during this time period (e.g., HHS: SR016f and SR015f, February 21, 2010). The fighting and its consequences led to a notable decrease in the agricultural life and production of the district, the research community, and in the economic security of both. Evidence of the intensity of the rocket attacks and fighting can still be seen in the community, as many houses destroyed remain untouched.

All of these threats also contributed to the general perception in the district that most children had been restricted from attending school during this time (e.g., FGD04mix, January 26, 2010).[286] This lack of access to schools was confirmed by the Directorate of Education in Khulm (KII: EX05, August 2, 2009).[287] He said that while officially there were students, teachers, and schools at this time, *de facto*, most teachers and students were too scared to attend because of the regular bombings and killings that were taking place within the district. The Directorate of Education said that even though the Soviets and the PDPA had brought education to other provinces and urban areas, that in Khulm, "They just brought bombing" (KII: EX05, August 2, 2009).

Overall, it can be seen that personal security was not considered to be strong in Khulm under the PDPA, though, as will be shown below, opinions and perceptions within Khulm varied on almost all topics and time periods discussed.

285 Two households described the loss of family members due to these mines: two brothers, a mother, an uncle, and a grandmother (HHS: SR005, February 2, 2010, SR012, February 10, 2010).

286 Within the household survey, a few respondents remembered children being able to attend school under the PDPA. One man said, "The gates of the school were always open during the regime for our children" (HHS: SR002, January 26, 2010). This viewpoint, however, was only stated by a small minority of respondents.

287 The Directorate of Education, though only in his position for eight months when he participated in the key informant interview, had been a principal for six years prior to that and a high school teacher for the 12 years proceeding his time as a principal. He, therefore has a unique, insider's perspective on education in Khulm District.

Also during this time period, the district faced another blow to its agricultural lifeblood beyond the fighting and mining of fields: the district was shifted administratively from Samangan to Balkh Province. This transfer was made in the mid-1980s according to a UNIDATA report (1992).[288] While this early date was confirmed by local residents of the district, it appears as though Khulm only formerly became a part of Balkh Province in 2002, when the "Decree on Separation of the First-Degree District of Khulm from Samangan Province and rejoining it to Balkh Province" was published in Gazette number 805, decree 179.[289]

The shift from one province to another, illustrated in Figure 21, created complications for Khulm District administratively. More importantly for the people of the Khulm Oasis, however, it also changed the terms of water access and usage rights that were vital to the irrigation system that fed the fields and gardens of the area, which in turn had a negative impact on the economic security of the people of the district who were highly dependent upon agriculture for their livelihoods.[290]

The relationship between Samangan and Khulm regarding water rights was long-standing, with water sharing arrangements going as far back as the 1880s, as noted by Adamec (1979: 568).[291]

---

288 The Directorate of Agriculture implied that the transfer was conducted due to political reasons (KII: EX13, January 31, 2010 and February 7, 2010). Khulm District was more strongly aligned with the Jamiat Mujahideen group, which had its troops stationed in Balkh Province and not in Samangan Province where the Jumbesh Mujahideen group had the support of the majority. He said that the Governor of Balkh decided that Khulm District would belong to his province instead of Samangan. It is probable that this transition was also motivated by the desire to collect taxes on the agricultural production in Khulm. One key informant told us that Khulm District had only belonged to Samangan for a short period of time, and that previously it had been a part of Balkh Province (KII: Com15, February 17, 2010). However, this was not put forward or confirmed by any other informants.

289 This decree went into effect on April 26, 2002 (AIA - MoC 2005).

290 The average farming income under the PDPA was about 48,500 Afghani per year, which is lower than the average income of farmers that would be seen under the Mujahideen Government (HHS 2010).

291 Twelve out of every 19 days the Tashqurghan River would be allowed to flow freely into Khulm, irrigating its lands, and the other seven days the water would run to Aybak of Samangan Province (Adamec 1979: 568).

Figure 21: Khulm District as a Part of Samangan and Balkh Provinces:

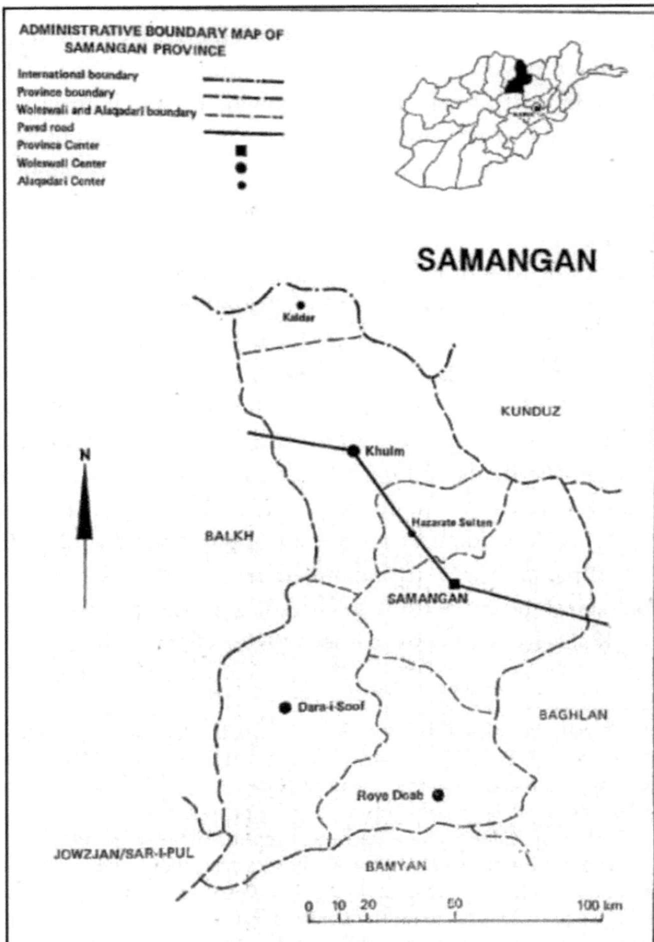

Khulm as a part of Samangan Province (left) under the previous administrative division, and Khulm as a part of Balkh Province (right) under the current administrative division. Source: (UNIDATA 1991: vii; 1992)

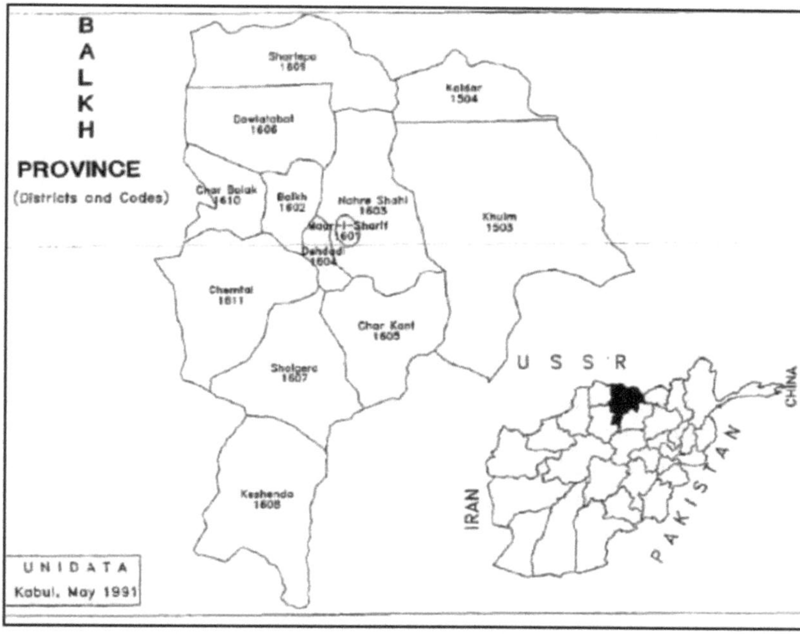

Once Khulm District joined Balkh, the water supply from Samangan was shut off through the implementation of a dam at Tang-i-Sara. This impacted very negatively on Khulm's agricultural system, as the only alternative source of water was a small spring in the Tang-i-Tashqurghan, known as the Hayat Spring (KII: Com31, March 9, 2010).[292] This spring produced smaller amounts of water, resulting in far less land being irrigated.

It was insinuated by the Director of Agriculture that it was because of this that the agricultural lands of Khulm began to be used on a rotational basis, a system that continues to the present day (KII: EX13, January 31, 2010).[293] Only one third of the arable land in Khulm is irrigated

292 One household survey respondent, when asked about this shift in borders, said, "We were upset, but what could we do?" (KII: Com18, February 23, 2010). One community elder noted that anyone who tried to turn on the water from Samangan under the PDPA would face bullets being shot at him (KII: Com31, March 9, 2010).

293 This rotation system is called *zamin-i-pejġule* in Dari (Favre and Kamal 2004:

each year, according to an established schedule (KII: Com03f, January 24, 2010, EX13, January 31, 2010). The Directorate of Agriculture also blamed Samangan for the seasonal floods in Khulm, claiming that the people of Samangan send the water into Khulm when they face flooding (KII: EX13, February 7, 2010).[294]

Overall, the household survey respondents in the research community did not agree on whether they felt that they had a good degree of economic security or not when the PDPA was in power. About half of the people said that the economic security situation was bad and the other half said it was good, making for a community almost evenly divided (HHS 2010).Those who said the economic security situation was good, appear to have concentrated on the time period after the Soviets entered Afghanistan prior to the Mujahideen uprising. Those who evaluated the economic security situation as bad focused on the situation after the Mujahideen troops had sprung into action, and therefore discussed the mining of the fields and continual fighting and its spillover into their gardens and fields, which prevented them from being able to actively pursue their agricultural practices. The Directorate of Agriculture in Khulm District, a lifetime resident, said that agriculture was the dominant source of livelihood for most households under the PDPA, with many in Tashqurghan's nearby suburbs nurturing fruit and nut tree gardens (KII: EX13, January 31, 2010).[295]

However, he also said that, "most of the people were too busy fighting to plant their lands" (KII: EX13, January 31, 2010). So while many people were dependent upon agriculture, its practice was rendered more challenging by the fighting that arose between the Soviet troops and the Mujahideen groups. Prior to the fighting, the market had

85).

294 It should be noted that the transferring of Khulm District to Balkh remains a controversial issue in the present day, with Uzbeks in the district wanting to belong to Samangan, the Tajiks wanting to remain a part of Balkh, and the Pashtuns being independent (KII: EX13, February 7, 2010). Some community members emphasized that people's preferences were connected to political party alignment, with those aligned with Jamiat wanting to remain a part of Balkh Province, whose governor is of the same political party (KII: EX07, February 23, 2010).

295 The main crops in the community of Shah Raheem were pomegranates and almonds, both of which are cash crops. The only other crop mentioned was wheat, and only by one household (HHS: SR013, February 14, 2010).

served as a place for the people of Shah Raheem to sell their harvest and to purchase other basic necessities, such as food supplies and fuel for heating and cooking.[296] However, when personal security started to be threatened, people were no longer able to safely walk to the district center and its market.

While farming was the main occupation of the Khulm households I spoke with, a number of other jobs also helped to sustain people in the area during this time. After farming, the most commonly mentioned occupation was working for the Mujahideen (HHS: SR010, February 9, 2010, SR012, February 10, 2010, YB004, February 24, 2010).[297] In general, those who earned their livelihoods outside of agriculture had higher incomes, which occurred not just during this time period, but holds true until the present day (HHS 2010).

Social Security under the PDPA

The experience of social security during the PDPA Government was also mixed, with about half of the household survey participants describing it as good and half describing it as not good, or bad (HHS 2010). For those who labeled this time period as one of weak or bad social security, the exodus of numerous families from the community was seen as resulting in fractured social relations. Households who were better off economically migrated, leaving behind mainly those who faced economic challenges (HHS 2010, KII: EX13, January 31, 2010). Therefore, with the more economically healthy families gone, the social security net was even less able to support those in need.

As mentioned previously, social cohesion in Khulm has historically been connected to the involvement of the community in the sadness

296 In 1982 the Friendship Bridge between Afghanistan and Uzbekistan was built at Hayratan, making trade even easier between the two countries (UNIDATA 1992: 69). One head of household said that, "Under the Soviets there was a good supply; the situation in Khulm was good. There were products imported from Uzbekistan and Iran" (KII: Com31, March 9, 2010). Generally, the shopkeepers remember this time as one of economic strength and security (Shopkeeper Questionnaire 2010).

297 Additionally there were shopkeepers (HHS: SR008, February 7, 2010, YB007, February 27, 2010), one man worked for the Soviet troops digging holes (HHS: SR015f, February 21, 2010), a teacher (HHS: YB003, February 23, 2010), and a few others whose livelihoods were non-agricultural in nature (HHS: YB006, February 27, 2010, YB007 and YB008, February 28, 2010).

and happiness ceremonies of other community members, as attendance confirms a family's perceived worth, goodness, and therefore their acceptance in the community. During the PDPA Regime, social security was ranked as the lowest of any time period covered by the household survey. A full quarter of all households explained that they could not attend the community members' ceremonies at all.[298] This was seen as resulting in greater isolation and the lessening of the sense of connection by those who were unable to attend the ceremonies (HHS 2010).

The survey respondents were not terribly positive in their descriptions of the relations that existed between community members at this time (HHS 2010). While, there were a number of households who said that the people of the community had strong connections with one another, the only other time period where this number was lower, was under the Taliban Regime (HHS 2010). The government, its Soviet supporters, and the Mujahideen groups were not seen as contributing a great deal to social security in a direct manner, though they did provide a small amount of assistance to people who worked directly for them.[299]

Empowerment under the PDPA

The fourth area of investigation of empowerment was not very easy to extract for this time period, though it is clear that the community members of Shah Raheem did not feel represented by the PDPA Government. While some people were empowered to relocate in order to better ensure their personal security, many could not. The household survey respondents did not indicate that they felt they were able to participate in the political realm or that they had a great deal of access to

298 The extreme fighting prevented people in the area from being able to help other community members, giving them assistance generally or economically (HHS: YB007, February 28, 2010). It is also possible that the constrained economic situation in some households prevented them from partaking in the ceremonies because they could not provide presents or contribute to the ceremony in some of the traditional ways. Fears of raising Soviet attention and suspicions about the purposes of such meetings, was also seen as preventing people from participating (HHS: YB003, February 23, 2010).

299 The Soviet troops were reported to have provided food for people who worked for the PDPA (HHS: SR001, January 20, 2010, SR002, January 26, 2010). The Mujahideen were said to have sometimes provided money to families who helped them through the provision of supplies or manpower, with money sent by the leaders of the different Mujahideen groups in Pakistan (HHS: SR012, February 10, 2010, YB004, February 24, 2010).

information. However, the involvement of members of the community in armed opposition to the government does show that some members of the population felt empowered to fight for what they wanted.

Development Efforts under the PDPA

As was discussed earlier, during this time period, the Soviet Government supported many of the development efforts in Afghanistan, however, in general, people in Khulm did not remember development efforts trickling down to them in the district.[300]

Some district residents spoke of government efforts to supply electricity to the district, noting that these attempts were not successful. One community member said that the Soviet troops stationed in Khulm built an electricity supply center that provided electricity for the people of Tashqurghan and some of the surrounding villages through a generator.[301] Another group of elders said that the Soviet troops had tried to extend an electricity line from Uzbekistan, but that this had failed (FGD11, July 10, 2011).

The Directorate of Agriculture also mentioned that the PDPA Government established an agricultural cooperative that trained people in up-to-date agricultural practices, and also provided some fertilizer and seeds to its members (KII: EX13, January 31, 2010).

The fighting that went on between the Soviet troops and the Mujahideen soldiers caused great damages to Khulm District, especially in and around the district center, resulting in increased development needs. The fighting destroyed many of the roads in the area (e.g., KII, FP08, July 14, 2009), along with buildings and other structures. "Almost all of the school buildings [were] damaged due to the war" (UNIDATA 1992: 114). An estimated 31% of all shelters in the district were destroyed by the fighting between the two actors, while a further 17%

300 Almost one third of the households that participated in the household survey remembered the PDPA Government and their Soviet supporters as having contributed nothing to their economic well-being and having brought them no development projects (HHS 2010). "The Russians [Soviets] did not bring any development project[s] to our community. They did not even give ten Afghanis to my family and there was fighting every day," said one survey respondent (HHS: SR005, February 2, 2010).

301 This generator was said to be stolen after the PDPA fell from power by a commander of the Mujahideen, who then sold it to the people of Shibirghan for his own profit (KII: Com02, January 20, 2010).

were damaged (UNIDATA 1991: 41). Additionally, four restaurants, eight mosques, and five shops were destroyed, and 22 additional shops were partially damaged (UNIDATA 1991: 41).[302] Some community members told us that their almond trees had been destroyed by the fighting, resulting in them having to replant the trees and wait years for them to become productive again (e.g., KII: Com02, February 2, 2010).

Dominated by the fighting that erupted between the Soviet supporters of the PDPA government and the Mujahideen groups, the time of the PDPA was a time of regular warfare that disrupted the personal security reportedly felt before the conflict erupted. The fighting resulted in physical threats to district citizens, preventing them from being able to freely go to the market or work in their own fields out of fear for their personal, physical well-being. This had a negative impact on the economic security of the population, which was also assaulted by the change in water rights and distribution experienced in the mid 1908s. The fighting led many households and families to flee the area, seeking safer lands, leaving behind mostly those who could not afford to make such a move themselves. This contributed to the divided remembrance of social security under the PDPA, with some reporting social security as good and others saying that it was bad. The small number of families that remained, their relative poverty, along with restrictions on movement, restrained social relations and social security. Only a few development projects reached Khulm during this time period and aside from those who picked up arms and joined the Mujahideen, most people in the district were disempowered, hiding in their homes with little capacity to bring about changes.

Unfortunately, the rise of the Mujahideen to power did not dramatically improve the perceptions of the areas of investigation in the district, as will be discussed below. The Soviet troops pulled out of Khulm District February 4, 1989 (FGD11, July 10, 2011). The Soviet withdrawal from the country as a whole was completed eleven days later (Cogan

302 When the Soviets were situated in Khulm, the famous covered bazaar was also destroyed and its foundations were not built upon again until the 21st century. The destruction of the bazaar remains a bit of a mystery. Some local shopkeepers say that it was badly affected by the earthquake which rocked the area in 1976 (Adamec 1979: 335), and others blame the demise of the bazaar on Soviet rockets.

1993: 77). As discussed above, the PDPA Government managed to stay in power until 1992, eventually succumbing to the Mujahideen forces when they proved too strong for the military to hold back.

5.1.4 Mujahideen Gain Political Power in Khulm District (1992-1996)

While the individual Mujahideen groups saw the Soviet troops and the PDPA as a common enemy, this did not make them allies after the Soviets retreated to the Soviet Union in 1989. Though the fighting and violence were reported to have diminished in the district of Khulm at this time, it was not entirely eliminated, as the Mujahideen groups began to wrestle one another for control.

However, for the general population of the district, the emergence of the Mujahideen over the PDPA in 1992 was initially seen as positive (HHS 2010). Immediately following the Soviet exit, community members from the research community remembered different Mujahideen groups helping to demine their agricultural lands (e.g., HHS: SR015f, February 21, 2010), allowing them to get back to their fields. This provided them with the opportunity to continue with their disrupted agricultural efforts, and to regain their livelihoods.

Personal Security under the Mujahideen

However, as the Mujahideen groups fought for control over the capital of Kabul and within the provinces, the population's sense of personal security decreased again, until it was evaluated to be only slightly better than that experienced under the PDPA (HHS 2010), as Mujahideen battles became more serious. Overall, the majority of those who participated in the household survey said that they felt the personal security situation was not good under the Mujahideen, but the fighting was perceived as less intense and damaging than they had experienced previously (HHS 2010).

In addition to the fighting, it was reported that sometimes members of the Mujahideen would force the community to give it money and would become physically violent if people did not concede to their demands (HHS 2010).[303] A few households reported that the Mujahideen

303 Some community members complained that those who did not give what was

forced community members to become soldiers for them (e.g., HHS: CS005, February 20, 2010).[304] Far fewer households talked about this when remembering the Mujahideen Regime, as had discussed it when remembering the PDPA Regime. One community member noted that the Mujahideen groups were just fighting about "who got the chair," in other words, which group would have more power (KII: Com18, February 23, 2010). A set of community elders described the time period as one of "internal war" (FGD11, July 10, 2011).

Educational opportunities were limited during this time period, with most former government schools having been shut down and with those that remained being of poor quality. Official records indicate that there were eight primary boys' schools with 1,545 students and 35 teachers (all male) and one boys' secondary school with 145 students and ten teachers in 1990, but that there were no schools for girls in the whole of Khulm District (UNIDATA 1992: 89-90).[305] This was a dramatic decrease from the records of 1978 (UNIDATA 1992: 91). Attendance of school was emphasized as evidence of good personal security, therefore, the lack of access to schooling for most of the district, with the exception of community schools, contributed to the more negative perceptions of personal security under the Mujahideen.[306]

---

requested were tortured or whipped (HHS: SR013, February 14, 2010, KII: Com38, August 20, 2011). Two households (HHS: CS005, February 20, 2010, YB001, February 21, 2010) said that they had to pay up to 10% of their harvest to the Mujahideen as a tax.

One woman described the cruelty of the Mujahideen through the story of her brothers' killings (KII: Com22f, February 27, 2010). When one brother tried to leave the Mujahideen group, he was killed. Her other brother was killed after he took back the two *jerib (ǧarib)* of land the Mujahideen had taken from his family without compensation.

304 Some said that the Mujahideen would accept payments in lieu of providing a soldier (e.g., HHS: YB002, February 23, 2010).

305 These numbers should be read with caution, as many times official documents and statistics did not reflect peoples' lived realities, as fighting and threats prevented people from sending their children to school.

The Mujahideen were said to have burned down the limited number of girls' schools that had existed in the district (HHS: SR007, February 3, 2010, KII: EX05, August 2, 2009).

306 Some community members said that there were no official schools operating in Khulm at this time, and that only community schools existed (KII: Com26, March 3, 2010).

The Mujahideen, however, established four schools in Khulm district, all of which were restricted to male students only, according to the Directorate of Education (KII: EX05, August 2, 2009).[307] This lack of easy access to education in the district was noted by the household survey respondents, with an increasing number of households that said their children could not attend school under the Mujahideen as compared to under the PDPA Government.

The reemergence of agricultural products from the villages in the district and the capacity of the population to get such products to market, indicate that personal security improved, in terms of mobility after the Mujahideen took control. The number of people in the household survey who said that they sold their harvests to the market increased four-fold from the time of the PDPA (HHS 2010).

Economic Security under the Mujahideen

However, almost all of the shopkeepers we spoke with remembered the Mujahideen Regime as one in which the unofficial costs of transportation were incredibly high (Shopkeeper Questionnaire 2010). Trucks moving products were often stopped on their way into or out of Khulm District and demands for payment by Mujahideen forces were made, be it in cash or in-kind: "They would take money from drivers when they were bringing products here. They had many checkpoints and they would threaten people" (KII: SKQ05, March 8, 2010).[308]

Perceptions of economic security under the Mujahideen, despite the taxes and payments demanded by them were more positive from the viewpoint of the people of the district than during any of the other four time periods covered in this research – PDPA, Mujahideen, Taliban, and post-Taliban.[309]

Those who said that they felt they had good economic security said that this could be in part attributed to good harvests (e.g., HHS: SR004,

307 These schools were established by Masoud's Shura Nazar, which was a supervisory council that included members from different political parties, ethnic groups, and Mujahideen groups established by Ahmad Shah Masoud of Jamiat.

308 Another shopkeeper said, "It was very difficult for me to get rice and meat from Mazar. If we brought these products, the Mujahideen would take half of my products by force" (KII: SKQ02, March 3, 2010).

309 Twice the number of household survey respondents remembered this period as one with good economic security than remembered it in negative terms (HHS 2010).

January 27, 2010, SR014, February 14, 2010, YB005, February 24, 2010). While the Mujahideen were in power, farming was again the most common form of livelihood attainment, with a strong increase, a doubling, in the number of households in the survey attesting to their household dependence on agriculture (HHS 2010). The reported average farming income also saw an increase during this time period, rising from 48,500 Afghani per year under the PDPA Government, to 57,785 Afghani per year under the Mujahideen (HHS 2010). This increase in income could reflect the ability of people to return to their agricultural lands after they had been demined.[310]

At the same time, there is evidence that market prices also increased during this time period (Shopkeeper Questionnaire 2010), which could have negatively impacted on households' economic securities. The closing of international borders to all neighboring countries except for Pakistan and the increased costs of transportation related to bribery and "taxes," were seen as the main factors in the increased prices (HHS and Shopkeeper Questionnaire 2010). The closing of the borders also negatively impacted on the variety of goods available in the market of Tashqurghan, as trade that had flowed from the Soviet Union under the PDPA was forced to a halt.

In general, however, there seems to have been an expansion in the number of shopkeepers in the district.[311] This indicates an increased employment capacity of the market.[312]

---

310 The main crops grown under the PDPA were pomegranates and almonds (HHS 2010). It is interesting to note that four households talked about growing wheat (HHS: SR009, February 7, 2010, SR014 and SR013, February 14, 2010, YB005, February 24, 2010), while only one household said they grew wheat under the PDPA (HHS: SR013, February 14, 2010). Some had farming lands outside of the village, away from the city, where the majority of wheat farming takes places, and they regained access to these lands due to a greater degree of freedom of movement. Other crops were also listed, such as clover, apricots, cotton, and tomatoes. This seems to show an increase in the diversity of land production in the area as compared to under the PDPA Government (cf. HHS 2010).

311 For example, the number of shopkeepers from the research community increased from only one under the PDPA to four under the Mujahideen (HHS 2010).

312 I would conjecture that a renewed market for construction materials and skills arose following the departure of the Soviet troops from the area, as many homes, schools, shops, etc. were destroyed during the fighting. This is supported by the increased number of households who reported having someone involved in the construction field as masons, carpenters, etc. (HHS 2010).

As in most time periods, multiple factors and forces were at play regarding economic security under the Mujahideen Regime. Some of these were positive and some were negative, but overall, the impression is of one where more people in the district of Khulm felt they had economic security than previously.

Social Security under the Mujahideen

Social security was seen positively in Khulm District under the Mujahideen, with an increasing number of people indicating that they felt that their social security had been good during this time (HHS 2010). They said that people had strong connections with one another and that this was shown through their participation in the sadness and happiness ceremonies of people in their village and community (e.g., HHS: SR012, February 10, 2010, CS003, February 17, 2010, YB002, February 21, 2010). No one said that they were unable to attend these ceremonies, as compared to quite a few people in the household survey attesting to facing such restrictions under the PDPA. There was even evidence that people in communities helped each other hold these ceremonies by providing food, dishes, mattresses, pillows to sit upon, pots, etc. to help other households (e.g., HHS: YB003, February 23, 2010, YB007 and YB010, February 28, 2010). Additionally, community members were increasingly able to assist one another economically during this time period as compared to during the PDPA (HHS 2010).[313]

It is interesting to note that there was little evidence of bridging social capital during this time period. This of course does not mean it did not exist, but rather that its importance was not strong enough to be mentioned by the people I spoke with in the district. Only one individual mentioned the government, noting that different Mujahideen commanders sometimes helped families of *Mujahid (Moǧāhad)* soldiers if

313 Almost double the number of household survey respondents noted that community members were able to support one another economically in times of need under the Mujahideen. Despite this notable increase, it is important to establish that an equal number of respondents testified that people were unable to help one another economically, saying that people were too poor to do so during this time (HHS 2010). Yet of these households, a fair number said that the social security situation was good, indicating to me that economic assistance was not the greatest indicator of perceived social security for these people, though there is some association between the two.

they had been killed in battle, a form of social security for those with such connections (HHS: YB004, February 24, 2010).

Empowerment under the Mujahideen

Understanding the sense of empowerment of the people during this time period was not easy and I am left with only the impression that people continued to feel dominated. The Mujahideen still threatened some community members, making them feel as though they did not have choices in life, but instead had to follow the rules provided to them. There is no evidence that the Mujahideen groups involved the community members in governance or in their decision-making process, though, as was noted above, some members of the community were a part of the Mujahideen. It appears, however, that the local *Mujahid* were not involved deeply in the decision-making at this time either, with most decisions being made in Kabul or in Pakistan.

The sense, in the end, is that people did not feel very well informed and that they were not able to participate, to hold the government accountable, or to take actions on their own to improve their situation, indicating a low level of empowerment.

Development Efforts under the Mujahideen

Even though the UNIDATA report (1992: 122) found that there were many development needs in Khulm in the early 1990s, it does not appear as though a great deal of development aid entered the district. Children were vaccinated and received some food hand outs, WHO and UNICEF were said to have built a 50 bed hospital between 1990 and 1994, but few other projects were noted by the UN or by people in the district (HHS 2010)(UNIDATA 1992: 124).[314]

The Directorate of Education added that four boys' schools, discussed above, were established while the Mujahideen were in power and could be considered a development effort. Overall, the people of the research community of Shah Raheem were unable to tell me about any other development efforts that took place in Khulm under the Mu-

314 The Chief of the district hospital in Tashqurghan stated that the Mujahideen established a second hospital in the district, which had no name, but was supported by the NGO Swedish Committee for Afghanistan (SCA) (KII: EX08, August 3, 2009). He was unable to provide more details about when this was built and how it was used.

jahideen, with one elder summing up the general opinion by saying, "In Mujahideen time[s] no fundamental or infrastructure projects were accommodated" (FGD11, July 10, 2011).[315]

At the same time, however, it is important to note that the population was also much less vocal about being frustrated with the government for not addressing their development needs, with fewer people lodging complaints against the Mujahideen Government than the previous Government (HHS 2010). Therefore, it is quite possible that there were additional development projects conducted during this time period that were not brought to my attention. Alternatively, it could be that there were different expectations of what the Mujahideen would bring to the people in terms of development and services.

Overall, it can be seen that the Mujahideen time period was one filled with conflicting perceptions of the major areas of analysis. The perceptions of personal security improved slightly under the PDPA, but continued fighting between the Mujahideen groups continued to threaten the well-being of the people of Khulm. There was a noted increase in freedom of movement, which when combined with good harvests, helped to lift economic security to the highest levels seen over the four periods investigated in this research. The increase in economic security, along with increased freedom of movement, also contributed to more people having a positive perception of social security in the district. However, there was still little evidence of empowerment in the district and even fewer development projects were cited as having been implemented under the Mujahideen. Most of these factors, however, would be evaluated even more negatively under the Taliban Regime, as is explored below.

315 One Khulm District elder said that he believes the people of Khulm preferred the Soviets and the PDPA Government to the Mujahideen because the Mujahideen groups did not provide anything for the people, and at least the PDPA and the Soviets provided some food assistance (HHS: SR002, January 26, 2010, SR001, February 9, 2010, KII: FP08, July 14, 2009).

5.1.5 Khulm District under the Taliban: Negative Perceptions of Securities (1996-2001)

The Taliban entered Khulm in April of 1998 (FGD11, July 10, 2011). Seen as the "real Taliban" or the "pure Taliban", these men brought a greater level of personal security to the area with their strict rules and regulations, bringing an end to the internal fighting and struggles of the Mujahideen in the district (FGD11, July 10, 2011, KII: FP08, February 7, 2010).[316] It was remembered fondly by a Khulm District elder, who said that it felt like, a "mother's hug," with everyone feeling incredibly safe (KII: FP08, February 7, 2010). This time period was remembered as one in which there were no checkpoints on the roads and no demands of money from the population, a dramatic shift from the previous years. This did not last, however, as control shifted to local commanders who abused their positions of power in the name of the Taliban (HHS 2010).

Personal Security under the Taliban

The severity and the threat felt from this second wave of the "Taliban" resulted in the people of Khulm evaluating this time period as one in which personal security was at its worst (HHS 2010). Almost all of the household survey respondents reported that personal security was not good.[317] Such unity amongst the household survey respondents is noteworthy for its rarity.

The Taliban were seen as threatening, with a dramatic increase in the number of households who felt threatened under the Taliban as compared to the Mujahideen Regime (HHS 2010).[318] These threats contributed to the lower evaluation of personal security that was reported for this time period (HHS 2010). One group of males from the community described the threats as "mental torture" (FGD11, July 10, 2011).

316 One district elder said that the Mujahideen left Khulm without fighting the Taliban, and only fought them outside of the district (KII: FP08, July 14, 2009).

317 The few households who were not negative about the personal security situation were also not positive about it, saying only that security was "a little good" (HHS: SR006, February 2, 2010, CS004, February 20, 2010, CS006, February 21, 2010).

318 However, it can be noted that the number of those reporting feeling threatened was almost equal under the Taliban as it was under the PDPA (HHS 2010).

As had occurred under other governments, the Taliban also required male Afghans to serve as soldiers on its behalf.[319] This requirement resulted in many families not allowing the young males of their household to leave their compounds for fear that they would be captured and forced to become Taliban soldiers (e.g., HHS: CS002, February 16, 2010, CS004, February 20, 2010, SR015f, February 21, 2010, YB009, February 28, 2010), as had similarly occurred under the PDPA. The number of households who discussed this increased greatly when compared to the number who said the same for the Mujahideen period, but is comparable to the number of households who mentioned forced soldiering when discussing the PDPA Regime (HHS 2010).

Some community members told us that they paid the Taliban weekly or monthly, in lieu of providing them with a soldier.[320] The average amount paid was equivalent to 2,500 Afghani per month (e.g., HHS: CS004, February 20, 2010, YB004, February 24, 2010, YB009, February 28, 2010), which is almost 50 USD, in a country which had a real GDP (PPP 1993 USD) of 800 USD in 1996 (UNDP 1996: 137).[321] These payments, therefore, were not insignificant to peoples' economic security.

During the Taliban regime, people's private lives were restricted in ways that were not mentioned under the other regimes. Minute aspects of personal lives, including personal appearance and expression of reli-

---

319 Some community members reported each family having to provide one male soldier (e.g., HHS: SR004, January 27, 2010) and some reported that it was one soldier per *gozār*, or small street (e.g., HHS: SR015f, February 21, 2010).
One woman revealed that her husband had been volunteered to represent their *gozār* because they were the poorest people living there. While he was away, the community provided her and her children with food (HHS: SR015f, February 21, 2010). This was confirmed by a set of brothers who said that people in the community "rented" their neighbors to give to the Taliban as soldiers (KII: Com10 and Com17, February 7, 2010). This often required families having to sell their livestock and other means of livelihoods in order pay the Taliban or their neighbor to avoid having to go to the front lines themselves.

320 People reported being physically assaulted by the Taliban if they failed to pay these "taxes" or provide a soldier (HHS: CS002, February 16, 2010, YB010, February 28, 2010).

321 People remembered having to pay varied amounts ranging from 100 Afghani (HHS: CS005, February 20, 2010) to 50,000 Afghani (FGD04mix, January 26, 2010).

gious devotion were outlined in Taliban edicts, which could result in severe physical recriminations if not followed.[322]

This also held true for women who did not adhere strictly to Taliban edicts about their attire or have a *mahram*, or male relative, by their side when they went into public.[323] One woman explained that the Taliban time period was "the worst time for women" because they had the least freedom of any of the periods discussed in this research (KII: Com20f, February 26, 2010).

A number of families, who had not migrated during the fighting between the Soviet troops and the Mujahideen, left the community of Shah Raheem during the Taliban time period, migrating to seek greater personal security and freedom.

Despite the media, and some academics, hypothesizing that Afghans have grown nostalgic for the order that the Taliban brought to Afghanistan (e.g., Constable 2008; Wildman 2010), albeit through repressive and extreme means, the people of Shah Raheem and Khulm in general expressed no longing for the Taliban times, and more often remembered the insecurities they had experienced under the Taliban than the securities (e.g., FGD11, July 10, 2011).[324] One elder explained, "There was no justice, for instance, if you had a problem there was nobody to listen to your problems" (HHS: SR006, February 2, 2010). Stealing was raised as an issue by more people than under the previous regimes (HHS 2010), including by "Taliban" commanders (FGD11, July 10, 2011).[325]

322 For example, all males were required to attend mosque, grow a beard, and wear a turban (e.g., HHS: YB002, February 21, 2010, YB004, February 24, 2010, YB006, February 27, 2010).

323 One district elder described how women who left the house on their own would be whipped by the Taliban if found (KII: FP08, February 7, 2010). A woman said that the Taliban would cut women's hands off if they had gone outside (and to obtain their rings), and that women were hit and killed for no reason at all (KII: Com20f, February 26, 2010). She went on to say that even the thought of seeing a Talib, "would almost give me a heart attack," she was so afraid of them (KII: Com20f, February 26, 2010).

324 There were a few exceptions to this generalization. One man said that the security situation was better under the Taliban than in the present, explaining that, "They [the Taliban] had good rules and regulations. If you left the pen on the ground no one would take it. If you leave it now, someone will take it from in front of you" (KII: Com25, March 3, 2010).

325 A few households expanded upon this topic and related how the Taliban had

The few people who expressed that they felt good about their personal security under the Taliban supported their opinion by saying that the Taliban did not threaten their family and that this was a sign of good personal security for them (HHS 2010). Why most people felt threatened by the Taliban and these families did not feel so, was not made clear.

In terms of access to school, officially the Taliban only banned school for girls. Boys were allowed to attend if they wore a turban, according to a local district government official (KII: EX05, August 2, 2009). Community members, however, reported there being no schools in the district that their children could attend (HHS: SR011, February 9, 2010, SR014, February 14, 2010, CS003, February 17, 2010; KII: FP08, February 7, 2010, Com25, March 3, 2010). I was unable to confirm with confidence that other government schools existed at the time, but the increase in the number of households who said that their children had no access to education, indicates that access was, at best, decidedly limited, which was seen as a negative factor in peoples' evaluation of personal security for this time period (HHS 2010). As occurred in other parts of the country, there is evidence that some teachers continued to teach students in their own homes, especially female teachers.[326]

Economic Security under the Taliban

Economically, the situation was not seen as having improved after the Taliban came into power. Prices of goods remained high in part because the Taliban continued to maintain closed borders, not allowing a regular flow of imports and exports into and out of Afghanistan.[327] At the same time, the second wave of Taliban appear to have taken over where the Mujahideen left off, stopping vehicles at checkpoints and requiring a payment from those transporting goods before allowing them

taken their farmlands from them by force. One man said that the Taliban threatened to kill him and his family if he did not sell his lands to them (HHS: SR010, February 9, 2010), and a second man explained that a commander took 175 *jerib* of his land after torturing him for it (HHS: SR001, February 9, 2010).

326 A female head of school, who lives in the research community, said that though the schools were closed under the Taliban she taught twelve students in her home (KII: Com27f, March 6, 2010).

327 Community members talked about how hard this was on them, as there were no imports of wheat, one of the staple foods of the community (FGD04mix, January 26, 2010).

to continue on their journey (e.g., KII: SKQ05, March 8, 2010). Mazar-i-Sharif was the only trading partner for Khulm shopkeepers and farmers as this time. The lack of secondary or tertiary roads between Khulm and Mazar-i-Sharif meant that goods were transported on the main Kabul-to-Mazar road, making trucks easy targets for those seeking bribes.

In addition to the reduced sources of goods, a massive drought drastically limited the agricultural production of the area (Shopkeeper Questionnaire 2010).[328] Fewer households reported having been able to sustain themselves at this time (HHS 2010).

The number of households who reported that they were farmers dropped by about half from its high under the Mujahideen Regime to just under half of the survey households, which was also accompanied by a slight downshift in the average annual reported income of farmers, from 48,500 Afghani per year under the Mujahideen Government to 47,563 Afghani per year (HHS 2010).[329] Additionally, the number of community members who reported selling their harvest to the market dropped markedly from the Mujahideen period (HHS 2010). This was in part reasoned to stem from the restrictions on movement established by the Taliban Regime and the fear some households expressed at the idea of leaving their household during this time, but all together it adds up to a picture of economic deterioration in terms of households economic security.

Many people in Khulm reported that under the Taliban, they lived through the worst poverty of their lives (HHS 2010). People talked about the destitution they felt they were living in, telling stories about having to sell their personal effects in order to survive, or having to go for days without food (e.g., HHS: SR001, January 20, 2010, YB003, February 23, 2010, YB004, February 24, 2010). In no other time period did

---

328 One shopkeeper (KII: SKQ12, March 10, 2010) explained that, "Under the Taliban there was a terrible drought too, and then people were jobless and could not buy anything."

329 The most commonly farmed products under the Taliban remained pomegranates and almonds (HHS 2010). A few households also mentioned that they grew wheat (e.g., KII: Com02, February 2, 2010).
The only two other agricultural products mentioned were apricots and poppy (HHS 2010). Poppy was only mentioned by one person, though it is possible that it was grown by others who did not want to admit to it due to current negative perceptions regarding its growth (HHS: CS004, February 20, 2010).

people talk so openly about the sacrifices they were forced to make due to economic insecurity.

The number of households who discussed the lack of jobs available under the Taliban Regime increased substantially when compared to the number who raised this issue when remembering the Mujahideen Regime (HHS 2010). However, the range of reported sources of income increased, with people saying that they had worked as a mason, a miller, a blacksmith, and a driver, all careers that were not mentioned previously (HHS 2010), which can likely be attributed to former farmers searching for other means of earning an income.[330]

The lack of money for basic foodstuffs, restrictions on freedom of movement, and the increased costs of doing business for the shopkeepers led to a reported contraction in the market economy of Khulm (Shopkeeper Survey 2010). Many shops closed their doors, with only one of the four shopkeepers in the household survey remaining open (FGD04mix, January 26, 2010, HHS 2010). Those who kept their shops open during this time agreed with one shopkeeper who said, "Only sometimes were we able to open our shops, most of the time we were closed because of the war" (KII: SKQ06, March 9, 2010).

Social Security under the Taliban

Social security also shifted negatively under the Taliban, according to testimony in the household survey (HHS 2010). The vast majority of the households we spoke with described the social security situation under the Taliban as bad, a much higher number than under the Mujahideen and slightly higher than under the PDPA Government. One district elder said, "People could not afford to offer one piece of bread," to those who were suffering because they themselves were so poor (HHS: SR004, January 27, 2010).[331]

---

330 The average annual income in the non-farming sector was reported to be 97,875 Afghani, almost double the farming income and an increase of the average non-farming income from the Mujahideen days (HHS 2010). Subtracting one large outlier, the non-farm income stood at 60,428 Afghani per year, which was still significantly more than the average made by farmers (HHS 2010).

331 This was further supported by another district elder who stated, "We turned to our neighbors for help, but they did not help us because everyone was on their own" (FGD04mix, January 26, 2010).

The dire view of social security under the Taliban was attributed to the fact that so many people in the community were jobless. They said their inability to assist one another economically, a sign of low social security for them, was due to the poor economy. More of the household survey participants said that they were unable to help others economically than under either the Mujahideen or the PDPA Governments (HHS 2010). As one man said, "Everyone was busy with their own lives and survival" (HHS: SR010, February 9, 2010). This sentiment was shared by others who participated in the household survey and the key informant interviews (HHS and KIIs 2010).

While a poor economic situation was the main reason given for the lower level of social security under the Taliban, I would evaluate that this was only one factor and that it was perhaps the easiest for people to discuss with outsiders, as no similar decrease in social security was mentioned in connection with the worsening economic security reported under the Mujahideen.

Even though most community members indicated that they were still able to attend the happiness and sadness ceremonies of the people in their area, a few respondents explained that they had been unable to do so because their economic situation was so bad (HHS: SR002, January 26, 2010, SR004, January 27, 2010). While it does not cost anything to attend a wedding or a funeral, per say, a lack of money can prevent one from attending such an event. Appearance is very important in Afghan society, as is the maintenance of "face," and one of the worst things a person can do is to bring shame upon their family. Therefore, for some, it is better to exclude one's self from community gatherings if they feel their attire or their gift will shame them, and, in turn, their family.

This kind of restriction from partaking in weddings and funerals would have increased the level of isolation of such members of the community. This would have been especially true for females, leaving them with few, if any, opportunities to create and maintain social connections with people from outside their immediate household or family. These ceremonies were one of the few events where women were allowed to interact with others from outside their household. This implies that worsening economic security can negatively impact on social security.

The number of heads of household who said that people had strong connections with one another under the Taliban dropped significantly from the number who said this during the Mujahideen Regime, which was already below what was said during PDPA times (HHS 2010). This shows that the sense of togetherness or cohesion had declined.

Empowerment under the Taliban

When the Taliban ruled, the people of the country were further disempowered. Previously, decisions about one's personal life, the status of the women in one's family, and one's religious observance were mostly left up to the males of a family or household, though with a great deal of social pressure to conform to established social norms.

When the Taliban came to power, these kinds of decisions were relegated to the government. This was explored by the household survey participants, who talked about the rules they were required to abide by, but did not necessarily agree with. One community member described the Taliban as, "cruel tyrants" (HHS: SR001, January 20, 2010), emphasizing their control and power. Community members also talked about there being distance between themselves and the local Taliban representatives, who were described as commanders more interested in the personal gains they could obtain than being a link between the government and the people (KII: FP08, February 7, 2010). People felt they did not have control over their lives and they had no say in politics (HHS 2010). In all, people of the district did not report feeling empowered under the Taliban Regime (HHS 2010, KII 2010).

Development Efforts under the Taliban

Even though the Taliban was active in dictating social norms, it was not reportedly active in the realm of development in Khulm District (FGDs and HHS 2010). Few, if any, of the buildings, roads, etc. that had been destroyed during the fighting between the Soviet troops and the Mujahideen, and then the fighting between the Mujahideen groups, were repaired while the Taliban were in control.

No one who participated in the household survey for this research spoke positively of development under the Taliban, and neither did FGD participants (HHS 2010). Multiple Khulm households openly expressed levels of dissatisfaction with the Taliban Government in this regard, saying that the government had provided the district with no development projects or assistance: "The Taliban did not pay any attention to our people" (HHS: SR010, February 9, 2010). An elder who participated in a FGD said, "Indeed there was no project, no progressive program, no school, no road, and no clinic" (FGD11, July 10, 2011).

All in all, the Taliban Regime was a difficult one for the people of Khulm District and the people of the community of Shah Raheem in terms of the major areas of analysis: personal security, economic security, social security, and empowerment. The economic situation was the most challenging seen during the time periods under discussion, the personal security situation was intense, and social security did not provide a great deal of comfort for those families who might have needed it. As for empowerment, the Taliban Government excluded the people in its decision-making, imposing external rules into the personal lives of the people, leaving them with little chance to become informed, to participate, to hold people accountable, or to organize themselves.

5.1.6 Khulm District in the Post-Taliban Time Period: Improving Perceptions of Securities (2001-present)

Personal Security in the Post-Taliban Time Period

In general, personal security has improved in Khulm in the post-Taliban period, and while there continue to be some acts of violence, these are no longer systematic or regular in nature. One Khulm District representative noted, "There is no stealing and you can walk around twenty-four hours per day and you will face no problems" in Khulm District (KII: FP08, February 7, 2010).[332] Residents of the district have the impression that there is greater personal security in the post-Taliban period than under any other time period discussed in this research,

332 He was supported in this perception by two household survey participants (HHS: SR001, January 20, 2010, YB008, February 28, 2010) and the district level representative of MRRD (KII: EX07, August 4, 2009).
Since this research was conducted, a foreigner was killed in Khulm District in December 2010 (AFP 2010) and four Taliban soldiers were killed in Khulm in 2012, but prior to that, the most talked about security incident involved the theft of a small number of sheep (KII: EX06, January 18, 2010). The German police trainers further elaborated on this story, informing me that the main problem was not that the sheep had stolen, but that the rifles of the sheep herders had also been stolen (FGD03, January 24, 2010).

as shown in Figure 22 (HHS 2010). This was seen as being especially true for women (KII: Com20f, February 26, 2010). There was almost universal support for this impression, which is noteworthy in itself, as the same cannot be said for the answers provided for other time periods. It is possible that these positive responses were, at least in part, motivated by the desire to say what they thought I, as a foreigner, hoped to hear, but in general, I believe that the voices that came through were genuine.

Figure 22: Number of Households with a Positive Perception of Personal Security in Shah Raheem Community During Different Government Regimes

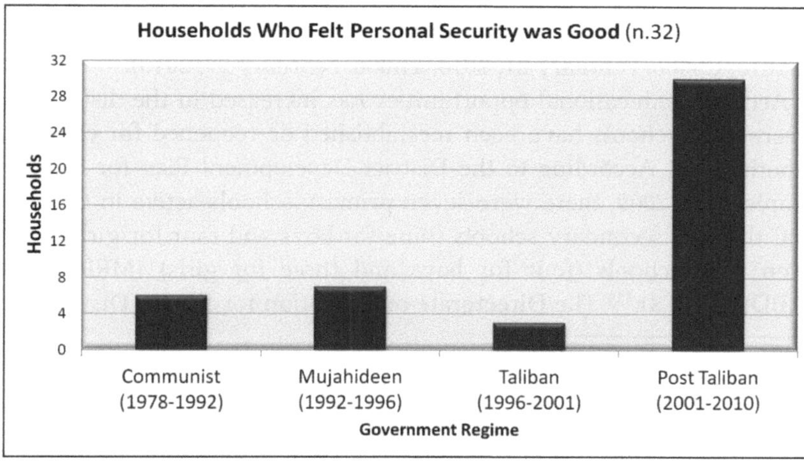

Source: own design based on (HHS 2010)

The end of the Taliban Regime marked an increase in the mobility of the people of Khulm, and a decrease in many other restrictions that had been imposed upon them through government decrees. Men in the district reported no longer being afraid to go to the market in Tashqurghan to buy their products and to conduct shopping for their household (HHS 2010, KIIs 2010). Even a few women are now allowed to conduct the shopping for their family, some with a *mahram* and some with their children or a female relative.[333]

333 Most women we spoke with told us that they were still not allowed to travel to

The increase in freedom of movement also extended to trucks providing Tashqurghan with access to imports and exports. All in all, the shopkeepers reported a revitalization of the market economy (Shopkeeper Questionnaire 2010).

The decrease in rules to abide by also relieved the population of Khulm of many of the threats that had stemmed from not obeying these rules. Many district citizens interviewed in the household survey said that the decrease in threats from the governments was one of the main reasons that they were more positive about their personal security than they had been for the other time periods (HHS 2010).[334] The decrease in threats contributed greatly to the increased sense of mobility, with people in the district reporting being able to work freely without threat, be that in the privacy of their own fields or gardens and in public spaces, such as the market (e.g., HHS: SR009, February 7, 2010, SR014, February 14, 2010, CS006, February 21, 2010, YB006, February 27, 2010).

Access to educational opportunities has increased in the district, as government schools have been reestablished or reopened for children of both sexes. According to the District Development Plan for Khulm, completed in 2009, there were seven primary schools active in the district, thirteen secondary schools (nine for boys and four for girls), and seven high schools (four for boys and three for girls) (MRRD and NABDP 2006: 3).[335] The Directorate of Education for Khulm District es-

---

the district market in order to conduct the shopping unless they had a *mahram* with them. At the same time, however, I ran into one such woman in the market one day without a *mahram*, so it is not clear how restrictive the norms are or how closely they are adhered to. I would venture to say that this varies greatly by household.

In some households concerns for personal security were put forward as an excuse for why women of a household are not allowed to go out in public on their own. One female *kalāntar* said that most of the women were not allowed to leave their household without permission from their husbands (KII: Com03f, February 10, 2010).

A few female teenagers said that they had faced some harassment when walking to and from school (FGD02f, January 20, 2010), but no other women reported being harassed or placed into a position of personal insecurity when walking or traveling to the district center.

334 In particular, some citizens discussed being free from threats of powerful commanders and thieves, and the threat of forced soldiering (e.g., HHS: SR002, January 26, 2010, SR007, February 3, 2010, SR013, February 14, 2010).

335 This is not yet equal to the numbers seen in the late 1970s, and the number of primary schools still remains below the eight primary schools recorded in 1990

timated that 90% of all school-aged children in the district were attending school, with a total of 16,930 students attending formal schools, though he still felt there was room for further improvements (KII: EX05, August 2, 2009).[336]

When people described their personal security in the post-Taliban time period a possible connection was made to the government and the police, which had not been made previously (HHS 2010). Almost half of the households expressed gratitude to, or satisfaction with, the government for the security it was seen as providing (HHS 2010), a sentiment that was also echoed in numerous key informant interviews (e.g., KII: Com25, March 4, 2010). The police were also given credit for the security by a number of households, with most of them indicating that the police regularly patrol their village during the day and at night (e.g., HHS: CS002, February 16, 2010, YB004, February 24, 2010, YB006, February 27, 2010, YB009, February 28, 2010).[337] The Governor of Khulm also credited the Germans and the Police Mentoring Teams (PMT) for helping maintain security in the district, claiming that their monthly patrols contributed positively to ensuring security in the district (KII: EX06, January 18, 2010).[338]

---

under the Mujahideen (UNIDATA 1992: 89-90). However, the schools have been confirmed to be active.

In the NSP community of Shah Raheem there are no government schools and students travel to the district center to attend school, however there is a *madrasa* in the community that attracts people who want to be a *mullah (mollā)* and students from a perimeter of 100 km (about 62 miles) (KII: Com05, January 31, 2010).

336 In the 2006 NABDP District Development Plan, education was seen as weak in the district. There was limited access to schools for students and not enough qualified teachers or official school buildings, leaving some schools operating out of people's homes or out in the open. A lack of secondary schools, especially for girls was also mentioned, as was the complaint that there were not enough *madrasa* schools in the district (cf. MRRD and NABDP 2006: 4).

337 One household in Chahar Sukh emphasized that the police are "always at our service" (HHS: CS002, February 16, 2010).
The Khulm District police force has ten cars, 45 police officers, 45 Afghan soldiers, and weapons, the number of which remained unsaid due to concerns about security. The district police partook in a two-month long training program conducted in Mazar-i-Sharif by German police trainers. The Police Chief told me, "Now people are very happy. Now the police are at the service of the people" (cf. KII: EX11, January 18, 2010).

338 The PMT are a part of the Focus District Development (FDD) program, a police reform program that is concentrated on increasing the operational and institu-

The research community's assessment of an increased level of personal security was supported by government leaders and the district level police force, along with other key informants (HHS: Com25, March 4, 2010, KII: EX11, January 18, 2010, EX06, January 18, 2010). This is a large shift in the historical picture provided in this research, as previously the government was seen as contributing to greater personal insecurity, not greater personal security (KIIs 2010).

On the whole, people in the district portrayed it as one with few security incidences and where threats were no longer prominent (HHS 2010). Even when asked about the new road that links Khulm to Kunduz, where there has been significant Taliban activity in recent years, people expressed no fear about the Taliban entering Khulm and disrupting the serenity of the area.[339] Unfortunately, the killing of four Taliban soldiers travelling on the Tashqurghan-Kunduz highway in May of 2012, indicates that perhaps the establishment of this road will lead to greater insecurity (Pak Tribune 2012). Community members, however, were enthusiastic that the road will be good for businessmen in Khulm (e.g., KII: Com31, March 9, 2010).

Economic Security in the Post-Taliban Time Period

Economic security in Khulm has increased in the post-Taliban period, though evidence of trials and hardship remain. This has led to a division in perceptions among the household survey respondents, with the

---

tional capabilities of the Afghan National Police (ANP). The program in Khulm included a district assessment, training of the district police in Mazar-i-Sharif, and a period of mentoring (cf. FGD03, January 24, 2010).

339 The Mayor of Tashqurghan assured me that the *Shir Baghali* desert would act to block insurgents from entering Khulm from Kunduz. He also said that the Pashtuns in Kunduz were connected to the Taliban, but that the small percentage of Pashtuns in Khulm District (he estimated at around 5%) do not support the Taliban and would be inclined to report on anything suspicious they saw or heard. He said, "Pashtun people will not allow the Taliban people to come here" (KII: EX10, January 18, 2010).
There is currently a checkpoint that has been built along that road in the desert between Kunduz and Khulm that consists of a small container surrounded by HESCO barriers, a type of military fortification, to protect against explosions and other threats (FGD03, January 24, 2010). No one is actually manning this station on a regular basis, however, because the police say they do not have enough personnel for the job and the Afghan National Army (ANA) says that it is not their job (FGD03, January 24, 2010).

majority seeing the economic security situation as good, but about one third of respondents being less positive (HHS 2010).[340] Even those with negative impressions of economic security testified that despite still feeling insecure economically, they felt economic security had improved over the previous time period (HHS 2010). So overall, there appears to have been a significant shift in people's perceptions of their economic security towards feeling more secure, even if for some, the shift has not been drastic enough.

In comparison to the other districts of Balkh Province, Khulm's families are statistically worse off economically, with 83% living below the poverty line, as compared to only 58% for the province as a whole (Kerry 2011: 9).[341] The people of the district of Khulm had a more conservative estimate of 59% of families in their communities seen as poor, 31% seen as in the middle, and 9% viewed as well-off (NSP Khulm Database 2009).[342]

Despite these poverty estimations, there is evidence of positive shifts in the economic environment of Khulm District. The economy of the district continues to be dominated by agriculture, and the number of farmers, or those who rely on agriculture as a main part of their livelihood strategy increased, as did the average farmer's salary.[343] The

---

340 Those with a negative perception of economic security stemmed from only one of the three villages that make up the NSP research community. The reasons for this were not clear, as the average reported income in this village was slightly higher than the other villages in the NSP community. One possible explanation is that it is the most exposed to NGOs and other development agencies because it has the same name as the NSP community and many NGOs in Afghanistan are now trying to work through the CDCs. This exposure could have created higher expectations or taught the people to emphasize their poverty in order to attract development programs.

341 However, the head of NSP in Balkh Province for the MRRD estimates that the NSP research community of Shah Raheem is right down the middle, in terms of being a rich or poor area (KII: EX04, January 25, 2010).

342 In the research community of Shah Raheem, 80% of families were categorized as poor in their 2009 Community Profile.

343 Almost every household either tends their own garden, tends someone else's garden, has fields outside the community limits, or in some other way is involved in the selling and purchasing of agricultural products.
The average salary for those involved in agriculture is 51,000 Afghani (HHS 2010). This is an improvement over the average farm salary under the Taliban, but still remains below the average salary reported under the Mujahideen (HHS 2010).

Directorate of Agriculture said that 144,000 *jerib* of land in the district was dedicated to agriculture (KII: EX13, January 24, 2010).[344] He estimates that each year about 50,000 *jerib* of land are cultivated, based on the rotation system established after the district was transferred to Balkh Province, with 16,000 *jerib* of that being covered by gardens (KII: EX13, January 24, 2010).[345]

Agricultural products from Khulm have been entering into the market and also being exported nationally and internationally, indicating a reviving market economy.[346] Pomegranates, in particular, have gained attention, and at least one international development project has focused on trying to export the fruit to places such as South Korea (Brown, T. 2007: 3).[347] The re-opening of the international borders to trade has helped to increase both imports and exports in Khulm, infusing greater life into the market of Tashqurghan. A shopkeeper said, "Now we can import any products without any problems, no security threats and no bribes" (KII: SKQ01, March 3, 2010).

The high level of dependence on agricultural production in Khulm District, however, means that water and agricultural issues continue to

---

344 Twenty percent of the land is dedicated to rain-fed agriculture, while 73% is recorded as being rangeland (Favre and Kamal 2004: 102).
The main crops continue to be pomegranates and almonds in the post-Taliban time period (HHS 2010). Additional crops mentioned include: wheat, barley, peas, vegetables, sesame, melons, cotton seeds, cherries, and apricots (KII: FP08, July 14, 2009).The household survey participants reported growing other products for household consumption, such as tomatoes, leeks, clover, turnips, onions, and spearmint (HHS 2010). There was a much greater variety of agricultural products mentioned by those interviewed than previously.

345 Only one third of the agricultural land is said to be cultivated in any given year, as each community only has water diverted to it off of the main canal about 100 days per year (KII: EX13, January 24, 2010).

346 Khulm agricultural products have been able to make their way to Kabul and Pakistan.

347 A Khulm Growers Association was established by pomegranate growers as of 2007, with the help of the Accelerating Sustainable Agriculture Project (ASAP) (Brown, T. 2007: 3). Unfortunately, this effort has not seen much success, with one district member saying, "They promised us, but they did not deliver" (KII: FP05, January 31, 2010).
In a year with good weather, Khulm District produces around 10,000 tons of pomegranates for export and enough for 2,000 people to eat locally (KII: EX13, January 31, 2010). In years with bad weather, the number of tons for export could drop to 3,000, as happened in 2009, due to unseasonal ice and to springtime flooding (KII: EX13, January 31, 2010).

dominate farmers' lists of concerns, and act as a threat to their economic security (HHS 2010). The need for access to clean water in the district has been noted by development organizations and companies, such as USAID contractor Chemonics International and the Asia Development Bank.[348] Both floods and droughts are common in Khulm District, and act as a threat to the economic security of those who reside there. Khulm experienced a major drought from 1998 through 2002 and again in 2008, as the Spring rains never arrived (Chemonics International and The Cadmus Group 2008: 49; Larsson 2008: 4). Floods affected the area in both 2005 and in 2009, destroying fields and sweeping away homes in some places (CARE 2009: 4; Chemonics International 2008: 32; OCHA 2009).

These droughts and floods resulted in many men migrating to Iran in order to find employment opportunities (Larsson 2008: 31; Pain, A. 2008: 1), highlighting just how vulnerable economic security is in the district to natural weather occurrences and how little preparation has been made in households to survive such threats.[349]

Migration was not just limited to being a reaction to drought or flooding, it is also seen as a coping mechanism for economic challenges that are being faced throughout the district even in times of good agricultural weather. Individuals migrated to Iran, and other locations, to look for work to help support the members of their household, as they believed there were no jobs for them in Khulm (KII: Com19, February 24, 2010, Com11, February 9, 2010, Com20f, February 26, 2010).[350] The Mayor of Tashqurghan estimated that 70% of the town's population was unemployed, which he said was leading to problems of stealing

348 Chemonics found that the lack of access to clean water and irrigation water had led to local armed landlords, militia commanders, and warlords abusing the *mirab (mir'āb)* water system in Khulm District (Chemonics International 2008:32), and the Asia Development Bank drafted a 10-year investment plan for farm water management and agricultural production in the Khulm River Basin (Asian Development Bank 2009).

349 The Khulm Directorate of Agriculture confirmed this, saying some migration was motivated by the failure of the harvests due to the drought in 2008-2009 (Afghan calendar year 1387) (KII: EX13, February 7, 2010).

350 Many households told us that their sons had travelled to Iran or another country in order to earn money, or that their sons hoped to because, "There are no jobs for people in Khulm" (KII: Com12f, February 14, 2010). Most migrants enter into Iran illegally, putting them at risk both in crossing the border and in Iran, but the gains are expected to outweigh the risks.

and smoking (KII: EX10, January 18, 2010).[351] However, when questioned whether this was an increase over previous time periods, he said it was not.

All migration in the research area since the fall of the Taliban has reportedly been economically motivated (HHS 2010).[352] The costs of migration are very high, and, therefore, the increase in migration during this time period seems to point to a true lack of employment opportunities.[353]

This lack of economic opportunities is blamed on the government (HHS 2010).[354] The government is seen as responsible for providing jobs for the citizens, with some demanding that the government build factories in which to employ the people (e.g., HHS: SR009, February 7, 2010, KII: Com26, March 3, 2010). They were also seen as responsible for people having to sell their personal effects in order to survive by those in the district who felt that economic security was not good in the post-Taliban period (HHS: SR009, February 7, 2010).

In addition to agriculture, district residents actively employed were also engaging in daily labor, tailoring, masonry, handicrafts, and for the first time people spoke of remittances contributing to their livelihoods (HHS 2010; KIIs 2010).[355] An increase in shopkeepers was also noted

351 This seems to be a rather large estimate considering that everyone we spoke with in the district was able to tell us what kind of work they did in order to make money, which would seem to indicate that while underemployment might be at 70% that unemployment was probably not at such an extreme high. Beath et al. (2011: 17) note that being involved in subsistence agriculture or animal husbandry in Afghanistan "is practically equivalent to being unemployed," which could help explain the high estimation provided.

352 In the post-Taliban period, migration is not a family or household affair, but usually consists of a male family member migrating on his own to look for work (HHS 2010). These men then send remittances back to their families.

353 One key informant estimated that it generally costs a person around 30,000 Afghani (about 600 USD) to get to Iran, and usually this money needs to be borrowed from a family member or money lender (KII: Com02, January 24, 2010). It often has to be paid back before the laborer can send money to their family or household as remittances (HHS 2010, KII: Com12f, February 14, 2010).

354 "Everyone in the community is jobless and the government does not pay enough attention to us," said one community member (HHS: SR006, February 2, 2010).

355 A number of district residents indicated that the women in their household were involved in creating handicrafts, such as embroidery, knitting, and bead-

(HHS 2010), indicating a rise in market power, and greater diversity of livelihoods was seen in the household survey (2010). At the same time, non-agricultural salaries also reportedly increased.[356]

There are currently over 325 active stores lining the main road in Khulm, with additional shops located on side streets and within courtyards (Market Survey 2010).[357] While a decrease from the 1970s when Charpentier and Centlivres conducted their research, this number of shops is still an increase from the number regularly operating under the Taliban (Shopkeeper Questionnaire 2010). The market is of vital importance to people from throughout the district who travel to Tashqurghan for the twice weekly market days.[358]

Though the Tashqurghan market is not likely to ever gain the same level of importance it had as one of the crossroad points of Asia, it remains a market of great importance in the area, and it is hoped that this importance will increase with the construction of the road connecting Khulm to Kunduz.[359] The Mayor of Tashqurghan said, "The road is a good opportunity for the people" (KII: EX10, January 18, 2010). He reasoned that the new road would mean that the flow of goods from Kunduz to Mazar would travel through Khulm, resulting in greater traffic

---

work to sell in the local and regional market (e.g., FDG05f, January 27, 2010). While this does not appear to have brought in a great deal of money, with an embroidered *šalwār kameez* selling for only 500 Afghani (about 10 USD) and a beaded bedspread that takes 20 days to make bringing in 300 Afghani (about 6 USD), its contribution to the household finances was mentioned, indicating that its existence holds importance at the household level.

356 A starting teacher's salary that was around 20,000 Afghani (about 400 USD) under the Taliban was 54,000 Afghani (about 1,080 USD) in the post-Taliban period, and a shopkeeper whose salary had been 15,000 Afghani (about 300 USD) per year under the Taliban was earning 49,000 Afghani (about 980 USD) in the current economic environment (HHS 2010).

357 The vast majority of the stores are general grocery stores that sell a selection of goods, from wheat, flour, fruits, and vegetables, to sweets, nuts, and soap (Market Survey 2010). Grain stores that sell flour and un-milled wheat and usually rice are also common, along with fruit and vegetable vendors. In general, the Market Survey (2010) of Tashqurghan found that almost half of the storefronts were oriented towards food, while the other half covered a much wider array of needs, including mechanic and auto parts stores.

358 Market days are Tuesdays and Thursdays. The market is open the other days of the week too, but the largest amount of traffic is seen on market days.

359 The road was funded by the German Government and KfW Entwicklungsbank (Jedelhauser 2009). Construction on the 112 km (about 70 miles) road began in January of 2010 and was still ongoing as of October 2012 (Devex 2012).

through the market and that it would also simultaneously connect the people of Badakhshan, Takhar, and Kunduz to Khulm.

Photo 6: Camels Being Led Through the Village of Shah Raheem to the Tashqurghan Bazaar on Market Day

Source: photograph MB Wilson, January 2010

A small number of households in the district raised the issue of development assistance when they were asked about economic security, indicating that they felt some of the projects had increased their economic security (HHS 2010). The main project mentioned was a hen project. This and other development projects will be discussed in more detail below.

Overall, it appears as though economic opportunities and therefore security began to increase in the post-Taliban period, though it still remains below that remembered under the Mujahideen time period, see Figure 23. This positive impression, however, also stands next to some very difficult economic threats, such as flooding and unemployment. It appears evident that though economic security is improving, there is still much to be done before the economic situation can be considered strong and secure for all of the people of Khulm.

Figure 23: Number of Households with Positive Perceptions of Economic Security in Shah Raheem Community During Different Government Regimes

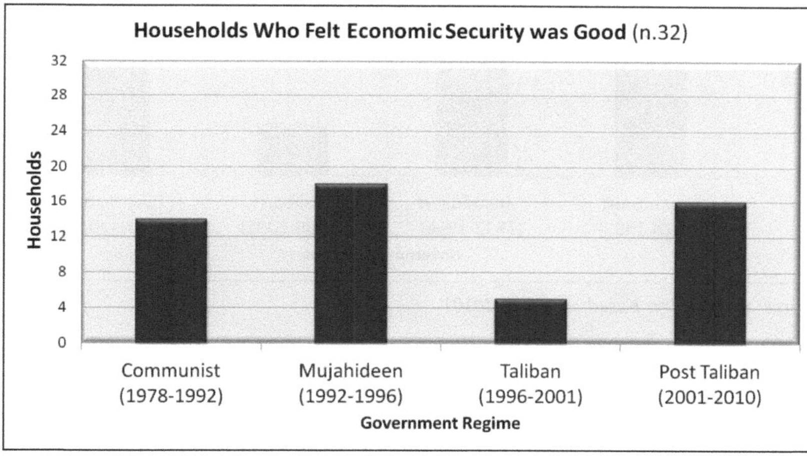

Source: own design based on (HHS 2010)

Social Security in the Post-Taliban Time Period

Perceptions of social security in Khulm, as with personal security, have reached a high in the post-Taliban period, as compared to the other three time periods under review, see Figure 24. More people in the household survey related that the social security was good than during any previous period, and no respondent said that the situation was bad (HHS 2010). In other time periods, survey respondents presented a divided front regarding the status of social security. While this seems to indicate an overwhelmingly positive perspective of the current social security situation, this should be tempered by the possibility that people often try to tell interviewers and foreigners what they think an outsider wants to hear.[360]

360 I have seen this dynamic in Afghanistan on multiple occasions, and believe it

Figure 24: Number of Households with Positive Perceptions of Social Security in Shah Raheem Community During Different Government Regimes

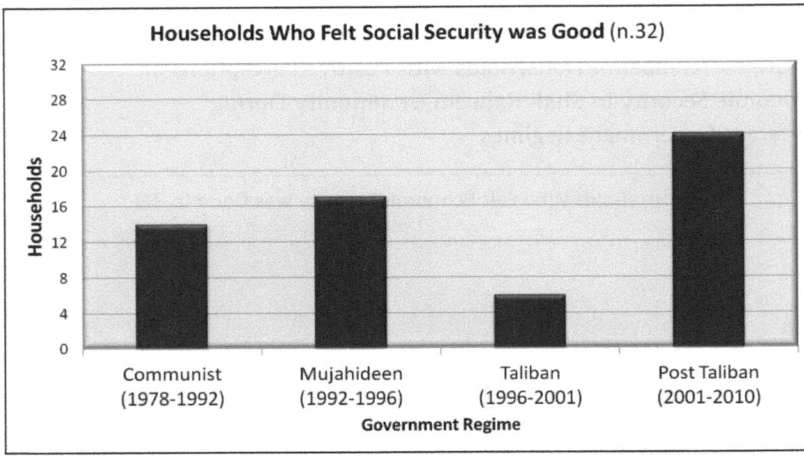

Source: own design based on (HHS 2010)

This existence of strong connections was reflected by an increase in the number of people who said they attended the sadness and happiness ceremonies of other community members (HHS 2010).[361] However, as in previous times, hosting such ceremonies can be quite expensive, with most economically well-off people reportedly spending 10,000 USD on a wedding, with smaller home weddings costing around 1,000 to 6,000 USD (FGD07f, February 9, 2010).[362] Therefore, ensuring some

---

could have played a small role in how household representatives responded regarding this question.

361 This is the first time in the survey that people talked about the reasoning behind their attendance of such ceremonies, with a couple of households saying that they attend the ceremonies so that the families do not "feel alone" (e.g., HHS: SR010, February 21, 2010). They also talked about how not attending such ceremonies implies that a family is not a "good family" (HHS: SR008, February 7, 2010, SR010, February 21, 2010). Reputation is incredibly important in Afghan society, which means that attending someone else's sadness and happiness ceremonies is a means of supporting that family's reputation and place in society, or can be viewed as a form of social security.

362 Some couples wait four to five years after becoming engaged to hold their wedding, in order to save for the event. The groom and his immediately family are not the only ones involved in paying for the wedding, the extended family is also generally involved (FGD07f, February 9, 2010, KII: Com06, January 27,

level of social interconnection and security can be counter-balanced by a loss of economic security in the form of costs.

However, more community members than before said that people assist one another with the large burden of holding such ceremonies through providing materials and helping the family prepare food (HHS 2010). The post-Taliban period also had the largest number of households saying that people in the community support one another economically when they face problems.[363] This aligns with the perception that economic security has improved.

In contrast to this, general observations, spending an extended period of time in the community, and then remaining connected to this community, leads me to evaluate that although the social security situation is generally perceived as quite good, there are still many weaknesses evident in the social fabric of the district and within the communities in the district. Internal to the communities, individuals are not obtaining the social and economic support that one would expect when community members indicate that the people have strong connections with one another.[364] There is also a notable lack of bridging capital.

Family and neighbors were seen allowing one woman to fend for herself, her children, and her mother on the 10,000 Afghani (about 200 USD) that her husband was able to send to her from Iran on a quarterly basis (HHS: SR015f, February 21, 2010). Their lack of support is magnified by the health issues faced by three of her children.[365] The family is unable to survive on this amount, but she has had to turn to relatives in Mazar-i-Sharif because the local social support networks have been un-

---

2010).

363 The number of households that said that they could not help each other economically in the post-Taliban time period decreased from what was seen under the Taliban by a signification amount (HHS 2010).
One example of this is the economic assistance said to have been given to a poor, older woman from the community who was quite ill. Community members gathered money to transport her to Mazar-i-Sharif for assistance. Unfortunately, she made it there too late and died of appendicitis. The community again collected money from the "sort of rich people" to help her family pay for the funeral (cf. KII: Com03f, February 10, 2010).

364 More than double the number of households said that community members have strong connections compared to under the Mujahideen time period (HHS 2010).

365 Three of the six children have required surgery to improve their eyesight as they had gone blind or nearly blind in their early childhood.

willing to help her, even though she has a relatively wealthy relative living within 50 meters of her home.[366]

While she had been able to borrow things from people in the community in the past, she shared that this had at times resulted in people talking badly about her and her family. There is a sense that one gets that most people in the community are focused on themselves and are survivalist in nature, with limited efforts to support others in the community. Only the very rich seem to be removed from this, sometimes extending a helping hand, as reported by household survey and KII respondents (HHS and KIIs 2010).

Additionally, the household survey unveiled differing levels of suspicion of one's fellow community members. When discussing sources of conflict, the most frequently mentioned one focused on the distribution of humanitarian aid and development assistance, with community members accusing one another of taking an unfair share of the benefits or trying to gain all of the benefits for themselves. These concerns and suspicions were not unfounded, as one family reported that they were able to obtain 50 chickens from the chicken distribution program despite only being eligible for 15.[367] One village had to replace their *kalāntar* because he was not fairly distributing information about development programs.[368] There were disputes over a quilt project and accusations of stealing in another development project (KII: Com33, Sep-

366 Excusing her relative she said, "Everyone is just trying to survive, and therefore he cannot help me" (HHS: SR015f, February 21, 2010). However, she also related that her brother does not even give her permission to search for firewood in the garden that he owns to heat her one room home, despite the fact that this would cost him nothing and the garden is next door to her household.

367 Each female head of household was supposed to be provided with 15 chickens, however in this household, each of the older daughters also received 15 chickens and the youngest daughter was provided with five chickens. The family spoke openly of their numerous chickens, but I noted that they did not report them in the official hen project documents that the female head of household was in charge of maintaining (KII: Com02, January 24, 2010, Com03f, January 24, 2010).

368 One key informant told us she did not know about a microfinance project run by Hand in Hand because only the *kalāntar's* relatives and neighbors had been invited to join the program. She said this also occurred for a tailoring project that was implemented in the village (KII: Com03f, February 10, 2010). This was confirmed through observations and a FGD conducted with the ten women who were beneficiaries of the microfinance project in the village of Shah Raheem who all came from one *gozār* (FGD08f, February 9, 2010).

tember 26, 2011).[369] People of the community also called into question how CDC members were running the NSP subprojects in Shah Raheem, with some thinking that they were taking money from the projects, highlighting the lack of trust in different communities. One man summed it up, saying that he felt that social relations were not strong in the community anymore and that "even the elders [had] become corrupt" (KII: Com25, March 4, 2010).

A low level of bridging social capital was also evident in Khulm District. A key informant from Khulm talked about the lack of trust many people of Afghanistan have in other people, in the government, in development organizations (both local and international), etc.[370] Another man said, "People in Afghanistan are just so confused about what is going on in their country that they do not trust anyone or anything, and this has destroyed the hope that they had previously" (KII: FP03, January 11, 2010).

The District Governor was seen as keeping things provided by different organizations meant for distribution to the people (KII: Com10, February 7, 2010).[371] In a focus group of males they explained that they were hoping to obtain some land from the government, but they all agreed with one man who said, "Without a bribe we will not get this

---

369 Women in one village said that they had been promised a quilt making development project. In the end the NGO running the program implemented it in another village and the women from the first village felt they had been cheated (KII: Com03f, February 10, 2010, Com22f, February 27, 2010). They were very angry, and said that they deserved the program more than the people of the other village, and that the only reason the people of the other village had received the program was because of nepotism. No consideration was given for the fact that there is greater economic hardship in the second village as compared to the first (cf. HHS 2010, KII: Com03f, February 10, 2010).

370 For example, when discussing the Ministry of Agriculture, one district key informant said, "They promised us, but they did not deliver," when referring to a development project for the Khulm market and the export of pomegranates (KII: FP05, January 31, 2010).

371 This was raised by one teacher whose salary for a literacy class was paid for by UNICEF through the governor's office. This teacher said that though UNICEF paid the equivalent of 88 USD per month for his salary, he only received half of that, with 44 USD remaining in the Governor's hands (KII: Com25, March 3, 2010). He also explained that this was not unusual and that the Governor had also taken the carpets and wood sent by UNICEF for the schools in the winter and stationary sent to the school at different points in time (KII: Com25, March 3, 2010).

land" (FGD08f, February 9, 2010). One woman said that the benefits of distributions only go to the rich and that nothing reaches the poor (KII: Com20f, February 26, 2010). Similar sentiments were common refrain when people were asked their opinion about NGOs and development efforts in Khulm (HHS and KIIs 2010).

Empowerment in the Post-Taliban Time Period

The end of the Taliban regime also meant the end of Taliban edicts ruling the everyday lives of Afghans, and this was also felt in Khulm. Since then, the men have been able to choose how long to grow their beard and if they are going to go to their mosque every Friday. The women can wear high-heeled shoes without fear of retribution from individuals connected to the government. Though limited in nature, these freedoms give the people of Khulm more control over their lives than they had before, leading to an increase in empowerment.

The people of Khulm also had some limited ability to participate in selecting their government representatives. They were able to participate in the Presidential elections of 2004 and 2009 and the Parliamentary elections of 2005 and 2010. There were, however, one third fewer votes cast in the 2010 Parliamentarian elections in the country as a whole than in the 2005 elections, indicating a decrease in involvement.[372] As was succinctly stated in a report on the elections, ". . . As a result, both male and female members of Parliament have extremely narrow electoral bases and thus not much incentive for broad representation while they are in office" (Worden and Sudhakar 2012: 3).[373] Additionally, as was noted earlier in this dissertation, the integrity of these elections has been called into question. As these elections were, unfortunately, one of the few ways in which the people have had an influence on their government, it becomes evident that the people of Khulm have not experienced great political empowerment.

In terms of local government, the people of Khulm have basically no influence in their provincial or district government. As with all districts, the governor of Khulm is appointed by the Provincial Governor, currently Atta Mohammad Noor, who was in turn appointed by the

372 I was unable to find voting records specifically relating to Khulm District.

373 Males who were elected to Parliament in 2010 had on average only 6,973 votes and females elected had only 3,257 (Worden and Sudhakar 2012: 3).

President of Afghanistan, which means that the chain of command goes directly to Kabul and not to the people. At the village level, however, villagers participate in the selection of their representative. In each village a government liaison is selected to act as a go-between for the government and the people of the village, known as the *kalāntar*. Most often the *kalāntar* is selected by the village, with the individual being chosen by the male members of the village.[374] The municipality then decides whether the candidate is acceptable, providing them with an official letter declaring them the *kalāntar* for their village, or alternatively the village's candidate can be rejected. However, once officially named as *kalāntar*, the community wields influence by virtue of paying the salary of the *kalāntar* (cf. KII: Com25, March 4, 2010).

An example of this power can be found in one of the research villages, where the people decided it was time to replace their *kalāntar* with a new one, as they did not trust him, believing that he served only himself and his family. The people selected a new man to act as *kalāntar* and forced the former one out of the position, evidence of villagers being empowered to hold some people accountable (cf. KII: Com03f, February 10, 2010).[375] This was one of the few stories shared with me that showed the community or its individuals holding someone accountable. When it comes to accountability in general, however, there is evidence that the people in Khulm District remain less than empowered.

One example of the lack of accountability centers on a program manager of one development program, which included a distribution of wheat, oil, and peas. He was accused of stealing from the program (KII: Com03f, February 10, 2010). The women who participated in the program did not report this program manager to anyone, nor did they speak of their suspicions when the man's supervisor visited them to conduct monitoring. The women I spoke with lied to the supervisor and

374 Once an individual is selected, usually based upon a consensus, their name is submitted to the municipality who gathers together a group of 14 elders from the community to act as character witnesses for the candidate. If the elders' testimony is convincing, the individual is approved by the municipality (KII: Com25, March 4, 2010).

375 Please note, women were excluded from this process entirely. A female *kalāntar* is selected either by virtue of being the wife of a male *kalāntar* or based upon their age, with the most elder being selected.

only said positive things about the program and the program manager because they feared that telling the supervisor the truth would mean creating problems with the man who lives near to their community and that they would receive no more projects from the program implementer in the future (KII: Com03f, February 10, 2010).

These stories and many more that were relayed to me highlight how, in some situations, the people of the community are not empowered to hold others accountable for their actions, for fear of the possible consequences.

Even the most respected elder in one village was unable to influence others to take advantage for their own gains. After water wells and taps were installed in the village by DACAAR, it was found that people were removing the taps in order to get their water faster. This resulted in people further down the line not being able to get enough water. Despite talking to the community multiple times, he told me that these actions continued to occur, noting: "In all countries there are people who do not do their responsibilities well, they do not respect each other and only want their own profit" (KII: Com31, March 9, 2010). He felt that he had no way to hold these people accountable.

The women of the research community of Shah Raheem, and of Khulm District in general, face greater constraints on their empowerment than do their male counterparts because of the social norms associated with the genders in Afghanistan. Most women I spoke with in the district, except for those who had grown up outside of the district and had some formal education, said they were reliant upon gaining the permission of their husbands or male relatives in order to leave the house, including obtaining permission to gather in order to speak to me for this research (KII: Com03f, February 10, 2010, Com22f, February 27, 2010, Com33, September 26, 2011).[376] Women are, therefore, strongly

376 One woman clarified that because she was from Mazar-i-Sharif, she had a greater degree of personal freedom herself (KII: Com03f, February 10, 2010). The women who participated in one FGD explained that most of them were able to walk to the market in Tashqurghan with permission of their husband or other male family member, but only if they wear the fully enveloping *chaderi* (also known as the *burqa*). While I do not ascribe to the theory that whether one wears a *burqa* or not implies something about the empowerment of the individual who has chosen to wear or not wear this piece of clothing, it is interesting to note that the social restrictions on women's behavior is currently restrictive and removes their ability to make such decisions for themselves, even

connected to their household and do not have great social networks of support, be that emotional support, economic support, or otherwise. Even within the household, women are not always allowed to play a strong role in household level decisions, such as school education, the marriage of their children, household spending, etc. (FGD07f, February 9, 2010). These restrictions on women's participation in decision-making are detrimental to their empowerment.

Development Efforts in the Post-Taliban Time Period

The affects of the decades of combat that were visited on Khulm District left their toll. Many houses and compounds were destroyed during the multiple rounds of fighting (UNIDATA 1992: 113). "The town of Kholm [Tashqurghan] has been severely affected by the war and the road network around the town has been badly damaged" (UNIDATA 1992: 114). In the post-Taliban time period billions of development dollars have flowed into Afghanistan.[377] These funds have supported development projects of all types and sizes throughout the country, though this is not always visible in the rural areas. One Khulm District resident said that when his brother came back to Khulm after having moved abroad he asked where all of the foreign aid had gone because he could not see many improvements in Khulm (KII: FP08, February 14, 2010).

Despite his brother's negative impression, Khulm has not been exempted from development efforts in the post-Taliban period. The Afghan Government, NGOs (both local and international), and development companies have been involved in bringing development projects to the district, some more successfully than others.

Immediately following the exodus of the Taliban, many development organizations went to Khulm and conducted surveys, saying they would start development projects. The people of Khulm expected these NGOs and other groups to return to Khulm to implement projects, but

---

if it is to a lesser degree than reported in previous time periods. Women from the research community explained that they thought that women should wear a full *chaderi* (*burqa*) when in public and that they thought the Shia women who wore only a black *abāyah* were not covered enough (FG02f, January 20, 2010).

377 As was noted in the historical analysis of development efforts in Afghanistan, the Afghan Government estimated that it had received 57 billion USD in development assistance since the start of the US-led invasion, as of 2011 (Provost and Cummins 2011).

one elder estimated that only 10% of the organizations actually returned (KII: FP08, February 14, 2010).[378] This tainted people's views of NGOs, outsiders, and development projects, causing a strong level of distrust that was also felt and expressed upon the introduction of the National Solidarity Programme (NSP) to the area (KII: FP08, February 14, 2010). Despite these misgivings, a number of development projects have been implemented in the district, most of which can be placed into one of four categories: infrastructure, including electricity; economic support, including agricultural projects; health care; and education. District members, however, still feel there is a great deal of development work to be done.[379]

Decades of fighting in the district, as noted previously, have had a devastating impact on the infrastructure of Khulm. Roads, culverts, bridges, buildings, and more were destroyed, most of which had not yet been rebuilt as of the beginning of 2010. Most roads in the district are graveled at best but are mostly dirt, with only the access road to Tashqurghan from the Mazar-i-Sharif to Kabul highway previously being paved. This is currently changing, as the access road is being repaved and extended 112 km to Kunduz (KfW Entwicklungsbank 2011: 2),[380] and the World Food Programme (WFP) has implemented a "Food for Work" program to hire local laborers to gravel secondary roads in

378 A district elder, who is currently employed by an international NGO, said that 20 different organizations had asked him to make lists of beneficiaries for them, but they never returned with the promised distributions (KII: FP08, February 14, 2010).

379 The Chief of Police listed the problems of Khulm as including: the lack of drinking and irrigation water, professional doctors, and enough schools (KII: EX11, January 18, 2010). Community representatives from the whole district prioritized the lack of electricity, agricultural water for irrigating the land, clean drinking water, health clinics, schools, and employment as their most pressing development needs during the District Development Planning Session conducted in Tashqurghan (July 26-27, 2009).
The District Development Planning Session was run by individuals from the UNDP. It brought together representatives from each of the 91 communities in Khulm District, with the representatives mostly consisting of NSP community development council (CDC) members.

380 As stated in an earlier footnote, the road was funded by the German Government and KfW Entwicklungsbank (Jedelhauser 2009). Construction on the 112 km (about 70 miles) road began in January of 2010 and was still ongoing as of October 2012 (Devex 2012).

the district (KII: FP08, February 14, 2010, EX10, February 18, 2010).[381] This project was generally seen in a very positive light, as it helped to address multiple needs of the people at once: better roads, the means to feed their families, and employment.[382] Some NSP communities additionally committed their NSP funding to constructing or reconstructing roads or other pieces of infrastructure in their community.[383]

Other infrastructure projects have included the construction of a flood protection wall along the river by GIZ, with USAID funding. The HALO Trust has worked on demining the district. DACAAR operated within the district and within the NSP research community, establishing water taps and pipelines to provide safe drinking water.[384]

Finally, one of the main development needs expressed in the district, in terms of infrastructure, was the need for electricity. The vast majority of the district remained without access to electricity from the grid at the time of the field research, though it arrived in Tashqurghan towards the end of 2010.[385] The household survey respondents said that they had no access to electricity. However, during the course of the research and in a number of KIIs, it was discovered that many of the community members did have access to electricity through a private generator, albeit not very regularly.[386]

---

381 Each worker was paid 3 kg of beans, 2 kg of oil, 1 kg of salt, 4 kg of peas, and a 17 *ser* bag of wheat (KII: Com19, February 24, 2010).

382 While the majority of those who spoke of the WFP roads project said that it was a "food for work" program, a minority labeled it as a "cash for work" program (e.g., KII: Com40, February 26, 2010), indicating some level of confusion.

383 The most common NSP projects in the district included clean water projects and roads projects (KII: FP08, July 9, 2009).

384 Water taps were established in five villages of Khulm District in 2009, including one of the three villages that make up the NSP community of Shah Raheem (KII: Com31, March 9, 2010).

385 Khulm previously had access to electricity for at least a few years from the electricity plant that was built in 1971, mentioned in the section on development under the PDPA. Evidence of this could be seen in the old and unused electricity poles situated throughout the district, including in the NSP community of Shah Raheem.

386 Many families paid a local man with a large generator to provide them with a very limited amount of electricity on a monthly basis. Generally, the nights on which they were connected to this electricity source, households had electricity for three hours (19:00-22:00). Some of the richer community members had their own generators, but most could not afford this luxury (cf. KII: Com20f, February 26, 2010).

There have been a number of economic development projects in Khulm in the post-Taliban period. Most of these have either focused on agriculture, as the main source of livelihoods for most of the families, or on helping women to generate income.

One program provided families in Khulm with live chickens to rear.[387] Hand in Hand implemented a microfinance project within the district.[388] FireUp was implemented by the NGO Mercy Corps in the district of Khulm, focusing on women's income generation,[389] and the Afghanistan Small and Medium Enterprise Development (ASMED) program helped to establish a Khulm leather producers' organization in 2009.

Non-governmental sources of agricultural assistance include the Horticulture, Livestock Program (HLP), which worked with women to increase their agricultural practices and knowledge.[390] BRAC, an NGO, also had an agricultural program that focused on women in 15 different villages of the district, including the research community of Shah Raheem.[391] This focus on women and agriculture was seen as unusual by

387 Twenty-five families in the NSP research community of Shah Raheem were included in the program. They were provided with 15 chickens, with a total of 1,000 chickens being disbursed in the district. Women were trained in chicken keeping and healthcare, and provided the necessary medicines (KII: EX13, August 15, 2009).

388 Hand in Hand started in Khulm in 2008. It gave each person 5,000 Afghani (about 100 USD) to buy animals and feed. The people were expected to pay back the loan within 10 months by fattening up animals and selling them or their products for cash (KII: Com11, February 9, 2010). In one village of the NSP community women were targeted beneficiaries of this program and in another village men were the target beneficiaries.

389 FireUp had six projects in Khulm District with 33 women to produce handicrafts, including embroidered products, quilts, and prayer rugs. Each project lasted 50 days, during which the women were paid for their work and were allowed to keep what they produced to sell in the market (KII: FP12f, February 4, 2010).

390 HLP began in Khulm in 2007. Funded by the World Bank and supervised by the Ministry of Agriculture, the program had established over 200 gardens in the district of Khulm by the end of 2009. The program provided women with gardening tools, fruit plants, and regular training and information sessions (cf. KII: EX13, January 24, 2010). The program office was still active July 2012 (personal communication, district informant, July 24, 2012).

391 BRAC distributed almond and pomegranate plants to women in the district and provided them with trainings on how to maintain the plants and harvesting. Only those who had a garden were invited to partake in this program (cf. KII:

the Directorate of Agriculture who at first said that, "Most women are not involved in agriculture" (KII: EX13, January 24, 2010). However, he later clarified that women are often the ones who are in charge of collecting the harvest, not the men. This was especially true for the harvesting of almonds. Additionally, a seed and fertilizer provision program was started in Khulm in 2009 by International Relief and Development, Inc. (IRD), with funding from USAID.[392] A similar program was implemented by CARE International, which provided seeds and fertilizer to district farmers (KII: FP10, July 21, 2009).[393]

The head of the Khulm District hospital said that BRAC, with funding from the World Bank, runs all seven of the medical facilities available in Khulm District (KII: EX08, August 3, 2009).[394] The Chief of the Khulm District Hospital in Tashqurghan said that the services are provided free to all who seek assistance, including medicines (KII: EX08, August 3, 2009, Com02, March 10, 2010).[395] This was contradicted by a few people in the district who said that despite the fact that these things should be free, they often were forced to pay bribes or to purchase medicine from local pharmacies because it was not available at one of the health service points (e.g., KII: Com10, February 7, 2010, Com02, March 10, 2010).[396]

EX13, January 24, 2010 and January 31, 2010).

392 Through it, 1,663 farmers received wheat seeds that were planted in the winter and were ready for harvest by June (KII: EX13, January 24, 2010).

393 The program, which started in 2008, provided 3,685 to 4,548 families with wheat seeds and fertilizer, depending on which FP staff member was spoken with (KII: FP10, July 21, 2009, FP06, January 14, 2010).

394 The district hospital based in Tashqurghan has a catchment area of 15,000 people. It has 50 beds, out- and in-patient departments, and offers vaccinations, x-rays, internal medicine, laboratories, obstetrics, a pharmacy, and more. There are two comprehensive health centers. They provide vaccinations, have a pharmacy, a male and female doctor each, conduct x-rays, and provide laboratory services. There are also two basic health centers. These have no regular doctor, but they have a midwife, a nurse, and are able to provide vaccinations. Additionally, there are two health-sub centers, which offer only primary health care and are staffed by a nurse and a midwife (cf. KII: EX08, August 3, 2009).

395 The Chief of the hospital has lived in Khulm his whole life, except for a period of study to gain his medical degree at Kabul University. He has worked in the district hospital for 18 years, and therefore was able to discuss healthcare in the district over the different time periods focused on in this research (KII: EX08, August 3, 2009).

396 One community member said that the hospital had asked him for 5,000

Educational opportunities in Khulm have increased in the post-Taliban period. The Directorate of Education for the district provided evidence of the government serving 16,239 students,[397] which he estimates to cover about 90% of all school-aged children (KII: EX05, August 2, 2009), with the other ten percent too poor to attend.[398] This is especially true for boys who help contribute to the household economy. However, the Directorate of Education of Khulm feels that the Balkh Line Ministry of Education does not pay enough attention to the educational needs of Khulm District (KII: EX05, August 2, 2009).

Despite the high enrollment rates presented by the Directorate of Education, evidence from the surrounding areas indicate that more girls than boys are unable to attend school, especially as they reach older ages when social norms and conventions that view women as belonging out of the public eye are considered more relevant. Girls who are deemed to be too old to be going to school often attend the girl's UNESCO literacy courses that are held in communities throughout the district.[399] There are also additionally Koran classes for girls that are held by an educated female from within the community (FGD02f, January 20, 2010, FGD10f, March 6, 2010, KII: Com03f, January 24, 2010). Males in the district attend Koran classes at the *madrasa* in Chahar Sukh Village or at their local mosque.

Overall, the people of Khulm are seen as relatively well educated, with an estimated literacy rate of 40% (25% male and 15% female) (KII: EX05, August 2, 2009), above the national rate of 26% (ICON-Institute

---

Afghani (about 100 USD) to operate on his hernia (KII: Com17, February 7, 2010).

One district elder, who is also an FP staff member, estimated that 90% of the population of Khulm has access to the clinics and uses them, and that the remaining 10% are not able to afford to visit except on rare occasions (cf. KII: FP08, July 14, 2009).

397 The District Development Plan for Khulm from 2009 found there to be 17,023 students in total, 44% of which were female (MRRD and NABDP 2006: 3). These kinds of conflicting numbers are quite common in Afghanistan.

398 In the research community, 201 boys and 81 girls were said to be attending school, which accounts for almost all children of school-age (NSP Documents, KII: Com03f, January 20, 2010).

399 These courses provide lessons for groups of 20-25 girls per village, six days a week for two hours each day (FGD02f, January 20, 2010). There is one course held in Shah Raheem Village and a second held in the village of Yaqubai in the research community.

2009: 65). The Directorate of Education believes that this high level of education can be attributed to the relative security of the district and its proximity to Mazar-i-Sharif (KII: FP08, July 15, 2009).

Despite all of these development efforts, people in the district and the research community, often do not feel like their development needs are being met. A male community member in the research area said, "When we watch TV, most countries give assistance to Afghanistan, but we have not seen any of it" (KII: Com30, March 7, 2010). One can feel the sense of disappointment in the district that they have not received more development projects.

5.2 Extracting Meaning: Understanding the Research Area

This historical overview is intended to establish an understanding of Khulm District and the environment in which the empirical data can be embedded. At the same time, it provides interesting perspectives on the four main areas of investigation – personal security, economic security, social security, and empowerment – along with the additional topic of development. These insights help provide localized definitions and perspectives on each of these topics, insights that help make analyzing the impact of a participatory development program easier and more adapted to the environment in which the program was implemented.

When looking at personal security, there is a strong sense that it has been in large part determined by who holds political power in Afghanistan, i.e., the national, provincial, and district government. When the national government is battling against opposition groups within the country for power, it increases personal insecurity in areas because of the fighting, however, this is not the only way in which the government is connected to personal security. Government and opposition forces have imposed restrictive laws and norms based upon threats. This has also increased personal insecurity in Khulm District.

When the PDPA came into power, personal security was still seen positively by many of the people I spoke with, however, when the Mujahideen rose in opposition, this sense of personal security diminished due to the fighting and threats that stemmed from both actors. When the Mujahideen came into power, the people of Khulm continued to see fighting and increased political pressure to conform to Mujahideen ideals, contributing to people feeling more negatively about their personal security. This sense was deepened even further when the Taliban came to power. Despite the end of regular fighting in the area, the political restrictions and the threats made against those who went against the rules were very oppressive, maintaining a low sense of personal security. In the post-Taliban period, personal security has increased as the threat of fighting has been kept at bay within Khulm, for the most part. In addition to this, the lifting of rules stemming from the government about the details of how to live one's life and how to behave, etc. have also led to increased perceptions of personal security, which indicates that personal freedoms and the ability to make choices for one's self are also seen as related to personal security.

Economic security is associated with personal security in the minds of many people in Khulm District. When people feel their own or their family's physical well-being is threatened they are more likely to feel economically insecure.

The strong dependence upon agriculture as a main source of livelihoods in the district means that anything that threatens the agricultural balance of the district, such as the decrease in water available for irrigation when Khulm District shifted from Samangan Province to Balkh Province, floods, and droughts, threaten people's economic security.

Examples of risks stemming from human interactions include the mining of agricultural fields and limited mobility due to ongoing conflicts. This not only limited people's capacity to grow produce and sell it, but also prevented people from being able to travel freely to the market to make sales and purchases. In terms of natural constraints or disasters, Khulm is vulnerable to issues related to water, be that irrigation water shortages, flooding, or drought. Water is a recurrent theme and concern of the people of the district. This can be seen in the large number of communities that prioritized NSP subprojects focused on water issues.[400]

For many people in Khulm, economic security is also seen in connection to the government and what it is not providing. The govern-

400 Three quarters of all NSP communities in Khulm, as of 2010, had a water focused project as their first priority in their Community Development Plans (CDP) (NSP Database).

ment was often blamed for not providing people with jobs or other means of support, especially in the post-Taliban time period. This is important to note, as the participatory development program this dissertation looks at is seen as a means of creating stronger connections between the people and the government in order to create a stronger connection and understanding between the two.

Social security in Khulm District is often seen as stemming from both personal security and economic security, though not only from these two. The most common means of demonstrating social security involved participating in the happiness and sadness ceremonies of the local population.[401] In general, we can say that during times of low economic security, such as under the Taliban, social security was also seen as low, as people were unable to assist one another in any but emotional terms. At the same time, however, the relationship is not direct and other variables are involved, including personal security. During times of personal insecurity, it is harder for communities to gather together, more difficult for them to extend a helping hand economically as they do not know what the future will bring, and it also leads to out migration, which disrupts social networks. Migration leads to fewer sources of social capital or security, leaving those who remain behind with a smaller and more economically challenged network on which to rely.

Additionally, there is evidence that economic insecurity has contributed to an increase in competition throughout Khulm district, which has decreased levels of social capital and social security. This can be most clearly seen in discussions regarding the benefits and deliverables of development programs. People accuse one another of taking an unfair share of humanitarian and development aid distributions, such as food, seeds, and fertilizer, and even water from newly established water sources. As people are seen as taking advantage of the development programs for themselves or for their families, roots of distrust have been created that contribute to creating social cleavages. An environment of accusations and mistrust has emerged that stands in contrast to

401 Sadness and happiness ceremonies were described earlier, and refer to landmark moments in an individual's or household's life, such as funerals, births, weddings, etc.

stereotypes of Afghan rural communities as having strong social bonds that protect all members of the community.

Traditional social norms and restrictions remain strong in the district, however, and that continues to negatively impact on the empowerment of certain segments of the population, especially women.

One of the important things to take away from this historical review is the idea that empowerment should be looked at from an internal viewpoint, as much as possible, in alignment with the very fundamentals of participatory development approaches. The local perspective, local voices, and indigenous knowledge are as vital (Sillitoe 2000: 4; Woolcock 2009: 8).

This can require a recalibration of what empowerment looks like and how it should be measured. In the case of Khulm District, I believe that when looking at empowerment one must evaluate it based upon an understanding of the historical lack of empowerment that has existed. It is the steps, however small, towards having access to more information, participation, the ability to hold people accountable, and the capacity to organize, that are important. Such steps must be measured against the local baseline, not against an externalized image or imagination of what empowerment looks like. Following the Taliban regime, empowerment could even be seen in the seemingly simple, everyday act of people making their own decisions about what to wear or how to groom themselves.

Development efforts in Khulm District have also been highly dependent upon the government of Afghanistan at the time and how open the government was to external assistance. Soviet assistance under the PDPA Regime in Khulm aligned with development efforts throughout Afghanistan, and was focused on larger infrastructural projects that were selected and determined by the government or by foreign funders, a very top-down approach stemming from Modernization theories. During the Mujahideen and Taliban time periods little was accomplished in Khulm in terms of rebuilding the infrastructure that had been destroyed by the years of fighting or helping to build up the economic strength of the district, amongst other efforts. As billions of aid dollars entered Afghanistan in the post-Taliban time period, some of that money made its way to Khulm. An increase in the number of development projects implemented in Khulm is in evidence. Most of these projects have imitated development projects from the 1950s and 1960s and stemmed from a top-down approach, however, a few projects have attempted to integrate a participatory approach into their implementation, with the NSP standing out as the most prominent. The impacts of this type of approach will be explored in the empirical data on the impacts of the NSP on the community of Shah Raheem.

Overall, this historical presentation helps to emphasize that the different areas of investigation have shifted over time, noting that they do not remain static, but instead are fluid and changing. It also highlights the diversity of opinions, memories, experiences, and impressions within the district and within the research community, with some people finding a certain time period to be one of economic security, while their neighbor feels very differently.[402] This diversity of viewpoints and understandings will become very evident in the empirical research for this dissertation. In order to stay true to these different lived experiences, I present the multiple perspectives and then, through my own analysis, present my understanding of the situation. I believe that it is very important to treat the data that comes in from the field with respect, giving validation to the different people who took the time to present me with their memory of a situation, and I believe this is done not by pushing aside those opinions which do not match the majority, but by making note of them in the analysis. This oftentimes creates a more complicated story or picture, but one that I believe is closer to the complicated nature of lived experiences. It is also in alignment with the increasingly complex understanding of participatory development approaches and their impacts in the field that has been emerging.[403]

In the end, the importance of the three perspectives discussed at the beginning of this chapter – Khulm as an oasis, as a market, and as a battleground – have been explored. Agriculture supports a great deal of the population of Khulm District economically and is supported by the oasis. The market is also vital to the population. The battles that have occurred have left both physical and psychological scars on the district,

402 This vision of a community as diverse stands in contrast to more traditional understandings of community within the development framework, where a community was understood to be homogeneous (Mansuri and Rao 2007: 13).

403 This increasingly complex and contradictory understanding of participatory development approaches is discussed further in the section on participatory development literature.

both in terms of development and in terms of social capital and trust. The unique history and circumstances of Khulm District have helped to shape the communities that rest within its borders, and at the same time are representative of a larger Afghan story.

6 The Implementation of the National Solidarity Programme in Shah Raheem: Introducing the Independent Variable

Having explored the historical environment of the district, it is time to examine how this environment was impacted by the implementation of the NSP in the community of Shah Raheem. The program was first introduced into the community in November of 2006 during NSP II.[404] Over the three year lifespan of the NSP in Shah Raheem, a community development council was elected and three individual development infrastructure subprojects were constructed: a deep well, a shallow well, and a community center.

6.1 Community Mobilization in Shah Raheem

6.1.1 Three Villages Form One Community for the NSP

The story of the NSP in the community of Shah Raheem began with the official establishment of the "community" with the local government. Some communities were already recognized by the local government, and some, such as Shah Raheem, came together solely for the purpose of the NSP and were required to register themselves.[405] The community of Shah Raheem actually consists of three villages or hamlets that came together to form one community because they had been told by the Facilitating Partner (FP) staff that none of the communities were large enough on their own to partake in the NSP independently (FGD01, January 19, 2010, KII: Com18, February 23, 2010). While this information was not accurate, as two of the communities had over 25 families residing in them, it resulted in the three villages of Chahar Sukh, Shah Raheem, and Yaqubai coming together and registering themselves as one

404 There have been three separate, though overlapping, phases of the NSP: NSP I from May 2003 to March 2007, NSP II from April 2007 to September 2011, and NSP III from September 2010 to September 2015 (MRRD and NSP III 2012: 8).

405 As noted earlier, a community in NSP is defined as, "a rural settlement in Afghanistan consisting of at least 25 families" (NSP 2007: 5).

NSP community with the District Governor.[406] The coming together of multiple smaller villages to form one NSP community, or the separation of larger villages into multiple NSP communities has been noted by other researchers, such as Monsutti (2009: 29), Mielke and Schetter (2007: 82), and Brick (2008: v). Each of these authors is cautious about this approach of not utilizing or building upon already established rural groupings, with some concerned that it could increase local tensions (Monsutti 2009: 21) and some concerned about the establishment of new layers of categorization on top of the many that already exist at the local level (Mielke and Schetter 2007: 82).

The three villages that make up the community rest alongside one another along the graveled road that stretches north from where the old covered bazaar of Tashqurghan once stood. Chahar Sukh is the first village one encounters when driving from the city center of Tashqurghan. It is easily recognizable by the large *madrasa*, or religious school that sits at its southwestern edge. Just beyond Chahar Sukh, with no visible border, lies the village of Shah Raheem, and then a few hundred meters further down the road sits the village of Yaqubai. The community members said that these villages had been determined by previous governments many decades prior, during the rule of Mohammed Nadir Shah in the mid 1900s (KII: Com02, January 20, 2010, Com39, February 23, 2010).

The establishment of the NSP community of Shah Raheem is, therefore, an imagined community in which individuals identify with their village first and not with the larger NSP community. This could impact on the sustainability of the development council formed by the NSP, a concern also noted by Brick (2008: v) in her research on the program.

406 Chahar Sukh Village only has 15 families. One community member said that the number of families is so low because so many of the families fled the area during the fighting, and many of them had not yet returned because their homes were damaged or they had found somewhere else to live (KII: Com15, March 10, 2010).
Please note that whenever referring to the village of Shah Raheem I will use the word village, all other references to Shah Raheem relate to the NSP community as a whole.

Photo 7: Modes of Transport on the Main Street of the Community of Shah Raheem in Winter

Source: photograph MB Wilson, February 2010

For the people living in the NSP community of Shah Raheem, their understanding of "village" boundaries are not geographically defined, though there is geographic proximity, but are more clearly delineated by attendance at the village mosque (KII: Com02, January 20, 2010). This is in alignment with two of the local perceptions of rural categorization presented by Mielke and Schetter (2007: 75, 79) in their research on local perceptions of the concept of the village in Kunduz Province of Afghanistan.[407] In the NSP community of Shah Raheem, each village has its own mosque and their own separate irrigation wa-

407 Mielke and Schetter (2007: 74) found that the concept of village in Kunduz Province, a neighbor of Balkh Province, had multiple different translations into the Dari language that were not equal to the concept of village based in colonialism and modern Statehood. They found, as did I, that a territorialized notion of a village was not the primary local perception regarding "the loci of rural community life."

ter turn off, and these are seen by the people living in this area as the defining features of a "village" or of local organization (KII: Com39, February 23, 2010).[408] Community members who live within what could be perceived as the geographic boundaries of a village, but attend the mosque of a different village are considered to be members of the village in which they attend mosque. They, therefore, contribute to the salaries of the *mullah* and the *kalāntar* of the village of their mosque, and are involved in that village's decision-making processes.[409] This distinction is clear and logical for the Afghans in the area, even though it might be confusing to outside observers.

Brick (2008: v) noted that the divergence between "natural community boundaries" and NSP community boundaries are not uncommon, and that, "The drawing of CDC boundaries in a formalization process may be a future source of conflict." Fortunately, this has not led to conflict within the NSP community of Shah Raheem at this point. It did, however, alter the way the program was implemented overall from the program guidelines and framework, with each village voting for and implementing its own NSP subproject with money stemming from the NSP block grant instead of making those decisions as one entity.

---

408 The irrigation system runs through all three villages from Tashqurghan. A gate valve is opened for eight days at a time per community, resulting in each community having access to water only eight out of every 24 days, according to the *kalāntar* of Shah Raheem, who also acts as the *mir'āb* for water distribution in that village (KII: Com02, January 20, 2010).

409 In Yaqubai, each family was said to contribute 560 Afghani (about 13 USD) per year for both the *mullah* and the *kalāntar*. Of that money, 6,000 Afghani (about 120 USD) goes to the *kalāntar* and the remainder is distributed to the *mullah* (KII: Com39, February 23, 2010). The *kalāntar* of Yaqubai, said that each family must pay 540 Afghani (about 11 USD) for each male household member over the age of 18, and that the *kalāntar* gets paid 20 USD per month (KII: Com25, March 4, 2010).

6.1.2 Introduction of the NSP into the Community of Shah Raheem

CARE International was the Facilitating Partner (FP) responsible for the implementation of the NSP in the district of Khulm. As of March 2010 they had established functioning CDCs in 75 communities, all of whom had implemented NSP development subprojects. Only 16 communities in the district had not yet been covered by the NSP at the time of this research. These were mobilized under NSP III.[410]

The NSP was first introduced to the community of Shah Raheem on November 17, 2006 by the facilitating partner's Senior Community Mobilizer (NSP Documents). While the NSP was a new program in Khulm and in Shah Raheem, many community members had heard about it from family members and social contacts from other districts where it had already been implemented (KII: EX14, January 16, 2010). This was an advantage for the FP staff, as it meant that there was already some knowledge about, and interest in, the program when they first entered the community.

The FP staff described utilizing a standardized approach to introduce the NSP to Shah Raheem. First they held a meeting with the community elders (males), such as the *malik*, *khan*, *kalāntar*, etc. of each of the three villages. In this meeting the FP staff introduced the NSP to the community, providing them with detailed information about the NSP and how it was intended to function. It is during this meeting that the FP staff determined whether to move forward with the NSP in a particular community, based upon the level of support they received from the elders they spoke with. If the elders were interested in the NSP, the FP staff had the community sign a contract with them that outlined both parties' responsibilities (cf. KII: FP07, July 12, 2009). The elders of Shah Raheem agreed to partake in the NSP and a contract was signed by members of each of the villages in the NSP Shah Raheem community. It was also signed by the FP staff, see Figure 25.

The community members of Shah Raheem said that the FP staff notified the *kalāntar* of each of the three villages that they would hold their first NSP meeting and that everyone in the community was in-

410 CARE had started implementation of the NSP in the remaining 16 communities of Khulm District as of July 2012, with eleven having already completed their NSP block grant and five still ongoing (personal communication, July 26, 2012).

vited to join, with one community member saying that especially the men were invited (KII: Com37, May 31, 2011). The community leaders then communicated this to the village members via the local *mullah* in the mosque.

Figure 25: Reproduction of the Contract Signed between the Facilitating Partner and the Community of Shah Raheem

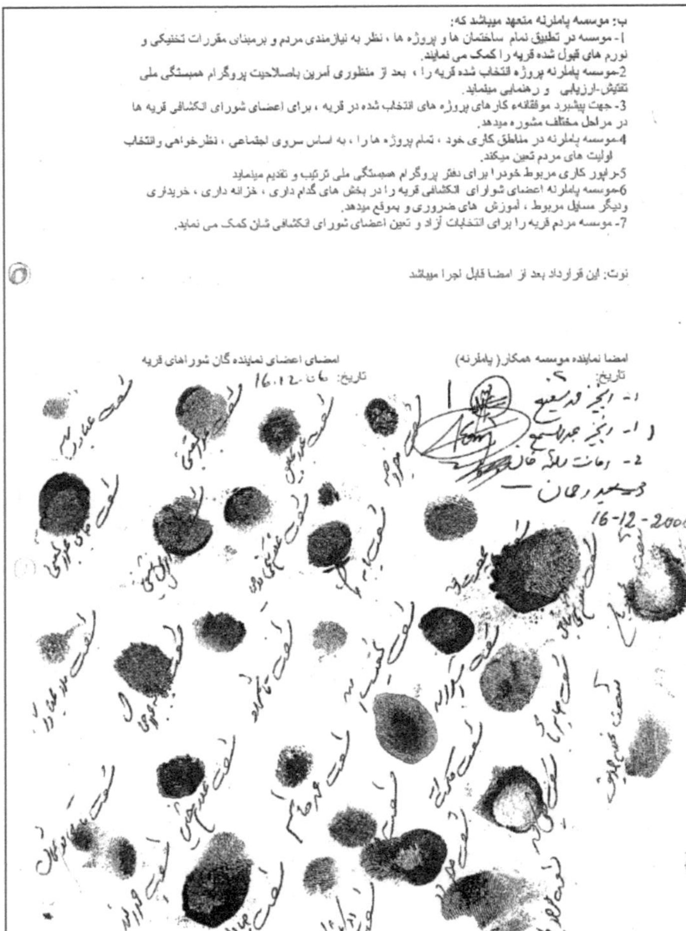

ب: موسسه پاملرنه متعهد میباشد که:
1- موسسه در تطبیق تمام ساختمان ها و پروژه ها ، نظر به نیازمندی مردم و برمبنای مقررات تخنیکی و نورم های قبول شده قریه را کمک می نمایند.
2-موسسه پاملرنه پروژه انتخاب شده قریه را ، بعد از منظوری آمرین باصلاحیت پروگرام همبستگی ملی تفتیش-ارزیابی و رهنمایی مینماید.
3- جهت پیشبرد موفقانهء کارهای پروژه های انتخاب شده در قریه ، برای اعضای شورای انکشافی قریه ها در مراحل مختلف مشوره میدهد.
4-موسسه پاملرنه در مناطق کاری خود ، تمام پروژه ها را ، به اساس سروی اجتماعی ، نظرخواهی وانتخاب اولیت های مردم تعین میکند.
5-راپور کاری مربوط خودرا برای دفتر پروگرام همبستگی ملی ترتیب و تقدیم مینماید
6-موسسه پاملرنه اعضای شورای انکشافی قریه را در بخش های گدام داری ، خزانه داری ، خریداری ودیگر مسایل مربوط ، آموزش های ضروری و بموقع میدهد.
7- موسسه مردم قریه را برای انتخابات آزاد و تعین اعضای شورای انکشافی شان کمک می نماید.

نوت: این قرارداد بعد از امضا قابل اجرا میباشد

امضا نماینده موسسه همکار(پاملرنه)
تاریخ:

امضای اعضای نماینده گان شوراهای قریه
تاریخ: 16.12.

16-12-2006

Source: CARE International (NSP Documents)

Some community members reported having three meetings before the meeting where the CDC elections were held (e.g., KII: Com18, February 23, 2010), while others indicated that all mobilization activities occurred in one meeting (e.g., FGD06, February 3, 2010). The FP staff of Khulm District said that on average they spent two months working to mobilize the NSP communities (KII: FP06, July 6, 2009), though it appears as though this is not an indicator of the amount of time spent working with the community per say. Voting for CDC members took place two and a half months after the first NSP meeting was held in Shah Raheem, but at the most, three meetings were held, which would only be one meeting about every three weeks.[411]

Meetings were led by the FP and involved explaining the voting process and gathering data on the population of the community that was then put into the NSP Form 1: Community Profile.[412] While the male community members said that everyone was invited, they put special emphasis on the attendance of traditional leaders, including religious leaders, *khans*, and the educated (KII: Com37, May 31, 2011). It is important to note that "everyone" does not necessarily include the women of the community, as they are often not considered as important actors in the communities.

In some NSP communities a deep distrust of NGOs and outsiders was reported (Boesen 2004; Kakar 2005: 12-13; McCarthy 2006; Nixon 2008a: 19). This was also true in the research community of Shah Raheem, even though some of the FP staff was from the same district or province, and all were Afghans.[413] The NSP FP Community Mobilizer said that in some communities, including Shah Raheem, people's cyni-

411 Nixon (2008a: 21) found that mobilization in NSP communities throughout the country usually took about six months, though he also highlighted the fact that it was occurring much more quickly in certain communities, with a range of two to five meetings being the most common.

412 The community profile was filled in by elders in a community, in collaboration with the local religious leader. These men were seen as honest and trustworthy by the FP staff (KII: FP09, July 20, 2009). The FP then checked this information, with the same men who created it, making their triangulation efforts redundant.

413 In other districts of Balkh and other provinces, communities sometimes accused the FP staff of being "boys of Bush" (KII: FP07, July 12, 2009). This had negative connotations, as it meant that they were seen as working on the behalf of, and in the interest of, a foreign leader.

cism usually continued until at least one NSP subproject had been completed in their community, or even in a neighboring community (KII: FP07, July 12, 2009, FP14, July 15, 2009, FP08, February 14, 2010).[414] Only then did people begin to have faith that the program would actually deliver what it had said it would. The female FP Community Mobilizer in Shah Raheem said that the women she spoke with about the program were very skeptical because of what they described as false promises made to them previously by other NGOs (KII: FP12f, February 4, 2010).[415] This was commonly referred to in the KIIs (e.g., KII: FP12f, February 4, 2010, FP08, February 14, 2010).

While the male FP staff did not report facing severe problems regarding the inclusion of women in the NSP in Shah Raheem, the female FP Community Mobilizer and community members hinted that this was a source of conflict (KII: FP08, July 16, 2009, FP12f, August 5, 2009). In some NSP communities, the elders protested against the involvement of women in the program, based upon their social traditions and norms, which exclude women from being involved in decisions made at the community level (Kakar 2005: 23).[416] A number of studies that look at the NSP found that the participation of women was not universal by far, with one report indicating that only 38% of the community development council (CDC) members in the country were documented as female (World Bank 2010b: 12). Multiple other reports on the NSP have highlighted the tokenistic nature of female CDC members in many communities, indicating that even in communities where women are documented as participating (Azarbiajani-Moghaddam 2010: 18; Brick 2008: 13; Zakhilawal and Thomas 2008: 21) that this should be called into question.

Male community members in Shah Raheem said that the inclusion of women was not a problem and that really the NSP faced no opposition at any point (KII: Com38, August 20, 2011). However, the vast ma-

414 Nixon (2008a: 20) found that suspicions were only quieted upon receipt of the first installment of the NSP block grant.

415 I am cautious about these promises being labeled as such because I have seen that oftentimes no promises have been made by NGOs, but community members have interpreted their presence and discussions of problems as a promise to address those problems.

416 One community in Khulm District, Hajiistan, rejected the NSP completely because it required the participation of women (KII: EX04, July 29, 2009).

jority of the women we spoke with indicated that they had never been invited to, much less participated in, an introductory meeting for the NSP, such as the one the men described (e.g., FGD08f, February 9, 2010, KII: Com20f and Com21f, February 26, 2010), indicating that they had been excluded from the program for one reason or another. One can easily connect the men's lack of desire to have the women included with their actual exclusion. As was noted earlier, freedom of movement for women in this society is often determined by male family members with little to no input from the women.

Though both the female FP staff member and the community said that this issue was quickly resolved, evidence of the exclusion of women from the NSP in Shah Raheem at a meaningful level, support these questions regarding the inclusion of women. The women are documented as having participated in different aspects of the program in the NSP Documents, but as will be seen, there is evidence to the contrary that suggests that their involvement existed only on paper, something that has been found by other researchers in regards to the NSP (Azarbiajani-Moghaddam 2010: 35; Kakar 2005: 24).

6.1.3 CDC Elections in Shah Raheem: Conflicting Understandings

The community development council (CDC) elections are a vital part of the NSP framework and provide one of the fundamental differentiators between it and other, more top-down development projects. The CDCs consist of community members who are elected to take on the responsibilities of implementing and managing the NSP within the community, increasing the level of participation. The importance of the election of the CDC members was noted in the community of Shah Raheem, as evidenced by the fact that more people were able to speak about the elections than had been able to describe an introductory meeting to the NSP (KIIs 2010).

The men of the community said that the election meeting was held in the mosque of Shah Raheem Village, with men from each of the three villages attending (HHS and KIIs 2010).[417] The FP was said to have

417 Women are not allowed to enter the mosques in the NSP community of Shah Raheem, which means that all meetings held there exclude women automatically.

described the program to the general population and then helped hold elections to select the CDC members who would run the NSP in the community. Some male community members said that there were about 25 or 30 people in attendance, while NSP Documents stated that over 150 males attended (FGD06, February 3, 2010, KII: Com08, February 2, 2010, NSP Documents). The males of the community self-reported that at least one male from each family was in attendance at this meeting, though some individuals told us that no one from their household had been invited (e.g., KII: Com08, February 2, 2010). While the exact number of male attendees is disputed, it is not disputed that the males of Shah Raheem participated in NSP CDC elections.

Mixed information about the inclusion of women in Shah Raheem was, however, provided by those who participated in this research, with the FP staff and the men of the community saying that women had been involved, and most of the women saying that they had not been involved.[418] NSP Documents noted that 151 men and 154 women voted for CDC members in Shah Raheem, which would account for 96% of all eligible men and 82% of all eligible women. This makes it highly unlikely that so many women would have denied having been involved.

418 The female Community Mobilizer of the FP said that she held a meeting with the women of Shah Raheem to inform them about the program and the possible projects that it could bring (KII: FP12f, February 4, 2010).
Males in the household survey reported that the women of their household had voted and that all women had the right to vote (HHS 2010). Other men said, "We were informed by NSP that the women should vote, but it is not allowed for women to vote in Shah Raheem, it is our culture" (FGD06, February 3, 2010). Others said, "The families would not allow women to have a meeting," to select or elect CDC members, nor would they allow a woman from the community to stand as a representative for the other woman, as this was seen as against their religion and customs (KII: Com15, March 10, 2010).
Amongst the 80 women I spoke with from the community, I was unable to find more than two women who said they had attended a NSP meeting or voted for CDC members. One said she voted from her home when the female FP staff member came to speak with her (KII: CS003, February 17, 2010) and another woman said that six women from her village had gathered together in their village and voted (KII: Com14f, February 16, 2010).
Two groups of women who participated in an FGD stated bluntly that no women had voted in the community of Shah Raheem (FGD05f, January 27, 2010, FGD08f, February 9, 2010), and another woman said that the men were, "selecting their first wives to positions even though they could not read or handle the job" (KII: Com03f, January 20, 2010), indicating that no elections for the females had taken place.

However, one male voter told us that when the men voted, this was counted as a vote for their wife also (KII: Com08, February 2, 2010), which could explain some of the variances in perspectives, but still would mean that women had not in fact voted.

In an effort to unravel these conflicting stories, I returned to certain male community members and the FP staff to ask them to help me understand why such different versions of what had taken place existed. Why would the women tell me that they had not been involved, while most of the male community members and the FP staff insisted that they had been? The responses I got were quite similar. They explained that the women were simply too poor and too uneducated to remember having attended such a meeting (KII: FP12f, February 4, 2010, Com03f, February 10, 2010, FP08, February 14, 2010).[419]

These responses gave no credit to the women for being intelligent members of the community, which conflicts with my observations and experiences with them. Taking this into consideration, along with my understanding of just how rare such a large gathering of women would have been, I came to the conclusion that it is more likely that the women had not participated widely in a group introductory meeting to the NSP or CDC elections. This is in part based upon my belief that the women had little to gain from telling me they had not been involved if they had been and the fact that such meetings would have been easy for the women to remember due to its rareness in their social lives. Additionally, it is evident to me that the FP staff and the male community members could have felt that they had something to lose because it was understood that the program required female participation.

For the men who said they were able to vote, however, it does appear as though the majority felt that they were doing so of their own free will. They reported feeling no pressure from powerful people within or outside of the community such as the *kalāntar*, district or local leaders, etc. (HHS 2010).[420]

419 The male FP staff member noted that this was not limited to the women, but also included some of the men, saying, "These people have problems keeping information in their minds" (KII: FP08, February 14, 2010).

420 It is interesting to note, however, that some researchers have found many Afghans believe that, "a person should vote the way his or her community votes, not how they feel individually" (Tariq, Ayoubi et al. 2011: 117). This would imply that individual voters may be less independent than it might first

The elections were evaluated by the male community members as having been fair (KIIs 2010).

That being said, the NSP Form 2: CDC Election and Composition, and some of the male community members say that not only were the women of Shah Raheem included in such elections, but that six women were elected as CDC members along with six men. One provincial FP officer said that women were elected to a female subcommittee, but that, "If it was not compulsory, they would have found it impossible to select a female subcommittee" (KII: FP06, July 7, 2009), indicating what has been noted by other researchers, that inclusion of women was allowed because it was required not because it was seen as important (Kakar 2005: 24). This community member then went on to say that the female CDC or subcommittee was on paper only and that the women "did not have any more say in community decisions than before" (KII: FP06, July 7, 2009).[421] He explained that the FP staff and the male community members understood that unless they allowed women to participate in the NSP that they would not receive money from the MRRD. He said that this had been true in the community of Shah Raheem and also in the four other provinces where he had worked for the NSP (KII: FP06, July 7, 2009). For this FP staff member, the simple existence of a female CDC on paper was seen as a large step for the empowerment of women, regardless of whether it provided them with a voice or not, as it had never happened before (KII: FP06, January 14, 2010). This seems telling in terms of how much room there is for shifts in the empowerment of women in Khulm, with seemingly small and easily ignored changes from an external perspective seen as significant from a more internal viewpoint.

The female CDC members who were "elected" included: housewives, two tailors, and two women who said that they produced embroidery to contribute to their household incomes (NSP Documents). Their ages ranged from 35 to 60, with an average of 41.8 years of age. The women were not listed by their own names, but were instead listed under the names of a male relative, with their relationship to that male relative provided. It is said that this was required by the men of the

appear.

421 He was not alone in this assessment of the women's subcommittee existence being only for show, it was also supported by other FP staff and district citizens (KII: FP07, July 12, 2009, EX14, January 16, 2010).

community, as a means of "protecting" the women of their family from outsiders (KII: FP12f, February 4, 2010). This can be seen as disempowering, as the women of the community were unable to even use their own names, to have an identity of their own that was not related to a male relative.

Of these six women I spoke with three, and none of them had detailed knowledge about the NSP and none said that they were CDC members. Even if the election was not only just for show, it fundamentally became such, as the women who were "elected" wrote on their CDC registration form that they passed all of their power to the male CDC members of Shah Raheem (NSP Documents).[422] This effectively means that even if the NSP Documents were accurate, that women still did not truly participate actively in NSP decision-making.

The FP staff said that the women had given their rights away voluntarily because they did not want to be involved in the NSP process, with one saying clearly, "Women do not want to work on these projects" (KII: FP14, July 15, 2009). As noted earlier, Haward et al. (2004: 100) supports the idea that there are legitimate and valid reasons for people not to participate in participatory development programs, as there are costs as well as benefits involved in participating. For the women and families of Shah Raheem, one of the costs appears to have been disapproval from family members, in particular the men, and possibly the gossip of neighbors who believe that women should not participate in community activities.

Additionally, it was reported that the women were unable to partake in the procurement of materials for the subprojects and obtaining money from the bank because of restrictions on their movement in public (KII FP08, July 15, 2009).[423] This was confirmed by a CDC member who said that the women were involved in selecting the projects, but that, "they didn't have part directly in accommodating the project because in accordance with the usual rule, the women transferred their authority to the men" (KII: Com33, September 26, 2011).

---

422 The form was translated as saying, "The female *shura* of Shah Raheem Village don't want to participate in this *shura*. All of these females give the right to the males" (field notes, January 18, 2010).

423 Only two women from the community of Shah Raheem worked outside their home at the time of research (KII: Com22f, February 27, 2010).

This phenomenon was not an isolated incident, but occurred in almost all of the 75 NSP communities in Khulm District, according to NSP documents. Only two communities in Khulm District were exceptions (KII: EX07, August 4, 2009 and NSP Documents).[424] However, official NSP Documents indicate that each of these 75 communities had a female CDC, providing the illusion of participation, without actual participation.[425]

The near universality of women giving away their NSP responsibilities in the district, leads me to believe that the FP did not make strong attempts to prevent this from happening. This was confirmed by the FP staff themselves, who were unable to say that they had worked with the community to help them understand the importance of including women and/or other marginalized members of the community in the NSP (KIIs 2009 and 2010). When I spoke with the FP staff, they seemed to accept the community members' belief that women should not be involved in such efforts based upon their "culture" and "traditions" (KIIs 2009 and 2010). Whenever I raised the issue of female inclusion in the program with FP staff, they seemed unconcerned about the lower level, or lack, of female participation. Gender biases and cultural norms were also found to be barriers to the participation of women in the NSP by Kakar (2005: 27) in his evaluation of the NSP.

According to some community members, the FP staff moved from introducing the community to the program to holding elections very quickly (FGD06, February 3, 2010, KII: Com19, February 24, 2010). The quickness of the mobilization process could have constrained the amount of time they were able to spend in the community explaining the reasons why the NSP requires the participation of women. Mohan (2001b: 52) also noted that time constraints from external stakeholders can negatively impact on deep participation, i.e., active participation. The FP staff's acceptance of the male community members' opinion

---

424 Both the communities of Dedadi and Nahir Sharin reportedly had female CDC members who participated actively in the NSP (KII: EX07, August 4, 2009), though despite multiple requests, I was never able to meet with them to discuss their participation.

425 The female FP Mobilizer estimated that in 20% of the communities' women had reclaimed their NSP responsibilities and were involved as of 2009 (KII: FP12f, August 5, 2009). Unfortunately, I was not able to find any confirmation of this from any other FP staff, community members, or key informants.

could have stemmed from other reasons also, such as sharing in the same "culture" and "traditions."[426]

According to the NSP documents, six male community members were elected to serve as CDC members: two farmers, a teacher with a great deal of land, a mason, an electrician, and a man who supported himself through daily labor. It is interesting to note, that the CDC was not solely made up of elites from the community of Shah Raheem.[427] This was also the case in other CDCs in Khulm District: "CDC members are not the most powerful people in the community," and the powerful people are not always happy with this (KII: EX04, July 29, 2009).[428] The members ranged in age from 22 to 56, with an average of 43.8 years of age. The CDC had representatives from each of the three villages.

From the CDC members, four men were elected to the executive committee of the CDC. These positions bring with them specific responsibilities within the NSP framework, but are uncompensated. As the subprojects moved forward in Shah Raheem, it was these four men who took on the responsibility of managing the subprojects, with the other two CDC members playing no greater role than any of the other community members. In Shah Raheem the election of each of the four executive committee members – Chairman, Deputy Chairman, Treasurer, and Secretary – was said to be unanimous (NSP Documents).[429]

426 This raises a concern regarding hiring locally; FP staff from the locality are likely to share the beliefs and social norms that can possibly stand in conflict to the program they are implementing. At the same time, hiring locally also has many benefits.

427 Two other KIIs reaffirmed this for the district as whole, saying that in their experience working with the NSP in Khulm District, the *khan, kalāntar*, and *malik* were often not elected to the CDC. Although this did not always please these individuals, however, they would not often put up a fight about it (KII: FP07, July 12, 2009, EX14, January 16, 2010).

428 Other researchers have found that anywhere from a high estimation of 75% of CDC members, to a low of 5% were local elites in different NSP communities (Maynard 2007: 47). The World Bank sponsored randomized control trial found only 25% of the CDC members to be local elites (Beath, Christia et al. 2011: 11). In the research community of Shah Raheem, the percentage was around 25% for active CDC members (NSP Documents).

429 The Chairman that was originally elected in 2006 moved to Nimroz Province and then on to Iran after the completion of all three NSP subprojects. He left in 2010 to find work, and since that time the CDC Treasurer and the Deputy took on the Chairman's responsibilities (KII: Com37, May 31, 2011, Com38, August 20, 2011).

The Chairman selected for Shah Raheem was a young man of twenty-two years of age, which is quite noteworthy in a society where the social hierarchy is based to a large degree on age.[430]

Within the community of Shah Raheem, the CDC was not widely known about by the people. Even after it was described to community members to account for the possible use of different terminology, people often simply did not know what I was talking about (HHS 2010). The male Treasurer of the CDC was not even able to remember the names of the two other CDC members who did not serve on the executive committee just after completing their subprojects (KII: Com15: March 10, 2010).

Women who were listed as CDC members told me that they did not know that they were CDC members (e.g., Com27f, March 6, 2010). This was also seen in the district during the District Development Assembly for women from other communities.[431] Additionally, few people were able to describe for me what the role of the CDC was (HHS 2010), which is quite meaningful in terms of the sustainability of such a council.

6.1.4 Development Subproject Prioritization

Following the selection of CDC members in Shah Raheem, came the prioritization and selection of development projects by the NSP community. After the establishment of the male CDC (hereafter, simply CDC), and the passing on of their responsibilities by the female CDC, the male community members said that they gathered in the mosque of

Two CDC member elections have been held in Shah Raheem, the original in 2006 and a second one in 2009 (NSP Documents).

430 One community member said that originally the people had voted for the Treasurer to be Chairman, a powerful, older landowner, but that he had declined because he did not have enough time to take on the responsibilities associated with being the Chairman of the CDC, and instead took the position of Treasurer (KII: Com08, February 2, 2010).

431 One female participant in the District Development Assembly told me she did not know if she was a CDC member, then her husband interrupted and informed me that she was indeed a CDC member. I then asked her about the NSP development projects in her community and she was not able to tell me anything about them (personal communication, Khulm District husband and wife, July 26, 2009).

the village of Shah Raheem to discuss and prioritize their development needs (FGD06, February 3, 2010).[432]

The prioritization meeting is the opportunity within the NSP for the community members to voice their own opinions about their development needs. The FP staff said that after a list of development concerns was established, the community members present at the meeting were each provided with three votes to cast (KII: FP06, July 6, 2009). They were able to distribute their votes as they pleased. In total 900 votes were documented as being cast in Shah Raheem, 492 of which were said to stem from the women of the community (NSP Documents). Though the numbers can be called into question, it was evident in the KIIs and the HHS that those needs prioritized in the NSP documents were indeed major development concerns within the community (HHS and KIIs 2010). The number one community priority, according to the NSP Documents, was a desire for a clean water source to be established, with 294 votes. This was followed by a community center with 229 votes. The third ranking priority was to reconstruct the roads in the community, with 127 votes (cf. NSP Documents).

When community development priorities were discussed in the household survey, the first priority was electricity (n.28 out of 32), followed by clean drinking water (n.22 out of 32) and fixing the roads (n.20 out of 32) (HHS 2010). Trailing more distantly was the desire for a community center, which was mentioned by about one third of the respondents (n. 11 out of 32). As can be seen, the top three priorities listed in the NSP Documents align with some of the top household survey participant responses. While this appears to be quite tidy, it is complicated by the fact that the community members indicated that each village voted only for the project in their own village. Unfortunately, the NSP Documents do not provide a village by village breakdown of the prioritization voting.

Male community members remembered having taken part in this prioritization, and report that the females participated in a separate meeting (HHS 2010). However, the female community members we spoke with, including both those who are considered to be influential

432 As noted previously, due to restrictions that prevent women from entering the mosques in the community, it is evident that this meeting did not include any women.

within Shah Raheem Community and those who are not, did not remember having participated (HHS and KIIs 2010).[433] Women spoken to said, for the most part, that they had not been aware of a prioritization process and that they had not contributed their opinions (KIIs 2010). They also said that the men of their household did not consult with them regarding development priorities before they attended the meeting (e.g., KII: Com20f, February 26, 2010). Some women in the community told us that the men did not discuss development priorities at all with them, with one saying, "The men do not have to talk to the women about development projects" (FGD10f, March 6, 2010). When male community members were asked about this, they said that the women in their household had been represented through their vote (FGD06, February 3, 2010). The women, however, said that they would like to be involved in making their own decisions and not just have the men make them for them (FGD10f, March 6, 2010), indicating that the women's lack of involvement was not determined by a deficiency of interest on the part of the women.

What we found in the KIIs and in the HHS responses is that people believe, or choose to believe, there is only one opinion per household and therefore multiple people from one household at meetings are not needed. Individuals and their opinions and views are not the basic unit; it is the family or household. One man summed this up when he said, "As well, the people who attended the meetings were enough. So the rest did not attend the meetings because of this" (HHS: YB005, February 24, 2010).

I was unable to reconcile the information provided in the NSP Documents with the information obtained from the women who partook in this research. However, I would again lean towards believing that the women did not play a significant role in the NSP development project prioritization process, in contrast to what was stated in the NSP Documents, due to the women's negative responses to questions regarding their involvement.

There is a requirement that 60% of adults in any NSP Community participate in the prioritization process (MRRD and NSP III 2012: 17).[434]

433 This distinction is made to highlight that both educated and uneducated women were excluded.

434 One FP staff member said that 70% of eligible voters were required to vote (KII: FP07, July 12, 2009).

The intention behind this was to insure that more members of the community were being engaged in the NSP, especially women. This could provide a motivation for the FP staff and males to indicate that women had participated, even if they did not.

Once the development priorities of Shah Raheem Community were established, the community was supposed to write up a Community Development Plan as a part of their capacity building and empowerment. In Shah Raheem this was done by the FP staff. One FP staff member said, "The writing and filling of the form is done by CARE staff," and "the Technical Facilitator writes up the subproject proposal" (KII: FP07, July 12, 2009).[435] This means that the people of the community were not in charge of guiding the subprojects in Shah Raheem. This was also the case for most of the communities of Khulm District (KII: FP06, July 6, 2009).

From the Community Development Plan, a subproject proposal was written for each of the three different subprojects that were implemented through the NSP block grant in Shah Raheem, which are discussed in more detail below. When self-identified CDC members from the village of Shah Raheem were asked about these proposals, they had no information about them, and said that only the four executive officers of the CDC would have this kind of information (FGD06, February 3, 2010). This just highlights that the community was not as involved in the NSP process as is laid out in the NSP Operational Manual. This was not limited to the research community of Shah Raheem, but was also discussed by FP staff and MRRD staff in many of the NSP communities (KII: EX04, July 29, 2009, FP05, January 14, 2010).

435 This was supported by another FP staff member, who said that the FP staff had filled in the community profile and other NSP forms for the community of Shah Raheem (KII: FP06, July 7, 2009).

6.2 NSP Program Implementation and Management in Shah Raheem

6.2.1 NSP Meetings: Inclusion and Exclusion

NSP meetings for the male community members of Shah Raheem were reported to have occurred on a regular, monthly basis when the program first started in 2006 (HHS 2010). They were said to include representatives from each of the three villages (e.g., FGD11, July 20, 2011, KII: Com18, February 23, 2010).

The NSP meetings differed from previous community meetings, which were traditionally held on an *ad hoc* basis, as needed to resolve issues. The tradition of bringing elders together to address problems existed long before the implementation of the NSP in the form of a *shura*, which is one of the foundational traditions upon which the NSP is based (MRRD and NSP III 2012: 11).[436] Additionally, meetings in the past only rarely required villagers from all three villages to get together, but instead usually consisted of villagers from only one village meeting to discuss their village's problems (KII: Com38, August 20, 2011). They were most often conducted only with the male elders of the village and middle aged males with power and influence, such as commanders or *khans* (KII: Com38, August 20, 2011). The NSP meetings were said by some to be more inclusive in nature, with younger males also being included (HHS 2010). Traditionally, young people in Shah Raheem did not have a say in community decisions due to the age-based hierarchical nature of the society.[437]

The first NSP meetings focused on getting the program up and running in Shah Raheem, and those that followed were dedicated to the management and implementation of the three different subprojects that were prioritized. While there is again contradictory information about such meetings, it appears clear that at least some male community members gathered to discuss the NSP and continue to meet today, according to some of the household survey respondents and key infor-

436 In the Operational Manual, *jirga* and *shura* are used interchangeably (MRRD and NSP III 2012: 7).

437 For example, the female *kalāntar* of Yaqubai was selected to be the *kalāntar* by the oldest male in the community only because she was the oldest female (KII: Com23f, February 26, 2010).

mant interview participants, though whether they meet regularly or *ad hoc* is not entirely clear.

In the HHS, 56% of men reported attending NSP meetings (2010). A lower percentage of around 20-25% of the adult male community were observed gathering in the mosque to discuss issues of importance, including the NSP.[438]

Conflicting evidence about the involvement of women in NSP meetings was again discovered. The FP staff and the males said that women held regular development *shura* meetings separately from the men (e.g., KII: FP12f, February 4, 2010).[439] This was supported by the female FP Community Mobilizer who said that regular monthly NSP meetings were held amongst the women of the three villages of Shah Raheem (KII: FP12f, February 4, 2010).[440]

This information was contradicted by the women of the community, who said that there were no regular, monthly NSP meetings held for the women of the community of Shah Raheem, and that they did not attend any NSP meetings (e.g., FGD10f, March 6, 2010, KII: Com03f, January 20, 2010).[441] Even one woman, who self-reported as a CDC member said that, "There were never any meetings. We just wrote our names on the sheet, but we have not done anything more than vote for NSP" (KII:

438 This is a very loose estimate based upon the fact that multiple meetings were held in the community that I observed where around 30 to 40 men from the community were in attendance. There were a total of 157 eligible male voters in the community (i.e., those over the age of 18).

439 These meetings were described as happening on a monthly basis, and included representatives from each of the three villages that make up the NSP community of Shah Raheem (e.g., FGD06, February 3, 2010, KII: Com15, February 17, 2010, Com18, February 23, 2010). As women do not have access to the mosque, their meetings were said to have been held in community members' homes (KII: Com14f, February 16, 2010, Com20f, February 26, 2010).

440 She said that these meetings were actually supposed to be conducted twice a month, but that they were only taking place once a month (KII: FP12f, February 4, 2010). She said that she was only able to attend the meetings irregularly.

441 Men in the household survey indicated that 30% of the adult females in their households participated in NSP meetings on a regular basis, which again disagrees with the information provided by women of the community (HHS 2010). I would suggest that this is a very optimistic evaluation of female attendance, especially as a number of the men who said that women from their household participated in the meetings also said that women attending such meetings would be seen as shameful (HHS: SR008, February 7, 2010, CS003, February 17, 2010, CS005, February 20, 2010).

Com03f, January 20, 2010).[442] I would like to reiterate that women from different economic statuses, social standings, etc. were included in this research, which helps to eliminate the possibility that a certain subset of women were included in the NSP meetings and the rest were excluded or left out.

When I followed up on this contradiction with the female FP Community Mobilizer, she provided the same explanation as was provided for why the women did not remember voting for CDC members; the women are too uneducated and poor to remember attending meetings (KII: FP12f, February 4, 2010). As I am unable to discover a reason for why the women would have hidden participating in such meetings, I am more prone to believe that if NSP meetings for women took place, they were on a very small scale and included only a very limited number of women from the community. Even this possibility seems dim, as I spoke with a wide variety of women in the community, including a few who were listed as CDC members, who would be assumed to have been included. They said they did not attend meetings where the NSP or development efforts were discussed.

Additionally, the male *kalāntar* of Yaqubai noted that no female staff from the FP had ever visited Yaqubai to talk with the women or hold meetings, and having male FP staff work with women from the community would not have been possible due to the separation of the sexes that is generally maintained (KII: Com18, February 23, 2010). He said that the men of the community would say things such as, "Women do not have a role to play in these things. Just the men make decisions."

In the household survey, the participants were asked to explain what barriers existed that prevented adult community members from attending NSP meetings.[443] The main reason provided for why men in Shah Raheem were not able to attend NSP meetings was restrictions on their time (HHS 2010).[444] A number of males were living outside of the district and therefore did not attend the meetings (HHS: SR011, Febru-

442 She is one of the two females I spoke with who said that she had voted, but when I sought out more details about the voting, she was unable to answer my questions.

443 According to the NSP framework, all adults over the age of 18 should be included in NSP meetings even after the prioritization has been accomplished.

444 This could be because they were busy with work or because they were students.

ary 9, 2010, SR013 and SR014, February 14, 2010). Additionally, two households said that they had economic problems that prevented them from attending the meetings, though they were unable to clarify how the two were connected (HHS: SR002, January 26, 2010, SR004, January 27, 2010). Those who were too elderly to walk to the meetings also did not attend (HHS: SR005, February 2, 2010, CS006, February 21, 2010). Finally, one man said that he did not attend the meetings because he is illiterate and would not know what to do in a NSP meeting (HHS: SR002, January 26, 2010). While this was only a single person, it does speak to an understanding that I heard multiple times in the community: that those without education do not have anything to add to a discussion, conversation, or the decision making process. A person's lack of education is equated with them having no useful information and being less intelligent.[445] Additionally, it points to a lack of information about the NSP that should have been addressed via the FP, as the NSP is supposed to encourage the participation of all community members regardless of their education level.

The most common reason given for why women did not attend the meetings, almost all of which were shared with me by men, was that women were busy making food for their families, and busy with other housework and the production of handicrafts, such as embroidery (HHS 2010). This was then followed by what appears to be a possibly more deeply embedded barrier; a significant minority said that according to their religion women could not attend the meetings (HHS 2010).[446] They expanded upon this by saying that women attending such meetings would be considered shameful for them and their families (HHS: CS003, February 17, 2010, CS005, February 20, 2010, YB001, February 21, 2010, KII: FP12f, August 5, 2009).[447] Boesen (2004: 51) found that a lack

445 I strongly disagree with this sentiment, and believe people can be very intelligent without having been formally educated.

446 One man said that his daughter was not attending the meetings because her husband would not allow her to attend them, illustrating the restrictions on some women's choices by male family members (HHS: SR005, February 24, 2010).

447 One community member said that the shame was connected to the fact that local commanders would talk to the men badly if they allowed their women to attend such meetings: "It is more a fear of these men that motivates them and what the community will think about them than any belief that women are inferior" (KII: FP12f, August 5, 2009). She also raised the idea that some women

of permission from male family members was the most common reason cited for women to not attend NSP meetings, and that a lack of access to information about the NSP also played a role. Kakar (2005: 27) agrees with Boesen, but also believes that gender biases of FP staff, along with limited leadership experience, a lack of education, and cultural norms are factors in why women did not attend NSP meetings.[448] The women, however, clarified that they did not attend such meetings because they did not exist (e.g., HHS: SR016f, February 21, 2010).

The *kalāntar* of Yaqubai told us that the NSP meetings were more exclusive in nature than others in the community implied (KII: Com18, February 23, 2010). While he said that there was a representative from each of the villages of the NSP community, the general community was not included. The CDC Treasurer supported this, saying that for regular NSP meetings, only the four executive committee members of the CDC would attend, but that for meetings where other issues were discussed and needed to be resolved, the elders would also be invited (KII: Com33, September 26, 2011). He never mentioned the inclusion of male youth. Females were also excluded from these meetings, as will be explored in more detail below.

According to the NSP framework, all adults of Shah Raheem Community should have been included in such meetings, as they were also supposed to serve the purpose of creating greater transparency in the program, providing the CDC members a platform from which to inform the community of the progress of different subprojects and to discuss issues such as procurement. This was, therefore, supposed to provide the community with information that would allow them to hold their CDC members accountable. The accounting books for the projects were said to have been shared with the community during these meetings, along with the receipts for all purchases (KII: Com37, May 31, 2011). However, we met community members who told us that they had never

---

were ashamed to be seen in public because they could not afford good clothes or shoes (KII: FP12f, August 5, 2009).

A number of households also said that some of the women were too old to physically attend the meeting (HHS: CS004, February 20, 2010).

448 Two other male respondents said that they already had enough family members attending, and therefore the women did not need to attend, which goes back to the belief that there is only one opinion per family (HHS: SR006, February 2, 2010, YB005, February 24, 2010).

been informed about such meetings, much less invited to them (e.g., KII: Com08, February 2, 2010, Com10 and Com17, February 7, 2010), resulting in reduced transparency community-wide.

While some community members say that the NSP meetings continue to be held monthly today, others implied that the meetings are held less often, such as every third month (e.g., KII: FP07, July 12, 2009). These meetings now focus on other subjects, such as conflicts between farmers and landowners, inheritance issues, plant diseases, and security concerns (KII: Com37, May 31, 2011). Others said that once the subprojects were completed, they stopped holding meetings altogether (e.g., KII: Com15, February 17, 2010).

In terms of whether those who attended the meetings felt they were able to actively contribute to them, there is some evidence that the men who attended felt they were able to express themselves and have a say (HHS 2010). This feeling was not universal, however, though it was difficult to gauge this, as the meetings had taken place prior to this research and I was unable to observe them directly. I can say that during my FGDs with both men and women, only a limited number of individuals actively participated despite efforts to extract answers from all in attendance. This could indicate that perhaps all who attended did not actively participate in the NSP meetings, as was found to have occurred in many other participatory development projects (Cohen and Uphoff 1980: 223-224; Fonchingong 2006: 144-146; Narayanan 2003: 2485; Resurreccion, Real et al. 2004: 531).

In my mind, this indicates that one cannot rely on program documents to provide insights into actual participation. Attendance lists, sign in sheets, etc. should not be taken at face value for two reasons: (i) the attendance lists could have been altered to include community members who did not attend, and (ii) even if the attendance list can be relied upon, this does not automatically translate into active participation by those attendees. This second reason has been noted by NSP scholars (Enikolopov, Christia et al. 2009: 3) and development scholars, such as Cohen and Uphoff (1980: 220) who highlight three dimensions of participation including the "kind" of participation. In this, they refer to concerns about whether participation consists of only being in attendance or if it involves further engagement, sharing of thoughts and opinions, making decisions, benefitting from a program, etc. (Cohen and Uphoff 1980: 220).

I would extend similar skepticism to quantitative and qualitative studies in which researchers spend only a day or two in the research area due to the limited scope and nature of their data collection techniques and generally the lack of time to build the kind of trust that is sometimes necessary for people to feel they can tell the researcher their version of events instead of the sanitized or fabricated version often reserved for outsiders. Limited time frames also commonly result in a limited breadth of respondents, referred to by Chambers (1983: 18) in his six biases of research as a bias of person, which means that people often conduct research with men, people with power, and people who participated or benefited from a program, as they are naturally the most easily accessible. With the understanding that communities are not homogeneous, in agreement with a number of participatory development scholars (Cleaver 2001: 44; Guijt and Shah 1998b: 7-9; Mosse 2001: 32; Pain, R. and Francis 2003: 51; Strand, Toje et al. 2003: 1), I believe that time constraints and limits that are common for most research projects act as a disservice to understanding this diversity and can result in a very biased picture of what happened, or what the respondents want researchers to believe happened.[449] Based upon my first week in the community of Shah Raheem, I would have concluded, in good conscience, that women had participated in different aspects of the NSP process, however, in the second week I was able to speak with more women and a different story emerged. As the weeks turned into months, the second version of the story gained greater and greater support, until it became evident that it was likely the more accurate version of past events.

6.2.2 NSP Trainings: Limited Information and Impact

In order to help CDC members with their planning, implementation, management, and monitoring responsibilities, the FP is supposed to provide them with trainings. According to the NSP Documents, quite a

449 The heterogeneity of responses in the empirical work for this thesis provides ample evidence of the diversity that can exist in opinions, experiences, and even in perceptions of what occurred in the distant and recent past.

few trainings were conducted for both the men and women in the NSP community of Shah Raheem. Four to five male CDC members were recorded as having attended trainings on procurement, accounting, legal issues, gender, transparency, and logistics. Refresher courses were said to have been provided later in the program for procurement, accounting, legal issues, and logistics. As happened with other aspects of the program, the NSP Documents indicate that six female CDC members also attended the first round of six trainings. Only two of the refresher courses were said to have been provided for the women – procurement and legal issues – and only three women were recorded as attending these meetings.

While none of the women interviewed were able to remember partaking in the NSP trainings, the female FP Community Mobilizer stated that six of the women listed as CDC members had participated in procurement, gender, conflict management, logistics, transparency, and accounting trainings (KII: FP12f, February 4, 2010). A male FP Community Mobilizer also talked about the females partaking in the NSP trainings, saying that he thought they were more committed to the trainings than the men were (KII: FP14, July 15, 2009). He did say, however, that he thought the women were partaking in the trainings to increase their standing in society and not to actually work on the NSP in their community (KII: FP14, July 15, 2009). However, I believe that if the women had participated in any training events they would have readily spoken about them with me, especially if it was seen as something that could raise their social status. The representative of RRD for NSP in Khulm District added support to this understanding when he said, "It is better for their future to hold training courses for females because it helps inform them of CARE projects" (KII: EX07, August 4, 2009), implying that no NSP trainings were held for women in the district. Azarbiajani-Moghaddam (2010: 19) and Barakat (2006: 190) also found that women were not being involved in NSP trainings, and in the case of Azarbiajani-Moghaddam, this was also found despite NSP documentation that attested to them having been trained.

One third of the male household survey respondents said that the NSP had benefited them through its training courses (HHS 2010). While this is very positive, it must be taken with a grain of salt, as many of the trainings listed by these community members as NSP trainings, were not a part of the NSP, such as farmer trainings, literacy classes, embroidery, and tailoring (HHS 2010). Not one of the trainings mentioned by the HHS participants was one of the official NSP trainings (HHS 2010). Even in the KIIs, trainings were not often spoken of and when they were spoken of, it was in vague terms (2010). One male community member had memories of participating in bookkeeping and accounting training in addition to trainings in gender issues, procurement, and transparency (KII: Com38, August 20, 2011). However, he was the exception, which seems to indicate that if NSP trainings did take place for the CDC and community members of Shah Raheem, they did not have a great impact on the people who participated in them.

These findings are not isolated to Shah Raheem. Barakat (2006: 73) evaluated that only half of all the CDCs he interviewed had gained enough from their trainings or their past experiences to handle managing NSP subprojects. Zakhilawal and Thomas (2008: 21) found that NSP communities had not received much training and that CDCs, in general, did not have the capacity needed to manage development projects in the future.

6.2.3 NSP Development Projects in Shah Raheem: Infrastructure Built and Mostly Utilized

Three separate subprojects were implemented in Shah Raheem, one in each of the three villages. They are each seen as successfully completed NSP subprojects, according to NSP Documents.[450] The Balk MRRD Provincial Manager told me that 45 different subprojects had failed in the province, indicating that he was not uncomfortable talking about failed projects (KII: EX04, July 29, 2010). Yet, he also categorized all three of the Shah Raheem Community subprojects as successful. Each village was said to have selected their own development project, with the block grant money from the NSP divided between the villages according to population (KII: Com19, February 24, 2010). In alignment with the NSP framework, the community members were required to contribute ten percent of the project costs. In Shah Raheem this was done

450 Data on the number of unsuccessful projects nation-wide was not available, however, the existence of a "negative list" of NSP projects that were no longer allowed because they were not successful earlier, is evidence that a number of NSP subprojects have been implemented unsuccessfully.

through the donation of land, through in-kind services such as labor, and through cash payments, though cash payments were only limited to the wealthiest members of the community (e.g., KII: Com10, February 7, 2010).[451]

The four CDC executive committee members worked on all three subprojects in the different villages within the community. The FP staff was also involved in each of the subprojects, especially in terms of writing the proposal and helping to guide the implementation process. Additionally, they provided engineers to work with the community in order to provide and inform any necessary designs and to verify the feasibility of project proposals in the locations selected.

A deep well was implemented in the village of Shah Raheem, which is situated in between the other two villages of Chahar Sukh and Yagubai. A shallow well was implemented in Chahar Sukh Village, the first village of the community coming from the district center of Tashqurghan, and a community center was built in Yaqubai, the village furthest away from the district center.

Deep Well and Generator

The division of the NSP block grant according to village population meant that Shah Raheem, with its 51 families, had the largest share of the grant to work with.[452] The village decided to prioritize the construction of a deep well to provide a source of clean drinking water for the people (NSP Documents), as the water they were using came from the open streams that run through the community alongside the road and through irrigation canals. The *kalāntar* of the village of Shah Raheem said, "In the past, when NSP was not here, we did not have drinking water" (KII: Com02, January 20, 2010). Back then, community members gathered flood water from the stream and then let it sit in barrels to allow the sediment to settle, after which they used it as drinking water without any kind of treatment (KII: Com02, January 20, 2010). This was

451 It was explained that those who could afford to provided their contribution in cash, while those who could not afford this volunteered their time and labor to the program.

452 Shah Raheem Village families account for about 48% of the NSP families of the community of Shah Raheem (NSP Documents).

said to have caused numerous illnesses amongst the children and the adults in the community.[453]

The deep well project in Shah Raheem consisted of three aspects: the excavation of the deep well and a reservoir, the installation of a generator, and the construction of a building to house the generator. The project was implemented on land donated by the community, which covered 100 m2 (NSP Documents).[454] The proposal was submitted January 27, 2007 and was approved by the Project Management Unit (PMU) of the NSP April 28, 2007. The community received its block grant installment August 9, 2007, a little over six months after the community had submitted its proposal. The project was certified as completed on July 24, 2008, almost a full year after it had commenced, according to NSP Documents.

Photo 8: NSP Subproject: Water Spout and Covering of the NSP Deep Well and Storage

Source: photograph MB Wilson, February 2010

The village received 560,000 Afghani (about 11,200 USD) from the NSP block grant and each family in the village contributed the equivalent of 1,243 Afghani (about 25 USD) (NSP Documents). One community member reported that a male member from each of the households usually worked five days on the subproject as a part of their contribution (KII: Com08, February 2, 2010). Another community member said that people could provide either 1,000 Afghani (about 20 USD) in cash to the project or a male member of the family could work for ten days

453 One community member said, "We faced different sicknesses" (KII: Com10, February 7, 2010).

454 This land is situated in the center of the village, next to the mosque.

on the project (KII: Com10, February 7, 2010). Widows would often send a male child or grandchild to contribute labor, though some were given exemptions and were not required to contribute (KII: Com17, February 7, 2010).[455]

In their NSP subproject proposal, the villagers stated that they would provide the funds to buy the fuel to operate the generator, which is necessary to extract the water from the deep well (NSP Documents). Those pledges, however, were perhaps a bit optimistic (KII: FP06, January 14, 2010), or were perhaps made without consultation with the community members, who seemed to play little to no role in writing the subproject proposal.

About two months out of the year, towards the end of summer, the streams runs dry, and only then do the villagers of Shah Raheem contribute the estimated ten to 50 Afghani (about 1 USD) needed per family every two weeks to run the generator (FGD05f, January 27, 2010, KII: Com08, February 2, 2010, Com15, March 10, 2010, Com38, August 20, 2011). The summer is seen as the time of year when people are more likely to get sick from drinking stream water (FGD06, February 3, 2010). The *kalāntar* of Shah Raheem said, "The generator is needed for the well and people cannot always give funds or money for it because of their bad economic state" (KII: Com02, January 20, 2010). Claiming poverty, the community members said that they continue to use the local stream to obtain their water during most months of the year (FGD05f, January 27, 2010, KII: Com02, January 20, 2010). During the FGD with males from the village of Shah Raheem, one man said, "We expected NSP to pay for the generator during the winter" (FGD06, February 3, 2010), showing a sense of dependence on an outsider or an outside organization to help them with development projects and efforts.

The community members said that sometimes rich people from the village contribute money to run the generator, but that when this happened, the rich people would keep the water for themselves (e.g., KII: Com15, March 10, 2010). Some villagers had reached out to their eco-

455 There were around 20 widows living in the community of Shah Raheem (KII: Com03f, January 20, 2010)(Widows List), though the NSP Community Profile indicates that there were only eleven households that were headed by a female. Many widows had been taken into the households of relatives, moving in with a daughter and her husband, a son, or a brother (KII: Com03f, February 10, 2010).

nomically better off relatives living in Mazar-i-Sharif and obtained money from them for the generator (e.g., KII: Com02, January 20, 2010).

The Provincial Manager of the NSP for RRD, whose family lives in Chahar Sukh Village of the NSP community, was suspicious about this assertion. He evaluates Shah Raheem to be neither rich nor poor, and said that he thought the monetary contribution necessary to run the generator was not out of their reach (KII: EX04, January 25, 2010).

This lack of maintenance of the project has a precedent according to Brick's research (2008) on the NSP. She found that projects that required regular contributions, such as the deep well project in Shah Raheem, were not being maintained (Brick 2008: 22, 23). This was supported by Barakat (2006: 182) who discovered that communities were not planning for the maintenance and upkeep of their projects.

One FP staff hypothesized that the village of Shah Raheem was not really concerned with the level of cleanliness of their water when they prioritized the deep well, but instead were concerned with finding a source of water for the two months a year when the stream ran dry (KII: FP06, January 14, 2010). He said that when he was in Shah Raheem and saw people using the stream instead of the well, he asked them why, and they responded that there was no difference between the water in the well and in the stream, which implies that they see both as equally healthy. This indicates a lack of knowledge transfer from the FP staff who said they discussed the importance of clean water sources within the community (KIIs 2010).

One community elder said that because they were using the well instead of the stream for their water, there was more water for those communities downstream, which resulted in them being able to irrigate more of their land (KII: Com05, January 26, 2010).[456] While we know that this well is only used intermittently, this does indicate that if the

456 The *kalāntar* and other community members said that villagers from up to six different villages nearby use the clean water coming from the NSP well when it is operational, and that they are welcome to do so (KII: Com02, January 20, 2010, Com05, January 26, 2010). This was confirmed by a Shah Raheem elder who said that sometimes whole buses would stop at the well and people would fill 15-20 jerry cans with water (KII: Com05, January 26, 2010). When asked if this had caused conflicts, he responded by saying, "We would never fight over this" (KII: Com05, January 26, 2010).

well was used more often, it could have positive economic impacts for communities further downstream.

Shallow Well and Hand Pump

Chahar Sukh Village, with only 15 families had the smallest proportion of the NSP block grant money to work with.[457] While the villagers of Chahar Sukh talked about originally wanting to build a bridge, they changed their minds and jointly decided to address their water needs first (personal communication, July 2009). They too had hoped to install a deep well with a generator, as Shah Raheem Village had done, but the funds that they had were not enough for this and they instead selected a shallow well (HHS: Com15, March 10, 2010). The proposal called for the excavation of a shallow well, the installation of a hand pump, and hygiene education (NSP Documents).

Photo 9: NSP Subproject: Shallow Well with Hand Pump

Source: photograph MB Wilson, February 2010

While the community members were disheartened that they had only been able to install a shallow well with a hand pump, their lack of funds

457 Chahar Sukh accounts for 14% of the families in the NSP community of Shah Raheem (NSP Documents).

might actually have been to their advantage as they benefitted from the clean water that flows from the shallow well throughout the year. I was able to observe people obtaining water from this shallow well on a daily basis during my research, the same of which cannot be said for the deep well that the villagers of Shah Raheem had installed.[458] The CDC Treasurer said, "People use this for all their drinking needs" (KII: Com15, February 17, 2010).

The subproject proposal for the shallow well was submitted June 19, 2008, just prior to the completion of the village of Shah Raheem's well project, and it was approved only three months later by the PMU, on September 18, 2008. The CDC received the NSP block grant installment January 22, 2009. The project was certified as completed in just under a year, officially ending May 2, 2009 (cf. NSP Documents). A few delays affected this project, including the rejection of the first proposal submitted because their forms were not filled in correctly. The final report was also rejected as the CDC was seen as having spent above market price on the hand pump and they lacked a receipt from the company who was hired to excavate the well itself. In the end, all data was provided to the satisfaction of the PMU. These means of monitoring the subproject stemmed from the FP level, not from within the NSP community.

This shallow well project cost 188,800 Afghani (about 3,776 USD) in total, of which 150,000 (about 3,000 USD) was provided through the block grant, with the community contributing 2,587 Afghani (about 52 USD) per family in cash, labor, and land (KII: Com38, August 20, 2011, NSP Documents).[459] Those who contributed labor were required to work five days on the well, while those who contributed cash provided 1,000 Afghani (about 20 USD) (KII: Com15, February 17, 2010). The CDC Treasurer said that obtaining the 10% contribution from the community was simple in Chahar Sukh Village (KII: Com15, March 10, 2010). He noted that, "the people saw the project as something that would benefit everyone," and went on to say that the villagers even cooked a meal for the FP Engineer who came to help them in their planning.

458 I never once saw the NSP deep well in the village of Shah Raheem functioning during my fieldwork.

459 The shallow well was built on 15 m^2 of land donated by the villagers of Chahar Sukh (NSP Documents).

The village of Chahar Sukh seems to be satisfied with the NSP subproject that they implemented. One man from the village said, "The community is greatly changed for the positive because of the clean drinking water" (KII: Com15, February 17, 2010).

Unlike the other villages in the NSP community of Shah Raheem, the villagers of Chahar Sukh have made plans for how to maintain the shallow well they had built. In preparation for any possible problems with the well, each family has contributed 50 Afghani (about 1 USD) to a repair fund (KII: Com15, March 10, 2010).

Community Center

With its proportion of the NSP block grant, the people of the village of Yaqubai chose to construct a community center.[460] Community centers were described by FP staff as a, "house to hold traditional ceremonies" (KII: FP08, July 20, 2009). It is a place where weddings and other parties can be held, where conflicts can be resolved, where *shura* meetings and trainings can be had, and where development opportunities can be discussed. One staff member said, "People become united here" (KII: FP09, July 20, 2009). He explained that people could do anything in such buildings, even dance and sing, things which are not considered acceptable in the local mosque.

The subproject proposal for the community center was submitted June 19, 2008, the same day as the Chahar Sukh subproject proposal, and was approved by the PMU September 10, 2008. The block grant installment was received in the community January 22, 2009. The original plan was to build a five room community center of 160 m^2 by 200 m^2 of land donated by a community member (cf. NSP Documents). The project was completed just prior to my arrival in January of 2010.

460 Yaqubai is noted as having 41 families, accounting for 38% of the total families in the NSP Community of Shah Raheem (NSP Documents).

Photo 10: NSP Subproject: Community Center

Source: photograph MB Wilson, January 2010

One community member mentioned that there were some delays in the project because the community had not provided its 10% contribution, but this problem was said to have been solved in time, though no details were provided on how (KII: Com37, May 31, 2011).

The people of the community were generally very enthusiastic about the community center. One community elder said, "It will solve all of our problems" (KII: Com30, March 7, 2010). However, when asked how, he was unable to provide details. While this statement might have been an exaggeration, it does show the enthusiasm with which the community center was received in Yaqubai Village. People said that the guesthouse provided a place for everyone to gather under one roof, and that this would create strong connections amongst the people (e.g., HHS: YB003, February 23, 2010, YB006, February 27, 2010, YB007, February 28, 2010), indicating that the villagers believed the community center would help to create stronger social security. In addition to the sadness and happiness ceremonies that would be held in the guesthouse, the community center was also seen as a place in which they would be able to hold meetings, including development meetings with NGOs or the government, with the hopes of obtaining more development assistance and funding (e.g., HHS: YB003, February 23, 2010, YB008 and YB009, February 28, 2010).

The villagers told me that if people from the other villages of the NSP community of Shah Raheem wanted to use the community center for their sadness and happiness ceremonies, they would be allowed to (KII: Com19, February 24, 2010). However, when this research took place, it had only been used by people from Yaqubai, and only very limitedly.[461]

The villagers of Yaqubai were also clear that the community center was only used by men because it did not have a wall around it or toilets, both of which were seen as requirements for it to be usable by the women of the villages and community (HHS and KIIs 2010). This was confirmed by both the men and women of the community (KIIs 2010). It is said that the original plans for the project had included a toilet, but that there was not enough in the budget for this, so it was abandoned (KII: Com30, March 7, 2010). Budget discussions for this project do not appear to have included any females, or general community members, instead it included male FP staff, male community elders, and male CDC members. One community member said that they had made a request after the completion of the subproject to the FP to build the wall and the toilets, indicating that they had no plans for adding those things on their own or organizing themselves to find other sources of funding (KII: Com27f, March 6, 2010).

Despite the enthusiasm of the villagers for the community center, there were also a few concerns voiced. The quality of the community center was called into question by the *kalāntar* of Yaqubai, who said that the roof was already cracking two months after the building had been completed (KII: Com25, March 4, 2010). This was confirmed by a respected elder of Yaqubai (KII: Com31, March 9, 2010), though it was denied by a few other community members who were asked directly about it (KII: Com27f, March 6, 2010, Com30, March 7, 2010). The CDC treasurer acknowledged that there had been a problem with the quality

461 I only observed the community center being used once, for the funeral of one man's mother and during my initial meeting with community elders.

of the materials used to build the community center, but said that the problem was not big and had been solved, though he did not say how it had been solved (KII: Com33, September 26, 2011). In general, the villagers from Chahar Sukh, especially the men, were very happy to have the community center (HHS 2010).

Confusion about What Constitutes a NSP Subproject

The NSP in the community of Shah Raheem consisted of only three public infrastructure subprojects. This was not clear to the community members, many of whom listed a myriad of different projects as having been a part of the NSP. This confusion or misinformation was held by both male and female community members generally, along with CDC members and community members with influence and power who were directly involved in the management of the program.

When household survey participants were asked what benefits the NSP had brought to their community, they often listed development projects that had been done through other programs (HHS 2010). They regularly said that the NSP had brought them a repaired main road and that this in turn had helped them get to the market, the hospital, and other locations more quickly, and because it was graveled it had also cut down on the amount of dust in the air, making everyone feel healthier (HHS 2010). This road, however, was built through the WFP Food for Work Program mentioned earlier. Additionally some HHS respondents talked about how the NSP had brought jobs to their community (HHS: CS005, February 20, 2010, YB005, February 24, 2010, YB006, February 27, 2010, YB008, February 28, 2010), and that it provided them with seeds and fertilizer (HHS: CS004 and CS005, February 20, 2010), both of which did not happen under the guise of the NSP.

One CDC member said that the NSP had implemented between 50 and 100 different projects, indicating that the confusion regarding which projects were related to the NSP and which ones were not was even held amongst the CDC members of the program (KII: Com37, May 31, 2011). While he might have been exaggerating for dramatic effect, if he, as the treasurer of the CDC, is unable to distinguish between NSP and non-NSP projects, it implies a serious lapse in communication and clarifications, and a lack of accurate information.

6.2.4 Management and Maintenance of the Subprojects: A Male Endeavour

The management of the NSP subprojects was universally seen to have been done by the men of Shah Raheem. While there is confusion about women's involvement in the selection of CDC members, the prioritization of development projects, and NSP meetings, there is agreement that the women were not involved in managing and implementing the subprojects (HHS and KIIs 2010). One community member stated bluntly that, "In implementation of projects, women are not directly involved" (KII: Com37, May 31, 2011). This was reiterated as being true for almost all of the communities in the district by the RRD District Representative for the NSP (KII: EX07, August 4, 2009).

The community members had very little information to share with me regarding how the program was managed, mostly because they did not know. The CDC executive members clarified that they were in charge of the procurement of all necessary supplies for the three subprojects and were responsible for contracting with the companies involved in the construction of the wells and the community center (FGD06, February 3, 2010).

They were also in charge of the monetary aspects of the program, as the NSP block grant goes directly into the bank account established by the CDC members (KII: FP08, July 15, 2009). In order to help ensure that the CDC members do not take advantage of this arrangement, multiple CDC members are required to sign off on any withdrawals from the community account (KII: Com33, September 26, 2011). Despite this, there were concerns raised in Shah Raheem that the CDC members were stealing some of the NSP block grant funds for themselves (e.g., KII: Com25, March 4, 2010). This was not only a problem faced in Shah Raheem, as was noted by a former NSP FP staff member of Khulm, who said that he had seen multiple communities where CDC members had taken some cash for themselves during the procurement process (KII: EX14, January 16, 2010). He noted that while sometimes this was done for the personal gain of the CDC member, sometimes it was also done in order to give the community contribution back to the people.

The community is also responsible for the upkeep and maintenance of their subprojects, something they were expected to have planned for in their NSP block grant subproject proposal. In the three proposals written by the community of Shah Raheem such planning was included but not always followed.

Instead of creating their own plan or collaborating with the FP to create a plan, one male CDC member from Shah Raheem said that the FP provided them with a plan for the maintenance and operation of the shallow well in Chahar Sukh, and that the community members were following this plan (KII: Com37, May 31, 2011). Maintenance plans for the other two projects were not raised or known about by community members, even the CDC members from those villages. While it is positive to hear that at least a village has a plan, it is a bit disheartening to hear that the plan came from the FP and did not involve the community in its establishment, as the NSP is supposed to empower the people of the community to be able to draft and follow such plans themselves.

A key FP staff member said that he thought maintenance of the NSP subprojects was not very good (KII: FP10, July 21, 2009). He echoed Brick (2008: 22, 23), when he said that communities where subprojects demanded regular funding or upkeep from the community to continue to function, often did not maintain their subprojects very well, such as in the village of Shah Raheem. He found that having smaller groups responsible for the project, instead of the whole community, actually increased the level of maintenance because, as he saw it, "Large groups cannot even agree on one thing" (KII: FP10, July 21, 2009).

The CDC members told us that they, as the development *shura*, were responsible for the maintenance and protection of the NSP subprojects (e.g., KII: Com37, May 31, 2011). They said that the community also plays a role in this, albeit less directly. Some CDC members, however, implied that it was the job of the FP to maintain the projects and to monitor them (e.g., KII: Com31, March 9, 2010). One community member said that the people are very careful to preserve and maintain the projects because they address their needs (KII: Com38, August 20, 2011), however, this was not clearly in evidence in the villages of Shah Raheem or Yaqubai. As was noted earlier, while the village of Chahar Sukh established a repair fund and were prepared to contribute to the maintenance of their well, similar efforts had not been made in the villages of Shah Raheem and Yaqubai.

6.2.5 Sustainability of the NSP CDCs: Dependency on External Inputs

The next question that arises is: What is the sustainability of the NSP and the CDC? After the completion of their third project, Shah Raheem was to be handed over to the Balkh MRRD, as had been done with 66 other NSP communities in Khulm District by the start of 2010 (KII: FP06, January 14, 2010).[462] This handover was supposed to mark a transition from the FP supporting the communities' development efforts, including acting as a conduit for funding, to the MRRD line ministry in Balkh taking on these responsibilities. However, the FP in Khulm District found that they were still involved with their NSP communities following the handover, as RRD requested their assistance in providing data and organizing RRD monitoring trips to the communities (KII: FP05, January 14, 2010). I was also told that the communities did not fully understand the handover situation and continued to reach out to the Khulm FP when they needed money for new projects or when they encountered conflicts in their community (KII: FP06, January 14, 2010), a sign that there was a need for greater communication.

The CDC members said that they had been very active during the first three years of the NSP when the three different subprojects were being implemented (e.g., KII: Com33, September 26, 2011). The CDC Treasurer said that without subprojects to implement, the CDC was not really active: "We don't have any special activity or development project with money coming to our community account," implying that without this, the CDC was not able to function (KII: Com33, September 26, 2011). One CDC member expressed his belief that the role of the CDC is important and that they could do a good job in addressing poverty in the country if they were given more possibilities to implement projects (KII: Com37, May 31, 2011), but again he was looking for outside forces to provide him and his fellow CDC members with opportunities to serve the community, and did not put forward the idea of the CDC being able to accomplish things without external funding or support.

While the CDC members said they were confident that they had learned a great deal from participating in the NSP and that they were ready to take on development challenges and projects on their own

462 I attended the first handover ceremony on July 26, 2009.

(e.g., KII: Com33, September 26, 2011), I would tend to be much less optimistic. One FP staff said he thought the CDC of Shah Raheem would be unable to write development project proposals on their own, but that they could handle smaller projects that do not require technical advisors (KII: FP07, July 12, 2009), but even that seems hopeful to me, as the community appears to have taken on very little of the responsibility for the NSP subprojects in terms of writing the proposals, monitoring the work, etc.

Even though not operating on their own, there are a few ways in which the CDCs are currently being utilized. CDC representatives played an active role in the District Development Assembly that was conducted by the UNDP in Khulm in August 2009.[463] NGOs and others looking to implement development projects specifically for women approach the community through the women listed as CDC members (KII: EX14, January 16, 2010), and the same goes for men regarding gender non-specific development projects.

In an effort to build upon the NSP process, the Government of Japan decided to provide funding for a clustering project to bring together representatives from different communities and establish Cluster Community Development Councils (CCDCs) to implement development projects. Named the Inter-Communal Rural Development Project (IRDP), the clustering program was implemented in Khulm District starting in 2010 (KII: Com38, August 20, 2011) (UNHABITAT 2010).[464] It utilizes the CDCs that were established through the NSP to represent the development needs of their communities in cluster projects.[465]

---

463 Non-CDC attendees were even denied membership in the executive committee of the District Development Council (District Development Plan Observations, July 27, 2009).

464 The IRDP was implemented in three provinces in total: Balkh, Bamyan, and Kandahar. In the clustering project, each community was given 16,815 USD, independent of population (KII: Com38, August 20, 2011). Information learned during the three years of implementation of the IRDP was used to help inform a larger cluster project to be supported by the World Bank Japanese Social Development Fund (Murotani, Wakamatsu et al. 2010: 37).
This project was still ongoing as of July 2012 (personal communication, July 24, 2012).

465 Through this program Shah Raheem joined forces with seven other communities in the area to introduce development projects for the eight communities in the cluster. Three subprojects were implemented: (i) a deep well was excavated in one community, (ii) irrigation canals in six communities were repaired, and

Two representatives from each community were selected and worked on the clustering. It seems that having served on the CDC has helped the community representatives understand the process of the clustering and the role that they are to play. In the development projects, unfortunately, the participatory concept of the NSP does not seem to have transferred into the clustering process. Non-CDC members were unaware of the clustering project, which indicates that they were not included, informed, or consulted (KIIs 2010).

The Provincial Manager of the NSP for Balkh, reflecting on the cluster project, said that he thought, "The people are very happy about the program. They thought NSP was going to leave them, but now they believe they will be there" (KII: EX04, January 25, 2010). This fear of the NSP leaving indicates a desire for the NSP to continue its work within the community and it also highlights the idea that further support for the program could be necessary for the infrastructure established to survive.

The CDC members and the community generally agree that they would like the NSP to move forward, and that they would like the CDC to continue to play a role in future development projects (HHS and KIIs 2010). One CDC member focused on how he would like the CDC to be a manager that is inclusive, with the people participating and feeling responsible for their own development (KII: Com38, August 20, 2011). Time will tell how useful the NSP CDC structure is to the community and external development stakeholders such as national and international NGOs, international organizations, and the Afghan Government. Without external support, however, I do not foresee the CDC of Shah Raheem surviving and working for the development of the community, which will be discussed more when the impacts of the NSP on Shah Raheem on empowerment are analyzed.

External actors appear anxious for the NSP, or at least the structures created within the program, to continue into the future, and see its continuance as a means of developing the country. This can be seen in efforts that have been made to transform the CDCs into official govern-

(iii) a communal bathing house was being constructed (KII: Com38, August 20, 2011). This process has been led by Coordination of Humanitarian Assistance (CHA), an Afghan NGO (KII: Com38, August 20, 2011). They operate as the facilitating partner, in much the same manner as CARE International did for the NSP in Khulm District.

ment bodies.[466] Brick (2008: v) predicts that there are multiple trajectories for the future of CDCs in individual communities. She believes the one that will be the most common, is the slow dissolution of the CDC: "After project completion, CDCs communities were mostly likely to be winding down their activities. While communities were interested in CDCs continuing their work, they did not organize this work themselves and held the view that they could do so only if they received more project funds." Considering the NSP process in Shah Raheem, in which participation has been weak, especially from women, in which the community depended upon the FP to write its proposal and still is looking to the FP to provide it with further funds despite having completed their block grant, and where such a small part of the community is engaged and knowledgeable about the program and the CDC that it created, I believe that the CDC in the community will be not sustainable. Evidence of its dissolution could already be seen in the inability of one of its members to name those who were supposed to work on the council with him (KII: Com15, March 10, 2010). Brick (2008: v) also found that NSP communities with natural, pre-existing borders were more likely to be sustained, another strike against Shah Raheem.

6.2.6 Shah Raheem's Internal Evaluation of the NSP: Positive Perceptions

From a localized perspective, the NSP was seen in a positive light in Shah Raheem. In general, the community members who were familiar with the NSP were enthusiastic about it and felt that it was a good program (HHS 2010).

Many of the household survey participants were able to provide information on the benefits they experienced personally or as a community because of the NSP, even though often they were crediting the NSP with providing projects or services that had actually been funded by other programs and organizations (HHS 2010). When asked to com-

466 The MRRD proposed that the CDCs be institutionalized into the formal government, possibly placing them in the role of the village councils required by the Constitution. A CDC by-law signed by President Karzai in 2006, was intended to provide the CDCs with formal legitimacy, indicating that the CDCs are a part of local governance, though not actually the local government (Brick 2008: 13-15).

ment on any drawbacks they had experienced, over 80% of the respondents said that there had been "none" (HHS 2010). However, some did speak about the fact that they, or their family members, had contributed to the program by providing labor on the subproject in their village (HHS: SR008, February 7, 2010, SR016f, February 21, 2010).

The NSP was evaluated by the people as having been positive for their community (HHS 2010). It was seen as helping them solve their development problems, be that by providing the wells, the community center, or through the provision of myriad other development projects that were credited to the program (HHS 2010). The NSP was also seen as bringing people together and creating strong connections between them by over half of the HHS respondents (e.g., SR004, January 27, 2010, CS005, February 21, 2010, YB008, February 28, 2010), though this applied to the men only. One community member said, "NSP has a good and strong relationship with this community" (KII: Com18, February 23, 2010). A former FP staff member said that he thinks that the NSP is the most successful development program he has ever seen in Khulm District, though he was careful to note that there are still many ways in which it could be improved (KII: EX14, January 16, 2010).[467]

The community members were active in differentiating the NSP from other development projects during the HHS and the KIIs when asked directly about the process of implementation. They said that the major difference stemmed from their involvement in the program (HHS 2010):

> *The main differences between NSP and other organizations are that NSP has people active in the implementation of a project. Additionally we selected the projects ourselves and according to our desires and under the supervision of our representatives, 10% of the work physically was done by our village. Other organizations did not listen to our desires and implemented their projects. The selection of these projects was not in our control. They did not even get permission from us for implementing their projects, which were not effective for our people and our village as a whole. (KII: Com35, March 16, 2010)*

467 This man said the NSP has been 70% successful, but not 100% or 90% (KII: EX14, January 16, 2010).

The NSP was also viewed positively because it was not seen as having increased internal community tensions to the same extent that other development projects and programs had (HHS 2010): "For example, Hand in Hand has not done any good for us. They brought conflict amongst our people and the people did not have a part to play in the implementation of this project. Their project was not effective for our people and our village" (KII: Com35, March 16, 2010). The Mayor of Tashqurghan felt that the longer time frame of the NSP, about three years, was noteworthy and made it better than the other development projects in the district (KII: EX10, January 18, 2010), though from my perspective three years might not have been long enough.

Almost half of the HHS participants also felt that the NSP subprojects had been of a good quality, though it is interesting to note that only participants from Chahar Sukh and Yaqubai put this idea forward (HHS 2010). Perhaps this is because the Shah Raheem subproject is so rarely used, the quality of its construction is not as prevalent in the minds of the villagers living there. Additionally, despite the problems with the community center roof, almost half of the Yaqubai HHS respondents also talked about the high level of quality of the NSP projects (HHS 2010), which leads me to question whether they felt motivated to paint a more positive image than is reflected on the ground.

In the end, the community members were happy to have new pieces of infrastructure in their villages and therefore expressed satisfaction with the program, but what its implementation meant in terms of the four areas of investigation will be addressed in the next chapter.

6.3 Extracting Meaning: Evaluating the Implementation of the NSP in Shah Raheem

After becoming intimately acquainted with the story of the implementation of the NSP in Shah Raheem from multiple perspectives, it becomes evident that the process was not clear cut and that it was not strictly in alignment with the Operational Manual. A few of the most important ways in which the implementation process veered from expectations have been highlighted, including the exclusion of women from the process. It appears as though women were not allowed to participate in the voting for CDC members or community priorities, and that women were not involved in the management and oversight of the three subprojects implemented in Shah Raheem. In addition to this, there is evidence that this lack of participation contributed to a dearth of knowledge about the NSP in general amongst women in the community.

This lack of information about, and understanding of, the mechanisms of the NSP and NSP subprojects implemented in the community does not only apply to the women of Shah Raheem, however. Many men, even those that were serving as CDC members, also proved to be unclear about which projects had stemmed from the NSP and which had come through other programs and funders. This confusion points to what should be a deep concern for the NSP; that the process of the NSP does not appear to be providing the community members with information and knowledge about the program, despite its intention to utilize a participatory approach. The fact that many community members attributed other development projects to the NSP also emphasizes that the NSP has not differentiated itself greatly from other projects conducted in the area. The NSP subprojects are not remembered separately for the process by which they were delivered to the community, the democratic nature of the process, or any kind of empowerment achieved. Whether this is because of the nature of the program or because the community of Shah Raheem was not as engaged as they were made out to be through the NSP Documents is not entirely clear, but it is evident that, in general, the population of Shah Raheem was not as involved as planned for in the NSP framework.

General male community member participation in the program was mainly limited to voting for CDC members, voting for the development project prioritization, and providing labor for the different subprojects. The community development plan and the subproject proposals were written by the facilitating partner, with little to no corroboration with the CDC members or the people of Shah Raheem. The management and oversight of the subprojects appears to have been relatively limited to the four executive officers of the CDC, with little evidence of them having consulted with and informed other community members of the progress along the way.

All in all, the implementation of the NSP in Shah Raheem appears to have followed the mandated steps, but perhaps without the depth of participation, knowledge and skills transfer, and empowerment that is generally assumed to have happened by policy makers and development stakeholders who proclaim the program to be something akin to a saving grace for Afghanistan. For such individuals there is faith that, "The celebrated National Solidarity Program has provided block grants to Community Development Councils (CDC), on which democratically elected locals identify, manage, and implement development projects" (Lee, B. 2011). As has been shown, different aspects of this faith can be called into question in terms of the NSP in the community of Shah Raheem. As will be shown, the implementation process has strongly framed the impacts of the program on the four areas of investigation discussed in the analytical framework.

7 The Impact of the NSP on the Four Areas of Investigation in Shah Raheem

The implementation of the NSP in Shah Raheem and the response it has garnered from the community members living there and other stakeholders, combined with the historical understanding established regarding the district and the four areas of investigation, provide interesting insights into how this participatory program has impacted on these areas of investigation. As will be shown, the program has impacted on the personal security, economic security, social security, and empowerment in the community of Shah Raheem, though in much smaller or more nuanced ways than is usually reported for communities in larger research studies. In my eyes, this does not invalidate the program, but it does lead me to conclude that expectations of the NSP and other participatory development programs should be reconsidered and reframed to base them more solidly in the lived experience of the beneficiaries.

7.1 Personal Security: A Lack of Strong Positive or Negative Impacts

The people of the NSP community of Shah Raheem noted that they feel safer now than they have felt since the Saur Revolution in 1978; however, they did not attribute this increased sense of personal security to the NSP or the participatory approach utilized for its implementation in their community. Decreased physical fighting, decreased threats, and the fact that their children, both male and female, can go to school are seen as signs of increased personal security (HHS 2010). However, these changes mostly arrived before the introduction of the NSP to their community in 2006.

Evidence from the household survey indicates that personal security, until recently, had been a concern for the Afghans of the community of Shah Raheem for multiple decades. Over three quarters of those interviewed said that the personal security situation under the Soviets was bad.[468] Personal security saw a slight improvement when the Mu-

468 The Soviet troops embedded themselves in the district capital, while the Mu-

jahideen took power, with only 70% of the respondents feeling that the personal security situation was bad (HHS 2010). Yet, with the take-over of the Afghan government by the Taliban, the perception of personal security worsened to the point where nine out of ten people interviewed expressed that the security situation was "bad" or "very bad" (HHS 2010).

Since the fall of the Taliban, many people felt that most threats have been eliminated, as government led restrictions had become lesser in nature and the district had settled into relative peace. People no longer felt as though they were living in a battleground (HHS 2010). This has resulted in a drastic positive shift in perceptions of personal security in the post-Taliban period, where for the first time in the household survey, the majority of the people reported feeling positive about personal security (HHS 2010). This sense of security has only recently been called into question by the assassination of a German KfW employee working on a road project in the district center in December 2010 (The Local 2010) and the killing of four individuals labeled as Taliban on the newly constructed road that leads to Kunduz in May of 2012 (Pak Tribune 2012), noted earlier. The general feedback continues to be that Khulm is a relatively safe district, despite these most recent incidents (personal communication, June 24, 2012), but this could change if more incidents of violence take place.

The community members provided us with examples of what they perceived to have created a lack of personal security for them over the different time periods in the household survey. In general, threats to themselves or their families were a priority (HHS 2010). Almost all of those interviewed said that they faced no threats and that this made them feel secure. "Right now the security situation is really good. There are no security problems in our village and nobody threatens me or my family" (HHS: YB003, February 23, 2010). Previous time periods were seen as not good for personal security, in part because the children of the area could not attend school and people could not go to work. The opposite was talked about for the post-Taliban period: more than half of the households said that their children were able to go to school

jahideen groups who rose up to fight the Soviet troops situated themselves just on the edge of the community, placing the community members directly in between the two sides of the fighting, as was reviewed in the section on personal security under the PDPA.

(HHS 2010). Many households also talked about how people were able to work freely without facing any threats: "Fortunately the security situation is better and our boys and girls are able to attend school and we are also able to work freely without fear or threats" (HHS: SR002, January 24, 2010).

Despite some conjecture amongst participatory development scholars that such an approach would increase social capital, thereby increasing personal or physical security (Owen 2006: 9; Strand, Toje et al. 2003: 12), this was not evaluated to be the case in Shah Raheem (HHS 2010).

Physical fighting between the government and the opposition of the specific time period also played a role in how the people of the community of Shah Raheem perceived their personal security, and this was strongly linked to how both sides of the conflict treated the community members. Forced soldiering had a strongly negative impact on peoples' perceptions of their personal security, as did being forced to pay money to the government or anti-government forces in order to avoid having to actively serve as a soldier (cf. HHS 2010). The current Afghan Government has no mandatory military service requirement of its citizens and the taxation system is seen as occurring in alignment with government laws and not so strongly aligned with the desires of local commanders (e.g., FGD11, July 10, 2011).

In contrast to Shah Raheem, in some NSP communities it was reported that the program had increased the level of personal security. For example, in one district in Ghazni, where the Taliban are seen as very powerful, many newly constructed buildings built by the government in the last decade had been destroyed by the Taliban or their supporters (Glad 2009: 48-9). However, the NSP subprojects built in this district had not been destroyed, which one FP staff member attributed to the community being more invested in the survival of the subprojects because they contributed to them (KII: FP07, July 12, 2009). He believes that this has motivated the communities to actively protect their projects and to negotiate with the Taliban in order to ensure that the NSP projects remain safe. His final words on the matter were, "Communities with NSP are safer than communities without NSP" (KII: FP07, July 12, 2009). Similar sentiments were not expressed in Shah Raheem, with community members attributing the increased security to other entities and to changes in the political environment, as was discussed in the section on personal security in the post-Taliban time period. There were no reported changes in the number of threats or violent incidents occurring since the implementation of the NSP from community members. This was similarly found by Beath et al. (2011: 20), who stated: "The program has no significant effect on the number of security incidents in and around villages as reported by villagers."[469]

Despite the lack of significant evidence of the NSP having a positive impact on the personal security of the community of Shah Raheem, so far, there is also no evidence that the NSP has negatively impacted on the personal security of the community. Though the NSP does not intend to decrease personal security in the communities in which it is implemented, it can sometimes produce unintended effects, such as inspiring anti-government elements (AGE) to attack communities for their involvement with a government-led program. This is not often fully investigated or discussed, as the NSP is mostly viewed as a force of good that should not be openly maligned or criticized.[470] Nevertheless, there is such a potential risk.[471]

A limited number of NSP communities in other provinces have faced threats and violence for their involvement in the NSP. These stories were only relayed via one-on-one conversations with staff of facilitating partners due to sensitivity about speaking of a connection between the NSP and decreased physical security. The FP Provincial Manager of the NSP for Balkh talked about NSP communities having received *shabnahme (šabnāme)*, or night letters, in which the Taliban or other AGE threatened communities, individual members of communities, or FP staff, with the intention of scaring them into stopping their

469 Beath et al. (2011: 20-21) did, however, find that there was a long term positive impact on personal security in NSP communities and that even in the short term they were able to measure a difference in the number of security incidents recorded by ISAF (International Security Assistance Force) in and around treatment villages, but not in community reported incidents.

470 Personal communication has related that a research report that looked at the NSP was not published due to its critical findings of the program (personal communication, July 2012). This, however, could not be confirmed. Even if it was not published for a different reason, the rumor is indicative of a negative attitude towards critical assessments of the program.

471 A number of assassinations of NSP officials have been reported in newspapers, however, whether their involvement with the NSP contributed to their death or not, was not made clear in the articles (Frontier Post 2012)

work with the government and, therefore, the NSP. In some communities where these warnings were ignored, CDC members were killed, such as in Chahar Bolak District of Balkh Province (KII: EX04, July 29, 2009).[472] Such incidents are rarely if ever discussed in reports on the NSP, and are quickly brushed under the rug when they occur. Further research into increased insecurity related to development programs should be conducted, along with research into increased security in order to identify which factors influence the impact a participatory development program has on personal security.[473]

The people of Shah Raheem did not express any fear of anti-government elements, or fear of retribution for having participated in a government supported program, such as the NSP. They received no threatening night letters, nor had any of their projects been physically threatened. However, it should be highlighted that these types of threats were not experienced in the community of Shah Raheem prior to the implementation of the NSP in Khulm District due to the district's relative peacefulness, so it is difficult to know how this could have differed in a less secure environment, however, Beath et al. (2011: 20-21) did not find increased physical security in the two "insecure" provinces in which they conducted their research.

I raise the possibility of negative impacts on personal security to highlight that even though the expectation of the NSP, and other participatory development programs, is that it will improve personal security, such a positive impact is not predetermined. The potential of the NSP to impact on personal security can be negative or positive, and in some communities it could very well be a combination of the two. One FP staff member stated, however, that, "Khulm security is the same pre- and post-NSP" (KII: FP10, July 21, 2009).

---

472 Affolter (2009: 323) noted the killing of NSP staff in other areas as a sign of the program's success, as the Taliban target those programs that show the Government of Afghanistan in a positive light.

473 The concept that aid fuels insurgencies or conflicts is found in the "greed" theory and "bargaining models" of conflict. In the greed theory, an increase in the amount of resources that can be contested, results in an increase in conflict risk and creates greater incentives for fighting. From a bargaining model perspective, conflict is also contingent upon economic material resources, but conflict only arises when negotiations to determine a peaceful division of such resources fails (cf. Beath, Fotini et al. 2011: 6-7).

It is possible that Khulm will begin to face such personal security challenges in the future, however, especially as the road to Kunduz allows for greater ease of access to the district from the province of Kunduz which is facing problems with AGE. The death of the German bank worker in 2010 and the killing of Taliban members in the district in 2012 both indicate that the relative security that Khulm has experienced since the fall of the Taliban could soon be weakening, and perhaps the structures created through the NSP will help the communities to combat any influx of personal insecurity that threatens them or their NSP projects. However, there is no indication from the community members themselves or the stakeholders interviewed that they believe this will be the case. Follow-up research in a couple of year's time could be very informative in this regard.

A different aspect of personal security that is addressed through the NSP is that by working with the engineers of the FP, the NSP could have a positive impact on the personal security of those who collect water in both the villages of Chahar Sukh and Shah Raheem. The establishment of wells should eliminate the need of the community members to gather their water from running streams and irrigation canals, reducing the possibility of people falling into moving water or experiencing other accidents on the stream banks. Two households said that it was physically difficult for their children to gather water before (HHS: SR008, February 8, 2010, CS002, February 16, 2010), though they did not say if they felt their children had been truly at risk of injury or accident, so it is difficult to say if this does or will have an impact on personal security, but it is a possibility.

However, it is also possible that situating the NSP wells on the main road could lead to greater insecurity for those gathering water, as it places them nearer to the road, increasing their chances of becoming involved in a road accident. Again, time will provide insights into whether such things have had an impact, either positive or negative, on the personal security of Shah Raheem.

Additionally, the use of FP engineers to ensure the safety of the construction of the wells and the community center, should mean that these pieces of infrastructure are safe for general use by the public, unlike some other development projects that have been poorly constructed and collapsed (Nawa 2006: 1, 8, 11-12). Though this was not mentioned by those who partook in this research, it can be posited that it could positively impact on the personal security of those living in the community of Shah Raheem. As will be explored below, the NSP in Shah Raheem does appear to have had a positive, albeit limited, impact on the social security in the community of Shah Raheem, and it is possible that this could help create greater personal security in the community in the future, should the need arise.

7.2 Economic Security: Minimal Positive Movements

Economic security has moved in coordination with both personal security and social security in recent history, yet over the previous four governent regimes it has had slightly less divergance than the other two types of security. Half of the households we spoke with told us that the economic situtaion they live with today is good with an additional number saying that the economic security situation was an improvement upon what they had under the Taliban, but without necessarily labeling it as good (HHS 2010). About one third of the survey population told us that they still felt insecure economically and that they would evaluate the current economic security situation to be bad (HHS 2010). While overall there has been an increase in the positive perception of the economic security situation as compared to remembrances of the Taliban time period, it is still slightly lower than the population's perception of economic security under the Mujahideen, which had the greatest number of people telling us that they felt economically secure (HHS 2010). Though much of this shift in perception of economic security was connected to changes in the political situation in Afghanistan overall, the NSP was also seen as having contributed in some ways to increasing the economic security in the community of Shah Raheem by community members and other external stakeholders (HHS 2010).

The main indicators of economic security extracted from the household survey are quite basic. People discussed whether there were jobs available, what their incomes were, and what types of employment they had (HHS 2010). Additionally, the opportunity for community members to sell their production to the market of Tashqurghan and to be able to purchase food and other items there were seen as indicators of good economic security (HHS 2010). To these I would also add health factors, though it was not regularly raised by the HHS respondents or the other stakeholders interviewed.

In the community of Shah Raheem, the purpose of the three subprojects chosen did not solely focus on directly improving the economic security of the community members in the same way that other development projects or programs have, such as the hen project and the microfinance program. Therefore, it is not terribly surprising that the program was not seen as having strongly impacted on the economic situation in the community. That being said, there are a few ways in which the NSP has left a small footprint in the economic lives of the community and its population.

The first way in which the NSP impacted on the economic security of Shah Raheem was the influx of development funding into the community in the form of the NSP block grant. The program provided the whole community with 1,120,000 Afghani (about 22,400 USD) to implement its prioritized development subprojects (NSP Documents). At the same time, the program demanded that 10% of the cost of the subprojects be contributed by the community, which means that those who donated money or labor to the subprojects, sacrificed some of their economic security through either a direct debit of cash or through the loss of five to ten working days. During the HHS and the KIIs, this economic loss was rarely if ever mentioned, and usually when it was mentioned, it was in reference to allowing widows and other community members without monetary and physical resources to benefit from the subprojects without having to directly contribute themselves (e.g., KII: Com10, February 7, 2010).

The second way in which the NSP has impacted on the community is less direct and much less universal. Almost three quarters of all of the household respondents said that NSP brought clean drinking water to their family and their community (HHS 2010). This was seen as being closely associated with having fewer sicknesses. Community members said that the only source of water they had previously was the local stream and irrigation canal, and credited the NSP with providing them with clean water to consume (e.g., HHS: January 27, 2010, SR008, February 7, 2010, SR012, February 10, 2010, CS002, February 16, 2010, SR003). Twelve different households mentioned that having clean drinking water has meant that fewer people have been getting sick with cholera and typhoid, or not getting sick as often as they used to (HHS 2010).

This perception of increased health needs to be mitigated, however, with the understanding that only 15 of the 112 families in the NSP community of Shah Raheem were observed to be regularly obtaining their water from a NSP subproject and therefore taking advantage of this cleaner source of drinking water, those who live in the village of Chahar Sukh with the shallow well and hand pump. In the village of Shah Raheem, for around ten months of the year, they were still obtaining their water in the way they did previous to the implementation of their NSP deep well, via the local streams, which are no cleaner than they used to be. Finally, in Yaqubai there is a different source of clean water, which stems from the efforts of a Danish Committee for Aid to Afghan Refugees (DACAAR) water project, not from the NSP, though many of the community members told me that they believed that it had been implemented by the NSP (HHS 2010).

While not often immediately associated with increased economic security, the cost of taking care of someone who is sick, be it in monetary or temporal terms, is not insignificant. For adults, the lower incidence of illnesses in the family can mean a greater number of days working and less time taking care of people. Additionally, money would be saved on transportation to the hospital, fees to see the doctor, and the cost of medicine.[474] Often, families would take their children to private clinics for dehydration related to water borne illnesses. Each visit was estimated to cost between 1,000 and 2,000 Afghani (about 20 to 40 USD) (KII: Com10, February 7, 2010). One community member estimated that his family had to go to this type of clinic two to three times per year, costing them anywhere from 2,000 to 6,000 Afghani (about 40 to 120 USD) annually (KII: Com10, February 7, 2010). He implied that these costs were now being saved because of the well. A woman from Chahar Sukh also indicated that the NSP had decreased the number of times her children would get sick in a year from two to three times a year, to rarely if at all (KII: Com14f, February 16, 2010). While Chahar Sukh res-

474 Though the hospital in Khulm District is meant to provide medical services and medicines to patients for free, I was often told that this was rare, and that normally there were a number of fees associated with visiting the district hospital that were "unofficial" in nature as mentioned in the section on development in the post-Taliban time period.

idents benefitted from access to cleaner water because of the NSP year-round, the people of Shah Raheem Village had access to clean water through their NSP well only two to three months a year because they did not commit their economic resources to running the well in the other months. This seem to imply that for the people of Shah Raheem Village, the economic gains of having fewer sicknesses was not worth the cost of providing the necessary funds for the diesel generator.

Additionally, it was thought that the NSP had impacted positively on the economic security of the community because the two wells built were said to decrease the amount of time necessary for the daily task of gathering water (HHS 2010), dropping from around 20-45 minutes to around 5-10 minutes (HHS: SR014, YB006, February 27, 2010, YB007 and YB009, February 28, 2010). However, gathering water is mostly the domain of children and, because they are not reportedly involved in contributing to household livelihoods, these few minutes daily are unlikely to have a large economic impact.[475]

The other villages also said that this was an advantage for them, but we know that in the village of Shah Raheem water gathering is still being done in the same way as prior to the establishment of the NSP well for most of the year, and the water pumps in the village of Yaqubai can be attributed to another development program. One CDC member said he felt that the NSP had had a positive impact on economic security. He said that the NSP wells meant that people no longer had to walk and carry their water from such a distance, or pay others to bring water to them (KII: Com37, May 31, 2011). I was not able to find evidence that many, if any, households in the community had previously been paying to have their water delivered, however, indicating that any change in the economic situation due to an elimination of water delivery costs would be minimal at best.

While the vast majority of those who partook in the household survey indicated that they believe their family is better off economically this year as compared to last year (HHS 2010), there does not seem to be a strong connection between this change and participation in the

475 Only one family from the household survey indicated that one or more of their children contributed to the livelihood of the family (KII: YB006, February 27, 2010). Despite this testimony, I observed children working to contribute to their household's income, albeit mostly through handicrafts and other work within the household.

NSP. The three separate subprojects did not impact very strongly on the livelihoods of the households we spoke with and those interviewed were unable to establish a connection between their involvement with the NSP and an increase in economic security, due to gaining new skills in NSP trainings or through their increased social connections. The program also did not contribute to the main economic concern of the people in the post-Taliban time period: the lack of job opportunities available (e.g., HHS: SR001, January 20, 2010, and SR004, January 27, 2010), despite some community members and district officials falsely giving credit to the NSP for providing jobs.

7.3 Social Security: Small Positive Shifts for Some

The HHS, FGDs, and KIIs indicate that most community members in Shah Raheem are very positive about the social security in Shah Raheem today, and in the district in general (HHS 2010). One community member supported this by saying, "The social situation is better. People have strong connections with one another in our community" (HHS: February 7, 2010). The NSP is seen as contributing positively to this, albeit it in a more limited way than it possibly could have (FGDs, HHS, and KIIs 2010).

Social security in the community of Shah Raheem was negatively impacted by the Communist Government, the PDPA. It saw a slight uplift with the fall of the PDPA and take-over of the Mujahideen and was then followed by a decrease when the Taliban came into power (HHS 2010). The Taliban period is remembered as the lowest point for social security. After the Taliban was defeated, a stark rise in the sense of social security was discussed, with people indicating that on average they felt more social security in the post-Taliban period than under any previous regime, though the rise was not as drastic as was seen for personal security (HHS 2010). Eighty percent of those we spoke with in the household survey openly stated that the social security situation post-Taliban was good, with not one household conveying that they thought it was bad (HHS 2010).

Social security is evaluated by the community members mainly on how much support they provide to one another in times of need (HHS

2010). Support can come in both material and emotional terms. Despite attempts to get the community members to talk about what coping mechanisms households had for facing unexpected problems, such as a main breadwinner losing their job or the failure of a harvest due to freezing weather after the winter defrost, the community members remained committed to only discussing what happens when someone falls sick and how families deal with "sadness and happiness" events. These events include funerals, mourning ceremonies, and weddings, though they can also include the return of family members who moved away from Afghanistan, the birth of a child, etc.

When questioning community members about social security, one of the main things discussed was whether the people had strong relationships with one another, also described as strong connections (HHS 2010). I wanted to know who people would turn to in their hour of need, be it neighbors, family members, or the government. What I found is that very few people see the government as playing any role in providing social security to the people, and that there was an emphasis on the connections between people living in a village (HHS 2010). It should be noted that I observed many fault lines in intra- and inter-village relations, but I also witnessed many strong relationships. Most often these strong relationships were built upon kinship ties, though alternately, kinship ties do not guarantee a supportive relationship that extends to social support or security.

7.3.1 Bonding Social Capital: Increased Connections for the Males of Shah Raheem

The NSP was seen by many community members familiar with the program as a means of making intra-community social connections, known as bridging social capital, stronger within the community (HHS 2010). The social culture we observed in the NSP community of Shah Raheem was tightly knit, in terms of the social rules that everyone had to abide by, but it was less cohesive or warm in nature than survivalist. The NSP was seen for at least the male population as having contributed positively to their social security.

More than one half of the males who partook in the HHS (2010) made it clear that they appreciated the NSP and saw it as having brought about stronger connections between the people of their community: "It has made strong connections amongst our people" (e.g., HHS: CS003, February 17, 2010). A few even said that they now felt like they were all family members (e.g., HHS: CS004 and CS005, February 20, 2010, YB003, YB010, February 28, 2010, SR014, February 14, 2010). This was mainly seen as occurring due to the NSP meetings that brought the men of the three villages together more often and more regularly than they had been meeting before.

The participants in one FGD group said that they had previously had social relations with people from the three different villages, and that those relationships were considered to have been good, as, "nobody lives alone in the world," but that the NSP meetings were seen as helping to strengthen those ties even further (FGD01, January 19, 2010). Holding regular, monthly meetings to discuss the NSP and development issues gave the men the chance to get to know one another and build new levels of trust (HHS 2010, KII: Com31, March 9, 2010). The male community members I spoke with said they felt closer to others within the NSP community, noting that they had not had such close interactions with male individuals from the other villages before the NSP (e.g., HHS: SR001, January 20, 2010, CS002, February 16, 2010, CS006, February 21, 2010).

In the past, each village had its own meetings of male elders to resolve problems as they arose, separate from one another (KII: Com33, March 10, 2010, Com38, August 20, 2011). These meetings were exclusive in nature, with only male community elders and other men with power being included, leaving out younger and less powerful men, and all women. Only rarely did a problem arise that involved each of the three villages and, therefore, resulted in a gathering of representatives from each of them (KII: Com25, March 4, 2010). This changed under the NSP, with regular or semi-regular meetings being held that brought together community representatives from the villages of Chahar Sukh, Shah Raheem, and Yaqubai.[476] During one FGD with males of the community, it became apparent that they were even using the term NSP to mean, "relationships between the villages." They said that they had their own, unofficial NSP program before the formal NSP program, and

476 As shown in the last chapter, the regularity and attendance of these meetings is not entirely clear, but meetings were held.

went on to explain that they had social relations with one another (FGD01, January 19, 2010).

One community member said that while they had often attended one another's sadness and happiness ceremonies in the past, now they are much happier to do so because they know each other better (KII: Com15, February 17, 2010). He also said that they are more willing to bring food and other forms of assistance to help families during their times of happiness and sadness. This was reiterated by the *kalāntar* of Yaqubai who said, "We are close to each other," explaining that before they would only see people from the other villages at weddings and funerals, but "now we see each other to solve problems" (KII: Com18, February 23, 2010).

The lack of genuine participation of women in the NSP process that was outlined previously, results in women from Shah Raheem not feeling the same bonding social gains that were experienced by the male members of the community. The lack of meetings, or limited nature of such meetings, depending on which version of the past you believe, has not provided an environment in which all the women from the community were able to share their thoughts with one another and get to know each other as individuals, creating a possibility for stronger personal connections to develop. The women said that they felt no difference in their social relations since the implementation of the NSP (HHS and KIIs 2010).

Along with the NSP meetings, the community center subproject built in Yaqubai Village was seen by HHS participants from all three villages as a means of increasing social security (HHS 2010). By establishing a communal location where people can hold their happiness and sadness ceremonies, the community center was expected to support the social security that exists in the community and possibly expand it. The community center has room for people to invite more individuals to participate in their ceremonies than their houses normally allow for. It also provides a place for community members to hold meetings, such as NSP meetings or meetings to address development needs: "Actually, when something happens, for example a conflict or a discussion about development projects, the people come up with solutions to their problems in the community center" (HHS: YB009, February 28, 2010).

This was a statement of what was expected to happen in the future, not what had already happened, as the community center had only hosted two events when this research was conducted: a funeral and my introduction to my research. Despite its limited use thus far, the community center is seen as a positive factor that can bring greater social security to those who are included in its usage (HHS: YB006, February 27, 2010, YB009 and YB010, February 28, 2010).

Unfortunately, this does not extend to the women of the community, who are not included in the ceremonies or the meetings that take place in the community center. Social norms mean that women are restricted from attending because the community center lacks both a protection wall around the building and a toilet (HHS and KIIs 2010). These factors were not taken into consideration by those who wrote the subproject proposal or by the FP staff and community members who participated in making the plans for the construction of the community center. It is possible that had the women been included in the planning and implementation process that these deficiencies would have been addressed, and both genders could have benefitted from the community center, but it is also possible it would have changed nothing.

When asked about internal social safety nets and the provision of support to households or individuals who needed outside assistance, the picture painted was one of a community where households help each other by attending peoples' ceremonies and supporting those ceremonies through lending *toshaks (tošak)*, pots, and other household items.[477] When asked specifically if they would provide economic support to their neighbors if it was needed, there was a mixed response in the household survey. Some households said that they would help people economically, while others said that they were only able to support their fellow community members emotionally (HHS 2010).

Despite these positive testimonies of mutual support, I witnessed a great deal of distrust in the community, and the spreading of rumors, indicating that the bonding social capital in the community is not as strong as it could be. This distrust also emerged in connection with the NSP, as it has with other development projects and programs implemented in the area. It is possible that in time the stronger social ties

477 A *toshak* is a pillow which is commonly used to sit upon the floor in Afghanistan.

that have been developed through the NSP will also increase these levels of trust, but this has not been seen in the six years since the program was first implemented in the community of Shah Raheem.

This mistrust caused some problems in Shah Raheem, as those who were involved as CDC members were seen as corrupt by a few community members (e.g., KII: Com33, September 26, 2011), which led to tensions. The CDC members were suspected of benefitting personally from their involvement with the NSP through a secret payment from the FP (e.g., KII: Com03f, February 10, 2010) or through skimming funds directly from the NSP block grant (e.g., KII: Com25, March 4, 2010, EX14, January16, 2010). These kinds of suspicions were also found in NSP communities studied in Parwan (Echavez 2010: 17), and should not be viewed as isolated cases.

This distrust was also witnessed outside the confines of the NSP, with community members reporting that people were unfairly benefitting from different development projects, or ensuring that the benefits of such projects went first and foremost to their family members. Even my research raised such accusations, with community members spreading false rumors that I was paying people to speak with me.[478] Other sources of mistrust and conflict in the community included the distribution of aid and development projects discussed previously. Community members accused each other and government officials of not giving them what they had been promised, or taking what they felt belonged to the people of the community (e.g., KII: Com03f, February 10, 2010, Com33, September 26, 2011). This was not reported to have increased or decreased with the implementation of the NSP (HHS and KIIs 2010).

This lack of trust in other community members and a regular sense that someone was taking more than their fair share are signs of a tenuous social setting. Such unease and the sense of being cheated was reported by the women of the NSP community more regularly than amongst the men (HHS 2010), which could stem from their relative so-

478 No payments were offered or made to research participants. Prior to engaging with the people of the community, my research, its purpose, and the lack of funding attached to it were reviewed with the elders of the community. This was also reviewed prior to each KII, HHS, and FGD conducted during the research. Everyone was also clearly assured that they were free to participate or not in the research, without prejudice, making all participation completely voluntary.

cial isolation from non-household members and resultant lack of social connections and building of trust.

There is a sense of competition at times that can be seen in the NSP community, where people seem more concerned about getting as much as they can for themselves than ensuring that people in the community get an equal share. This breeds distrust and resentment, which lowers positive social capital.

It will be interesting to see if the social changes that were observed and reported in the male population of the community of Shah Raheem will last and if they will have a spillover impact on the relationships between the women in the community, or if the means of dealing with development and other issues will return to being even more exclusionary in manner, reverting back to the previous informal infrastructure for dealing with community-level decision-making.

7.3.2 Bridging Social Capital: Limited Changes in Relationships with Outsiders

The NSP is also invested in helping to encourage or establish bridging social capital between NSP communities and the Afghan Government, and other development stakeholders. As is common in many parts of Afghanistan, the villagers have had few connections with the local and central level government in the historical timeframe of this research. In the post-Taliban time period there has been an improvement in these relationships, though there is a long way to go before I would feel comfortable evaluating the relationship as one of mutual trust. A number of studies on the NSP have put forward evidence showing that the program has been improving relations between rural communities and the government in general (Barakat 2006: 38; Beath, Christia et al. 2010: 18; Enikolopov 2011: 1-2; Zakhilawal and Thomas 2008: 23). While I was unable to observe strong evidence of the NSP increasing these trust levels in the community of Shah Raheem in terms of actions and activities, community members and Afghan government officials both indicated that they felt the program had improved their relationship with one another (HHS and KIIs, 2009 and 2010). Therefore, even though concrete evidence was not easy to find, it does appear as though the NSP has, at the very least, had an impact on perceptions of the relationship that exists.

Many community members who participated in the HHS and the KIIs said that they thought that participating in the NSP had improved their relationship with the local government (HHS and KIIs 2010). One man even said that he feels like he is a member of the government because of the NSP (HHS: SR008, February 7, 2010).

While one might assume that this improvement would have stemmed from the fact that the NSP is an Afghan Government program, funded through the Ministry of Rural Rehabilitation and Development, the government funding of the NSP was never mentioned when community members were asked to provide details of why they believed the relationship had improved (HHS 2010). When they were asked who funded the NSP, most community members answered with the name of the facilitating partner NGO, CARE International (HHS 2010). Very few people mentioned the government when it came to funding, though a few did understand that the NSP was connected to the MRRD (HHS 2010). Barakat (2006: 171) and Boesen (2004: 22) had similar findings in their research, which showed that the community members in different areas were not fully aware of the government's support for the program. It appears that this was a missed opportunity for the NSP, in terms of garnering goodwill for the Afghan Government, but one that perhaps could be addressed by the program's facilitating partners or the MRRD. If the community members had more accurate and detailed information about the NSP, or had retained this information if it was provided to them, the impression or perception of the Afghan Government could have been better in terms of its ability to provide development opportunities.

When asked who served their development needs the best, more community members mentioned NGOs than any level of government or other development stakeholders (HHS 2010). The NGOs were seen as providing them with the largest number of development projects.[479]

479 Some of the development projects mentioned included: providing clean drinking water, fixing the main road that runs through each of the villages, distributing chickens to households, and implementing a microfinance project. Secondarily, NGOs were also credited with having provided trainings in agriculture, supported literacy courses, built the NSP community center, established the CDC, and provided wheat seeds and fertilizer. At the same time al-

The government, however, was the second most mentioned, indicating that the community is aware of some of the development efforts the government has made (HHS 2010). The government was cited positively for its promise to bring electricity to the community of Shah Raheem and the agricultural trainings it was seen as providing (e.g., KII Com34 and Com35 March 16, 2010).[480] The government was also credited with helping maintain the high level of personal security or physical security in the district, as discussed previously in the section on personal security in the post-Taliban time period (HHS: SR003, January 27, 2010, SR005, February 2, 2010, SR014, February 14, 2010, YB006, February 26, 2010, YB007, February 27, 2010).

With all that being said, however, one of the top eight benefits of the NSP, as listed by community members, was that the program had brought the community and the government closer (HHS 2010), with the district government being the focus of these comments. However, community members were only able to remember two or three interactions with their local district government in the previous three years (HHS 2010), indicating that any relationship that did exist was not based upon regular communications or interactions. One CDC member said that he thought the relationship between the community and the government was good, citing as evidence the fact that the governor attended the commencement ceremony for the deep well in the village of Shah Raheem (KII: Com15, March 10, 2010). As Brown (2012: 6) points out, these "one time endeavors" are not usually enough to establish lasting connections between government offices and the people. Another community member remembered the former District Governor providing the community with 200 empty bags to fill with sand when the area was facing the threat of serious flooding (KII: Com05, January 26, 2010), and saw this as proof of a strong relationship. Unfortunately,

most one third of the households who participated in the HHS said that they thought that despite these efforts the NGOs continued to not pay enough attention to them (cf. HHS 2010).

480 The Directorate of Agriculture did not mention any government sponsored agricultural trainings being conducted, though he has worked closely with different organizations that provide such training, such as HLP, and it is, therefore, possible that the government was being given credit for these trainings. The government contributed to bringing electricity to Tashqurghan, the district capital, in 2010 and the research community of Shah Raheem in May 2011 (personal communication, July 24, 2012).

this community member also related that the sand bags provided were not enough to stop the flooding, and the water took over agricultural lands in the community. The current District Governor only came into power after this flooding in 2009, and at the time of the second research fieldtrip (January – March 2010) the new District Governor had not yet visited the NSP community of Shah Raheem (KII: Com05, January 26, 2010).

The community was also interacting with the government in regards to their applications for land situated on the road to Hayratan in Khulm District. One community elder told us that the conduits to each village, the *kalāntar*, had conducted meetings with the District Governor to discuss the land distribution.[481]

Some villagers said that they now felt closer to the government (HHS: 2010, SR001, January 20, 2010, SR008 and SR009, February 7, 2010, SR011, February 9, 2010, SR014, February 14, 2010, CS002, February 16) than prior to the NSP, but I was not able to actually confirm an increased number of interactions, increased cooperation, or an increased confidence in the local government's capacity to honestly serve its constituents. People are still resolving problems at the local level and only if they cannot, do they turn to the District Governor, and this appears to be more about his power and influence as an individual than about him being a part of the centralized government. The fact that the government has promised electricity and started delivering this to neighboring villages has started increasing belief in the government as functioning for the people.

In general, however, few if any of the people we spoke with felt the government was doing something for them (HHS 2010). Two community members individually said, "The government does not pay enough attention to us" (HHS: SR006, February 2, 2010, SR009, February 7, 2010). The people talk about being happy their children can go to school, but they do not label this as something that is provided by the government. The villagers see the government as having the responsi-

481 It is said that Hazaras asked the Governor of Balkh Province to provide them with the land on the way to Hayratan, but he told them he had reserved this land for the people of Khulm District. The community members were working on their paperwork to claim individual pieces of this government land during the time of this research. Each family expected to get 450-500 m^2 of land (cf. FGD06, February 3, 2010, KII: Com39, February 23, 2010).

bility of creating jobs, factories, etc., and, therefore, still feel disappointed by the government on the whole. The government, however, is not seen as the enemy, which is a positive shift from previous time periods (HHS 2010).

From an external perspective, the local government officials made it clear that they much prefer working with NSP communities than non-NSP communities. The District Governor said the NSP is beneficial because it groups different villages together with one representative. He reported that people come to him to solve their problems, with even more coming from NSP communities (cf. KII: EX06, August 3, 2009). He said, "NSP changes the face of the communities in which it is implemented. NSP communities are stronger than non-NSP communities because they have been empowered and there has been a consolidation of the people with the government and with the community" (KII: EX06, August 3, 2009). This positive perception is very interesting, though whether it extends from his personal experiences and a true belief or a desire to try to obtain more development support for his district was not clear.[482] He was unable to provide detailed examples of how the NSP and non-NSP communities differed, though whether this was because he did not have an answer or due to translation issues could not be determined.

The Mayor of Tashqurghan also believes that the communities that have the NSP are empowered and that it has created a stronger relationship between the local government and the communities of Khulm District (KII: EX10, January 18, 2010). He said that those who have implemented the NSP come to speak to him about their problems more often than they did prior to implementing the NSP and more often than non-NSP communities. One FP staff member said that the bridging social connections between the community and the government had improved, noting that the community members were now familiar with the different departments of the government (KII: FP08, February 7, 2010). While it is important for the people to understand how their government functions, this does not convince me that the relationship has undergone a transformation.

482 During his interview, the District Governor was very positive about the NSP, though at times his lack of knowledge about the program seemed to indicate that he was not very well informed, which makes me question the validity of his evaluation of the NSP.

Bridging social capital is not solely the domain of the community and the government, it is also appropriate to speak of such social capital in terms of NGOs, other development stakeholders, and communities.

Though negative thoughts about NGOs are likely to continue to exist due to past experiences and spoiled expectations, the NSP was seen as being one development project that actually did what it said it would do in the community of Shah Raheem (HHS 2010). While this might not have shifted community members' opinions drastically about NGOs and development projects in general, it has provided them with an example of an NGO following through on what has been discussed, and perhaps has helped ease distrust in NGOs a fraction. This will be beneficial in allowing the community to create connections with other NGOs who might be able to provide them with further development projects.

The establishment and existence of the CDC in Shah Raheem has also helped the community to gain recognition and given other development providers a means of communicating with the community. By this, I mean that many NGOs and development agencies are trying to work with and through the CDCs established by the NSP. One such example is the Japanese Government funded clustering pilot project called the Inter-Communal Rural Development Project (IRDP) discussed in the section on the sustainability of the NSP infrastructure. Though not a part of the original NSP implemented in Shah Raheem, the IRDP utilizes the NSP CDCs, bringing communities together to jointly plan and manage development projects for the area. This has enhanced the connections that exist between communities, and the community members even reported feeling able to ask people in these other communities for their cooperation and guidance as demanded by different development projects (e.g., KII: Com38, August 20, 2011). The RRD representative of the NSP in Khulm agreed with this, saying that even though he believed that the communities being clustered together in Khulm were already close socially, that the actual creation of a cluster had brought them even closer (KII: EX07, August 4, 2009).[483]

---

483 Prior to participating in this clustering program, a community must have completed its first NSP block grant (KII: EX04, January 25, 2010), making the NSP a requirement to benefit from this project.

Another example stems from the UNDP, which brought together representatives from each of the 91 communities in Khulm District to draft a District Development Plan (July 26-27, 2009). While not all representatives were CDC members, most were, and only those who were CDC members were eligible to be elected as one of the executive officers of the District Development Council.[484] The Provincial Manager of the NSP in Balkh Province for MRRD also mentioned that CDC members used to implement the WFP's Food for Work Program within the province (KII: EX04, July 29, 2009).

Being a part of the NSP, therefore, can be seen to have had an impact on outside or external actors, in that it makes them believe that a community is easier to work with, which could lead to those communities having increased access to development projects and programs.

7.4 Empowerment: Only Slightly Impacted

Empowerment of the people is generally an ultimate or underlying goal of all participatory development projects, and this is no less true of the National Solidarity Programme (Craig and Mayo 1995; Gloekler and Seeley 2006; Narayan 2005). By involving the community members in the planning and implementation of a development program, it is hoped, and often assumed, that this will provide them with the tools and experiences needed to become empowered. When the NSP experience in the community of Shah Raheem is reviewed, however, one can see very little changes in the different aspects that make up empowerment, as outlined in the chapter on the analytical framework. These include: access to information, inclusion and participation, ability to hold others to account, and the capacity to organize (PREM and World Bank 2002: 14). The NSP has had only limited impacts on the empowerment of the people of Shah Raheem, as will be shown below.

484 During the Khulm District Development Assembly, one man was elected by blind ballot to be one of the executive officers for the District Development Council. When it was found out that he was not a CDC member the next day, a second round of elections had to be held to elect new executive officers, effectively meaning that those communities with a CDC were able to gain more power within the District Development Council (District Development Plan Observations, July 27, 2009).

Despite a lack of evidence of large shifts in empowerment in the community, a number of community members, especially those with a personal or professional connection to the NSP, were vocal in their support for the program and readily stated that they thought it had empowered the community members, though they were rarely able to provide examples of this (HHS and KIIs 2010). "The community can accommodate any kind of project accordant to the proposal . . . right now," said one CDC member (KII: Com33, September 26, 2011). Another said,

> *Yes we the members of [the] development shura see ourselves [as] enable[ed] because by implementing three projects in our village we learnt issues like purchasing and procurement, record[ing] and registering of cash, dealing [with] accounts and all the performances . . . Thanks God, now we are able to implement our projects with good preparedness and feeling responsibil[e]." (KII: Com37, May 31, 2011)*

Community members who were not a part of the CDC were less committed to the idea that the NSP had encouraged or improved their empowerment (HHS and KIIs 2010).

7.4.1 Access to Information Improved for Some

During the fieldwork, the lack of information about the NSP within the community of Shah Raheem was regularly evident, despite all of the best intentions of the program to be transparent and inclusive as outlined in the NSP Operational Manual (NSP 2009a). Some people within the community had never heard of the NSP and did not know that their fellow community members had been in charge of its implementation, calling into question the participatory nature of the program. Others, however, were able to provide basic information about the program, though not always accurate information. Only a limited number of community members proved themselves to be well informed about the NSP, and almost all of them were a part of the executive committee of the CDC, whose job it was to run the program within the community (cf. FGDs, HHS, and KIIs 2010).

In general, the NSP appears to have created greater access to information for men in the community of Shah Raheem, though not for women. While the men were not fully accurate in their descriptions and their knowledge of the NSP, they had at least almost all heard of the program and were aware of its existence in the community (HHS 2010). The men saw the NSP as a program that was intended to bring people together and to provide the community with development projects (e.g., HHS: CS005, February 20, 2010, YB002, February 21, 2010, SR004, January 27, 2010, SR013 and SR014, February 14, 2010).

When women in the community were asked about the NSP, quite a few of them said that they had no information about the program or that they had never heard of it (e.g., HHS: SR015f and SR016f, February 21, 2010, KII: Com20f and Com22f, February 27, 2010, Com01f, March 6, 2010).[485] Women who participated in the research had no information about the funding of the NSP (e.g., KII: Com20f and Com23f, February 26, 2010), though neither did many of the men (e.g., KII: Com19, February 24, 2010). This lack of knowledge is very concerning as it means that some of the most basic aspects of the NSP were not generally absorbed and understood by community members, indicating their exclusion from attaining information.

In general, women in Khulm District and in Afghanistan have very little access to information due to their confined social interactions, their low level of literacy, and the belief that women do not need access to information as the men can make decisions for them. The NSP could have addressed this lack of access to information through holding inclusive meetings and trainings with the women of Shah Raheem. While the men and the FP facilitators said that this occurred; the testimony of the women indicates that they did not participate (HHS and KIIs 2010). Therefore, this opportunity for the building up of social capital was lost, as was the opportunity to provide the women with information about the program and other development issues. Unfortunately, most women in the KIIs told us that their husbands did not share information with them about the NSP, which means yet another opportunity for

485 Again, different names for the program were used, as was a description of the program, in case the people we were speaking with were familiar with the program, but not with the terminology we were using.

women to gain information was lost (e.g., KII: Com20f, February 26, 2010).[486]

Community members also told me that information about the progress, procurement, and spending of the subprojects was only shared during the monthly NSP meetings (e.g., HHS: SR003, January 27, 2010), which women were not allowed to attend. As the majority of the men interviewed indicated that they attended at least some NSP meetings (HHS 2010),[487] this could explain their greater knowledge in regards to how the project functioned. Women's exclusion from all planning and implementation meetings contributed to their lack of knowledge in general about the program and about the specific details of the implementation process and the NSP subprojects in the community (KII: Com27f, March 6, 2010).

The majority of the men in the community of Shah Raheem were able to identify the two wells and the community center as having been built through the NSP (HHS 2010). They were much less clear on whether those were the only subprojects implemented with NSP funding. Many women did not create a connection between the NSP and the three subprojects implemented in their community, only noting that the subproject in their village had been constructed by foreigners (e.g., HHS: SR015F and SR016F, February 21, 2010, KIIs: Com20f and Com23f, February 26, 2010). They often did not even know that the three subprojects were part of one development program, and were only able to speak about the development subproject implemented in their own village (e.g., FGD05f, January 27, 2010, FGD10f, March 6, 2010).

In general, those familiar with the NSP, both men and women, were often unable to differentiate NSP projects from other development projects or programs. When asked what differences the NSP had made in their community, almost all of the HHS respondents said that it had helped bring clean drinking water and the community center (HHS 2010).[488] Additionally, however, almost half of the respondents said that

486 A few women reported that their husbands did share what happened at meetings with them, but that they did not seek their input (e.g., FGD08f, February 9, 2010).

487 As was stated earlier, the household survey participants indicated that 56% of the men in their households participated in NSP meetings, however, the regularity of their attendance was not verified (HHS 2010).

488 As was noted earlier, the claims of access to cleaner water are most applicable

the NSP had contributed to fixing their road (HHS 2010, KII: Com14f, February 16, 2010, Com18, February 23, 2010, Com19, February 24, 2010, Com25, March 4, 2010).). As noted earlier, the road was leveled and graveled through a WFP Food for Work Program, not the NSP. The community members were very grateful for the work done on the road, and credit for this was given inaccurately to the NSP.[489]

A female teacher, who was also recorded as being the Deputy Chairman of the female CDC, displayed confusion regarding which projects belonged to the NSP. She said that the playground equipment outside her school was built through the NSP, when in reality it was built by CARE International with UNICEF funding, and utilized a completely different development approach (KII: Com27f, March 6, 2010). One group of community women discussed a tailoring program that had been operating for five months at the time of research, which was supposed to train the women in tailoring and then supply them with their own sewing machines. This too was categorized as a NSP project, even though it was not a part of the NSP and was not a participatory development program (FGD05f, January 27, 2010, HHS: SR001, January 20, 2010). Some community members inaccurately credited the NSP with providing English courses (HHS: SR001, January 20, 2010 and SR002, January 26, 2010), literacy courses (HHS: SR003, January 27, 2010 and SR007, February 3, 2010, KII: Com18, February 23, 2010), and mosque classes (HHS: SR007, February 3, 2010).

Even the Mayor of Tashqurghan was misinformed about the NSP. He told me that he thought it was one of the best programs in the district because it provided people with jobs (KII: EX10, January 18, 2010). However, the NSP does not provide paying jobs to community members. Their physical work on the project is seen as their contribution to the costs of the project, and is voluntary, not paid. He was not alone in believing this, with community members also incorrectly mentioning

---

to the village of Chahar Sukh, as those in the village of Shah Raheem only rarely operationalized their deep well and the village of Yaqubai had obtained their clean water source through an alternative development project.

489 Household survey participants were clear that the newly improved road meant they were able to more easily get to the district center, saving them time and money. Additionally, about one quarter of the HHS households credited the road work with lowering the levels of pollution in their community (e.g., HHS: CS002, February 16, 2010, CS006, February 21, 2010, and YB007, February 28, 2010, KII: Com19, February 24, 2010).

the NSP as a source of employment (e.g., HHS: CS005, February 20, 2010, YB005, February 24, 2010, YB008, February 28, 2010, KII: Com18, February 23, 2010).

All of this credit was given mistakenly to the NSP. While this might seem positive, it is also an indicator that should be of concern to those who run the NSP, as negative results from alternative development projects could just as easily be attributed to the NSP as positive results. Communities that have participated in the NSP, or are participating in it currently, need more accurate information that is readily and easily available to all members of the community.

Accurate information about the source of funding for the NSP was also not easily accessible or generally retained in the community of Shah Raheem. Only just over half of the HHS participants were able to list MRRD as at least one of the funders of the program, most of the others thought the money was coming directly from the facilitating partner (HHS 2010). This was true for the women of the community, who often demurred when asked about NSP funding, indicating they had no information about it (e.g., Com20f and Com21f, February 26, 2010). It was also true for the men, who more often than not responded to questions about the funding of the program, but did so erroneously (KII: Com18, February 23, 2010).

Very few, if any, of the community members had a good understanding of the CDC, its purpose, and who its members were (HHS and KIIs 2010). Most often, community members, males included, were only able to tell me the name of one CDC member, if any, despite having said that they participated in the voting process in which the CDC members were elected (e.g., KII: Com10, February 7, 2010). Even the *kalāntar* of Yaqubai, who is highly informed about what goes on in his village said that he honestly did not know what the CDC members' responsibilities were (KII:Com25, March 4, 2010). He was able to list the four executive officers of the CDC and he knew that they were in charge of the finances of the project, but other than, he did not have any information. In ACTED NSP communities, a similar situation was found, with even some CDC members unable to describe what the other CDC member's responsibilities were (Liu 2007: 27).[490]

490 This lack of knowledge was blamed on illiteracy and a lack of education, however, it was noted that in some communities where illiteracy and a lack of edu-

One of the powerful men from Yaqubai, who is generally very involved in community decisions, said that the CDC members were actually just representatives of each village and that three men were responsible for procurement (KII: Com19, February 24, 2010). This confusion was further emphasized and confirmed when I conducted a CDC member focus group discussion (FGD) in the mosque of the village of Shah Raheem (FGD06, February 3, 2010). It was supposed to be an intimate FGD group consisting of the six male CDC members for the community, but when I entered the mosque there were over thirty men who showed up, all of whom claimed to be NSP CDC members and were supported in this by the other community members in attendance.[491]

The NSP, as a participatory development program, would normally be expected to have a high level of awareness amongst the population who supposedly participated in it. The lack of knowledge about the NSP and about which projects were implemented via the NSP and which were implemented through other programs, leads me to believe that in the minds of the community members there really were not great differences between how the NSP was implemented and how other top-down development projects were implemented. This raises a red flag regarding whether the program was truly participatory in nature or not.

7.4.2 Inclusion and Participation is Selective

As a participatory development program, the NSP is expected to increase the quantity and quality of participation of the people in the development process. This is often touted as one of the many ways in which the NSP has proven to be a success (e.g., Beath, Christia et al. 2010: v; Quarry 2008). This claim has been supported (Besley, Pande et al. 2005: 656; Haq, K. 2002: 5; Timsina 2003: 239-240) and disputed by empirical studies on participatory development programs in general

---

cation were common that there was a good understanding of the responsibilities of the CDC members. This was considered to stem from having FP staff that was highly engaged. That being said, however, it appears to invalidate Liu's assumption that literacy and education were the prime factors (Liu 2007: 27).

491 In the end, the FGD was conducted with only thirteen individuals, while the others observed.

(Agarwal 2001; Fonchingong 2006: 144-146; Narayanan 2003: 2485). While it does appear as though the NSP has influenced the community of Shah Raheem to open its decision-making process up to include more than just community elders, as it had in the past, it continues to be selective in nature, not realizing universal participation from all sectors of the community.

Women's participation in particular can be called into question in Shah Raheem. Throughout the research, I continued to press for a clearer picture of the involvement of women. After spending months working in the community; observations about gender relations; testimony from the women; and knowledge of previous levels of involvement, or lack thereof, of women in development affairs, lead me to believe that the women from Shah Raheem Community did not actively participate in the NSP. This was especially surprising to me because the community is located in a district and a province that are considered to be relatively less conservative than many other areas of Afghanistan. The lack of participation of the women of Shah Raheem in the NSP leads me to question the participation of women in the program in general. Perhaps there are many other communities where there is NSP documentation of female participation, as there is for Shah Raheem, but where this documentation cannot be triangulated with the testimony of the women it supposedly included.

Other studies, have also shown a lack of female inclusion in the NSP, including Kakar (2005: 23), AIRD (2006: 28), and Azarbiajani-Moghaddam (2010: 18), as discussed earlier in the section on the practicalities of the NSP and how it works. In Khulm District, where seventy-five communities had implemented the NSP at the time of research, only two had female CDC members who were actively involved in the implementation of the program, all of the other female CDCs that were documented in the facilitating partner's district office had signed their rights away to the men of the community, effectively meaning they did not participate in the planning and management of the NSP subprojects (NSP Documents). However, if one was not able to read Dari, one could quite easily conclude that the female CDC registration forms were evidence of female participation in the NSP throughout the district of Khulm, although the women had really abdicated their role to the men.

The women I spoke with indicated that they did not feel that the NSP had provided them with opportunities to participate, as they never felt they had the option to become involved (e.g., HHS: SR016f, February 21, 2010). They told me that the men of the community would not allow them to participate and that the will of their male family members dictated their actions (FGD05f, January 27, 2010, FGD10f, March 6, 2010, KII: Com03f, February 10, 2010, Com23f, February 26, 2010). This was confirmed by five male household survey participants (HHS: CS002, February 16, 2010, YB001 and YB002, February 21, 2010, YB008, February 28, 2010).

I observed during this research, in conversations with both men and women of the community, that women were quickly dismissed as not having anything important to add to conversations that address issues outside of the private sphere. Men often talked about the lack of education of their female family members in this regard, as did the women (HHS and KIIs 2010). For example, one community elders said, "Most of the women are illiterate, we will not discuss these issues with them" (KII: Com18, February 23, 2010).

The women felt that they had little to contribute to the conversation, and invalidated their own thoughts and opinions, saying the men of the family knew better (e.g., KII: EX07, August 4, 2009). This contributed to the lack of inclusion or participation of the women in the community of Shah Raheem. It should be noted, however, that uneducated men were not generally prejudiced against in the same manner as the women, and that they were mostly still involved in decision-making processes at the community level. One FP staff member said that he thought the number of illiterate people partaking in community decisions had actually increased because of the NSP (KII: FP07, July 12, 2009). He of course was referring to male community members. While I was unable to find further evidence of this, it is quite possibly true.

Despite the lack of female participation, it does appear that at least some of the males of the community were more involved in the planning, implementation, and monitoring of the NSP subprojects than they had been for previous development projects. The pool of people who were involved in the decision-making also expanded. Younger men, who were traditionally not involved in decision-making for the community, joined in the NSP meetings to elect CDC members and to prioritize community development needs (HHS 2010). How actively they were able to participate, however, should be questioned.[492] Only a very limited number of men told me that no one from their household had been invited to the NSP meetings, which was seen as a definite shift compared to the previous ways of addressing concerns in the community. These are all positive signs that the NSP did manage to increase participation in development planning and decision-making in Shah Raheem, though on a limited scale. Boesen (2004: 44) also found that the scope of people who participated in the NSP in other communities had widened in her study of the NSP.

Further evidence of this was seen by one FP staff member who said he observed a shift in power structures in the different NSP communities in Khulm District. He talked about how power had shifted from being solely in the hands of the *kalāntar* or *malik*, to being in the hands of the people, as development projects and questions were now being addressed to the CDCs instead of to individuals with power (KII: FP07, July 12, 2009). This would imply a more expansive and inclusive power structure than existed prior to the establishment of the NSP.

Another interesting aspect regarding participation in the NSP community of Shah Raheem was the make-up of the CDC members. The members themselves were a mixed group of men in terms of age, power, and economic status. The treasurer was a traditional power holder in the community, as an older *khan* in his 50s with a large amount of land covered in gardens in the village of Chahar Sukh. The Chairman, however, was a younger farmer from the village of Shah Raheem of only 22 years of age, who was not economically well off, but was seen as an active and honest man by many in the community (e.g., KII: Com38, August 20, 2011, Com33, September 26, 2011). The Treasurer was a wealthy man who was a mason from the village of Yaqubai, while the secretary was a poor working man from Yaqubai Village. The other two CDC members included a farmer and an electrician, one in his 30s and the other in his 50s, in the first round of elections, and a carpenter and a *kalāntar* in the second round of elections (cf. NSP Doc-

492 This is based upon my own observations of group dynamics during meetings, where only one or two of the older males spoke for the most part.

uments).[493] Therefore, the CDC shows a new level of diversity in terms of the decision-making body itself.

Not a single *kalāntar* from the three villages was elected, and neither was a *malik*, which is interesting to note, as these are some of the more common power holders in rural communities in Afghanistan. It is also interesting that the two most powerful men in the village of Yaqubai were not a part of the CDC and played little role in the implementation of the program. The same can be said for the village of Shah Raheem, indicating that elite capture was not obviously at play in this NSP community.

I discovered that in contrast to the expectation that power holders in a community see the NSP as an opportunity to consolidate or increase their power, in Shah Raheem, being a part of the CDC was seen less as a means of accessing power and more as a burden that should be shared. The *kalāntar* of Yaqubai described having a rotation system in place for the CDC members, to prevent select individuals from having to carry the responsibilities of being a CDC member for too long a time period, due to the fact that they were not compensated for serving as a CDC member (KII: Com19, February 24, 2010). This was not, however, recorded anywhere in the NSP Documents, nor known about by the FP staff (KIIs 2010).

7.4.3 Accountability is Weak

Accountability is not an easy thing to enforce in an insecure environment where historically threats, fighting, and other factors have rendered community members afraid of the consequences if they speak up against something they see as wrong. This was observed in the community of Shah Raheem, and the NSP has not yet managed to greatly influence this enough to free people to act.

This can be seen most notably in the fact that despite there being problems with the roof of the NSP community center in the village of Yaqubai, no one from the community confronted the CDC executive officer who was suspected of having purchased lesser quality materials for the subproject in order to keep the difference between what he paid

493 Two separate elections were held, the first in 2006 when the NSP was introduced to the community and again in 2009.

and what he recorded as having paid.[494] The *kalāntar* of Yaqubai said that some people in the village wanted to investigate what happened, but he said that they were too scared to ask the CDC members about the low quality of the construction and did not do so in the end (KII: Com25, March 4, 2010). The community members did not feel comfortable addressing this problem with the FP either: "Why should we complain against the people, maybe we will not get facilities?" (KII: Com25, March 4, 2010). In other words, he expressed a similar concern to the women who complained about a development program implementer stealing products meant to be distributed to them in different development programs: if they told the truth about the negative aspects of development projects, they would not be given such projects or receive benefits from such programs in the future. In the end, he said, "It is the responsibility of the people who gave the money to the people to ensure good quality," eliminating all sense of responsibility from the community or village members (KII: Com25, March 4, 2010).

It is interesting to note how difficult it was to obtain usage information from the villagers of Shah Raheem in regards to the NSP deep well. When asked how often they used their deep well, they would brush off my questions or say vaguely that they used the well regularly (HHS 2010). Only one household survey participant was forward enough to volunteer that the well was not being used because they did not have the electricity needed for the generator to work. She expressed that she found the project "not very useful" (HHS: SR015f, February 2, 2010). When I followed up with other villagers directly, they supported what she had said, but they had been unwilling to be the first to tell me (FGD06, February 3, 20120, KII: Com02, January 20, 2010). In general people said that they obtained their water from a communal well, even in the village of Shah Raheem, with some also noting that they utilized the streams because no one was holding their fellow community mem-

494 The NSP is supposed to help increase transparency. The FP staff in Khulm told me that each community had a poster on which they noted all of the procurement that was done for the NSP subprojects in that community. This poster was then supposed to be placed in an easily accessed location where all members of the community would be able to review it. I did not see this poster in Shah Raheem, nor did any of the community members or CDC members tell me about it. The FP staff member, however, did say that in other communities these posters had led to CDC members being questioned about their procurement (KII: FP08, July 16, 2009).

bers to account regarding their commitment to pay for the upkeep and maintenance of the NSP deep well (HHS 2010).[495]

Another community leader had a quite negative view of how the NSP had been run in the community of Shah Raheem, and expressed that it would have been better had foreigners run it:

> *NSP has done more projects than other organizations but it has not been good because the people are in charge and they steal some of the money. They buy bad materials for projects. Our community center is of bad quality and the ceiling is already broken and caving in. (KII: Com25, March 4, 2010)*

He praised the DACAAR project that had brought them water taps, saying, "They gave us a good project with good quality. If NSP was the same as DACAAR, we might not face problems" (KII: Com25, March 4, 2010). The DACAAR project had experts check on the quality of the work and for ensuring that everything was being well maintained. Throughout this interview, he was unable to provide any examples of people from the community holding people or organizations to account for buying bad materials or not doing a good job. He said that he did not believe the people of the community had the power to hold others accountable to produce quality projects the same way that foreigners could (KII: Com25, March 4, 2010). In the end, he did not believe that participating in the NSP had impacted on accountability at all.

There were also non-NSP examples of the inability of the community to hold people accountable. In 2009, two years after the start of the implementation of the NSP in Shah Raheem, the community did not go and speak to the District Governor after the 200 sand bags he supplied proved ineffective against the spring flooding. When asked why they did not speak with the District Governor, I was told that the Governor had too much paper in his office and he would just "throw away their paper" (KII: Com05, January 26, 2010), indicating an inability to hold the local government to account. This same community member said, "We need a foreigner to get things done for the local people" (KII: Com05, January 26, 2010).

495 Included in the subproject proposal were assurances from the community that they would pay for the maintenance and running of the deep well and its attached generator (NSP Documents).

There is one example of accountability that was shared with me. A company was held accountable by the community after failing to clean the stream from which Shah Raheem obtained its irrigation water. In 2009, a private company was hired to clean the spring. Instead of cleaning it, they covered it with stones and rocks. After discovering this, one of the community elders went and talked to the Provincial Governor about the problem and the Governor broke the contract that the government had with the private company. Male members from the community were then gathered to work together to clear the rocks and stones from the stream (cf. KII: Com31, March 9, 2010).

The lack of basic knowledge about the NSP, explored above, also contributed negatively to the community's ability to hold people accountable. Perhaps if the community members had been better informed and aware of whom was responsible for each aspect of the program, they would have felt more able to become engaged and demand that people do their jobs honestly and with good quality.

7.4.4 Capacity to Organize is Not Strongly in Evidence

The establishment of the CDC can be seen as creating a structure that allows the community of Shah Raheem to organize itself for action. The NSP is expected to encourage this capacity to organize by providing the CDC members and the community with the tools to implement development projects and the democratic processes with which to determine their priorities. Conflicting information was found during this research in terms of how the NSP impacted on the community's capacity to organize, with evidence of actions and activities being contradicted by community member's evaluations.

Both during the implementation of the NSP in the community of Shah Raheem and the community's maintenance and monitoring of the subprojects, the community did not display a great capacity to organize. Starting with the process of implementing NSP in the community, it is clear from the testimony of CDC members and the FP staff that many of the processes that should have been taken care of by the community, with only the assistance of the FP staff, were actually conducted by the FP staff without community assistance or input. The original assessment of the community was mostly conducted by the FP, the subproject proposals were written by the FP staff, budgeting for the projects was not organized and analyzed by the community or CDC members but instead by the FP staff, etc. (e.g., FGD06, February 3, 2010, KII: Com29, March 7, 2010, Com33, September 26, 2011). It appears as though the training that was said to have been provided for the community of Shah Raheem did not provide the CDC members with enough skills or information to handle these tasks on their own.

Despite this, the CDC members I spoke with were quite positive about the NSP's impact on empowerment, indicating an increased level of organizational skills. One described that the people felt that they could manage future development projects because they had been able to execute their plans for the three NSP subprojects implemented: "Yes we the members of [the]development *shura* see ourselves [as] enable[ed]" (KII: Com37, May 31, 2011).[496] He highlighted the fact that the participants were involved and had made their own budget, figured out procurement, etc. on their own, without outside help (KII: Com38, August 20, 2011). However, evidence from other informants gives me good reason to question this version of events.

Another CDC member also supported the notion that he and the other CDC members had been empowered through their positions in the NSP. He said that he had learned issues such as purchasing and procurement, accounting, and other aspects of running a development project (KII: Com37, May 31, 2011). In addition to this testimony of capacity, is the fact that the community was able to put some of this capacity into action. They managed to procure a 10% contribution for the NSP subprojects in each of the three villages via in-kind contributions (HHS: SR008, February 7, 2010 and SR016f, February 21, 2010). So, there is limited evidence of the community working together to accomplish a goal or objective.

However, this is one of the few actions that supports the claims put forward by the CDC members. It can be clearly seen that there was a lack of monitoring and maintenance of two of the NSP subprojects: Yaqubai Village's community center and the village of Shah Raheem's deep well. The community center in Yaqubai not only reportedly has a cracked roof, but also lacks a toilet and a protection wall, rendering the community center usable only by men. This goes against the participa-

496 It was quite common for the CDC to be referred to as the development *shura*.

tory nature of the program as it excludes half of the population (MRRD and NSP III 2012). While there is information about other NSP communities facing similar problems and addressing them internally by volunteering labor and collecting funds from all families in the community, the people of the village of Yaqubai have made no moves in this direction.[497] Instead, they simply told us that they expect the NSP to pay for these changes and they seemed to fault the NSP for not having included these items in the community drafted plan and budget for the community center, despite the fact that they were supposed to have led the planning and budgeting efforts. This kind of attitude does not project a strong sense of empowerment, but instead reveals a latent dependency.

In a similar vein, Shah Raheem Village, despite their written commitment to maintain the deep well and generator constructed with the NSP block grant, does not regularly collect or provide the money needed to purchase fuel for the generator. The only time community members are willing to pay for the fuel is when the nearby stream runs dry in the summer. This well, therefore, only serves the community about two months out of the year. When asked about this, villagers from Shah Raheem were quick to tell me that they hoped the NSP would pay for the fuel (FGD06, February 3, 2010). It is possible that the community did not know that it was agreeing to pay for this fuel in the subproject proposal because they had not been involved enough in the writing of it, or it is possible that they agreed to this but still hoped that the NSP would provide the funding. This was not easy to clarify, as it was seen as a sensitive matter to discuss the writing of the subproject proposals.

Additionally, when some men from the community were discussing the NSP deep well constructed in the village of Shah Raheem, it was talked about as something that had been provided for them by the FP, not something that they had accomplished (FGD06, February 3, 2010). This seems to indicate that the NSP project, despite its participatory approach, is still seen as a development project implemented and provided by an external entity, not as a homegrown, bottom-up development project.

497 One example was provided by an FP staff member who said that in a community he worked in, they managed to raise the money on their own to implement a hydro project after they had completed their NSP block grant (KII: FP07, July 12, 2009).

In general, there was a sense of helplessness often displayed in the NSP community. A community elder expressed the idea that external countries were needed in order to bring about development projects in the future: "Since Afghanistan saw so many years of war, it needs assistance from other countries" (KII: Com31, March 9, 2010). This seems to highlight an expectation that I encountered multiple times in the community, that foreigners and foreign funds are needed to bring about development, that without outsiders providing money or guidance there would be no development. Instead of trying to resolve problems within the community, community members continue to look for outside assistance: "The problems that we have such as electricity, water, literacy courses, it would be better to provide for us. And we can promote ourselves via these projects if NSP provides them for us" (HHS: SR005, February 2, 2010).

This view shows a lack of confidence in their own problem solving skills and in their ability to implement projects. While major infrastructure projects usually require outside assistance, be that from NGOs, international organizations, or the government, smaller projects that require much less money and technical expertise can still make large differences in people's life experiences, yet, even smaller projects were not being discussed in the community of Shah Raheem.

Alternatively, in some NSP communities, the CDC is seen as taking on a conflict management role, with the CDC members helping the communities to resolve differences that arise, showing a capacity to organize (Nixon 2008b: 48; Zakhilawal and Thomas 2008: 13). This has not proven to be the case in the community of Shah Raheem. In Shah Raheem almost all of the household survey respondents noted that the community elders were still relied upon to resolve problems in the community (HHS 2010), utilizing already existing capacity to organize, but in conflict with the ideals of allowing for the participation of all people that is supported in the NSP framework.

One of the ultimate symbols of empowerment is a community's self-initiated efforts to improve the lives of its members. Unfortunately, there is no confirmed evidence that the community members of Shah Raheem have moved forward with development projects on their own. While a few community members said that the CDC members had reached out to other organizations to find funding for further development projects (e.g., KII: Com33, September 26, 2011), this was contradicted by individuals who worked for these same NGOs who said that they had approached the community members to conduct development projects, not the other way around (e.g., KII: EX16f, January 31, 2010, FP12f, February 4, 2010). Similar findings emerged out of a study conducted on the NSP in 2007 (Liu 2007: 35). It found that in the 365 NSP communities in which it gathered data, finding alternative sources of funding outside of the NSP was not happening. It was evaluated that CDCs were not reaching out to alternative funders, and that this was an "undeveloped aspect of the program" (cf. Liu 2007: 35).

The community members of Shah Raheem are instead waiting for more funds to come through the NSP, which might not happen, or for NGOs and the government to come to them.[498] They seem dependent upon money and impetus from the outside. These are not the actions of a community that is motivated to organize itself and has the capacity to do so.

7.5 Summarizing the Findings: The NSP's Limited Impact on the Areas of Investigation

While the NSP has had some positive impacts on aspects of the four areas of investigation, overall, its impact has not been dramatic.

Personal security, while improved, was not seen as connected to the implementation of the NSP by the community members themselves, and there was no evidence of such a connection from an external perspective. The NSP did not have a strong impact on economic security, with only a few people indicating that the NSP had resulted in them saving money or time.

The NSP does appear to have increased social security for men of the community of Shah Raheem. The men believe that their interactions in connection with the NSP with men from their own and other villages have helped to strengthen their relationships with these men and helped to strengthen or deepen trust levels between them, increas-

498 NSP III was supposed to provide a second round of funding for 17,4000 NSP communities that completed their block grants under NSP I and II, but Shah Raheem does not appear to have been lucky enough to be one of the recipients (World Bank 2010b: v).

ing social security. However, this did not occur for women of the community.

While the NSP does appear to have shifted community member attitudes in regards to the local government, with them reporting an increased level of connection to the government, there was very little evidence that this was being replicated in actions between the actors. However, local government officials were also enthusiastic about the NSP and expressed the idea that it had helped bring them closer to the communities they work with.

Empowerment appears to have been slightly altered by the NSP, but in nuanced and minimal ways. Some community members appear to have been decently informed, but unfortunately their numbers are quite limited, with many community members having inaccurate information about the program amongst the men, and little to no information about the program amongst the women. There is evidence that the NSP resulted in a greater number of men participating in village and community level decisions and in the development process in the community of Shah Raheem, including men who traditionally were without a voice in Afghan society: young men and men without power. Whether they were able to actively contribute to that decision making is not clear. In terms of accountability, the community did not provide any evidence of the NSP having changed their ability or their willingness to confront people or organizations and hold them accountable. The program also does not appear to have made a strong impact on the ability of the community members of Shah Raheem to organize themselves and to take action.

Overall, it appears as though the NSP in Shah Raheem was just another development program that provided pieces of infrastructure, some more useful than others. Based upon the four areas of investigation included in this research, we can see that the program did not significantly change the social structure of the community of Shah Raheem, and it has not yet resulted in the people of the community moving forward and working together to address other development issues in the area. However, it is possible that some effects and impacts will be felt further down the road. While the NSP does not appear to be living up to the grand rhetoric being expounded by external stakeholders in the community of Shah Raheem, it is quite possible that it has laid the groundwork for future steps that will allow the community members to make positive changes in their community.

8 Bringing the Research into Perspective

This dissertation has presented a historical overview of development theories and approaches that have helped elucidate how participatory development has re-emerged as a favored development approach. By evaluating the literature on participatory development, the basis for the analytical framework was established, which helped guide the empirical and academic aspects of this research.

This analytical framework was then applied to the research location, exploring Afghanistan's development aid history, which was shown to mirror that of the larger development discourse. It too has experienced an expansion of participatory development programs implemented in the last decade. The National Solidarity Programme (NSP) was then introduced as the example through which the participatory development approach and its impacts were looked at.

In order to further ground the research in the research community of Shah Raheem, the history of the district in which the community sits was analyzed, focusing on the four areas of investigation of the analytical framework – personal security, economic security, social security, and empowerment. To this was added the additional topic of development efforts. This historical perspective of Khulm District helped to root the terminology and definitions used in the district setting and in the research community of Shah Raheem.

These efforts have made it possible to understand with more clarity how the NSP was implemented in Shah Raheem and how it impacted on the community. It also allowed the incremental nature of that impact to emerge.

8.1 Applicability of this Research

8.1.1 Shah Raheem as Representative of NSP Communities in Khulm District

Though this research focused on only a single National Solidarity Programme (NSP) community, and therefore is not necessarily representative of all NSP Communities, it does reveal that perhaps the NSP is having a lesser impact in terms of the four areas of investigation than might be suggested by the rhetoric of development stakeholders.[499] While it is true that the NSP has a larger geographic reach than any other development program in Afghanistan and that it has reached more people, these statistics are often used as evidence of a successful participatory development program that will have a lasting, significant, and positive impact on the communities in which it is implemented. This is coupled with the belief that these changes will trickle up from the community level, changing the country as a whole. I question these assumptions. My objections arise from this intensive case study, with support from secondary research on the NSP that evidences examples of exclusion, a lack of sustainability, and a lack of empowerment, resulting in minimal impacts with questionable sustainability.

The NSP in the community of Shah Raheem was selected for this research, in part, because it was evaluated by the FP staff as having an average level of success in the district regarding the NSP (KII: FP08, January 26, 2010). Shah Raheem was not seen as a paradigm of success and participation in comparison to the 74 other Khulm District communities in which the NSP had been implemented, nor was it seen as being significantly or even marginally less participatory or successful.[500] This positioning was supported by speaking with representatives from other CDCs in the district (personal communication, July 26, 2010), an examination of the NSP Documents for all NSP communities in Khulm District (NSP Database), testimony of FP staff, and through discussions with provincial and district staff of the Ministry of Rural Rehabilitation and Development who oversee the NSP (FP09, July 20, 2009, EX07, August 4, 2009, FP08, January 18, 2010, EX05, January 25, 2010, and personal communications, District Development Assembly July 26-28, 2009).

499 For example, US Congressman, Michael Honda (Honda 2009), said that, "NSP builds social capital by promoting good local governance empowering rural communities to take control over their lives and livelihoods. No additional government resources are needed."

500 There are 91 communities in Khulm District. At the time of the second research fieldtrip, only 75 had partaken in the NSP. The 16 other communities became involved with the NSP during NSP III.

This background knowledge makes it hard to argue that the "participatory" experience of the people of Shah Raheem is an exceptional case in Khulm District. The implementation of the subprojects and their usage were not strikingly worse or better in Shah Raheem than in other communities, the level of knowledge about the program was also poor in other Khulm communities, and the other NSP communities in the district were not seen as any more empowered by the NSP process than Shah Raheem (KIIs 2010, NSP Documents).[501] This intensive research into the NSP experience in Shah Raheem, therefore, is not one that looks at an extreme case, but is rather one that looks at what has been evaluated as "average" by those who implemented the NSP in Khulm District.

8.1.2 Shah Raheem as Representative of NSP Communities in General: Evidence of Inaccurate Data and Biases

Some will argue that the findings of this research should not be used to generalize about the National Solidarity Programme, and from a purely analytical perspective I might agree. At the same time, however, I feel that these qualitative findings are significant. I believe that the research here provides an in-depth perspective that is more closely aligned with lived reality in a NSP community than is represented in the NSP Documents or than is generally presented to the world. This is supported by the inaccuracy of the NSP Documents in Shah Raheem, and the very real possibility that more intensive research in other communities would unearth the same.

The methodology employed in this research, an intensive qualitative case study, allowed me to attain information over an extended period of time.[502] This helped me to identify conflicts in understandings of project implementation and contradictions between expressed ideas

501 The level of knowledge about the program was evaluated based upon my own observations and interactions with other CDC members and community representatives at the District Development Assembly (July 26-28, 2012) where community representatives from each of the 91 communities in Khulm District were in attendance, and conversations with the FP staff (KIIs 2010).

502 Research for this dissertation was conducted during three different rounds of fieldwork, two of which were concentrated over a period of five months in 2009 and 2010, and a third that consisted of multiple days of follow-up interviews conducted in 2011.

and thoughts about the program impacts, and lived experiences that were extracted from those who participated in the research. Such conflicts are not easily identified in short-term research; such as has been commonly employed for quantitative surveys and evaluations of the NSP.

I believe that only through spending extended periods of time in the field, listening to and working with a community, can a researcher start to understand how a participatory development program's implementation and impacts are viewed by the people who live there. If a researcher only has a few hours or a few days to spend in a community, it is much more difficult to get people in an Afghan community to talk about problems they might have encountered, or areas where a program may have fallen short. Problems in Afghanistan are seen as shameful, as I have experienced in this and other research projects in the country, and are something that should not be exposed to foreigners. Limited timeframes such as those employed by most research conducted on the NSP and other participatory development programs provide a researcher with only a summary or superficial understanding of a community and its experience, while more extended time periods allow inconsistencies to rise to the surface, providing the researcher with insights that go beyond the image the community would like to project.[503]

Many people have a stake in presenting a positive image of the NSP and its implementation in a community, a district, or in a province. The community members often believe that they will only be given future development projects if previous projects have had positive outcomes and followed the rules of the program. FP staff at the district level wants to project an image of competence and success in their facilitation in order to satisfy their funders. Provincial FP staff and MRRD staff also want to create a positive image of the NSP and the work they have done to please the funders and those they report to within their organizations. This can often result in inaccurate information being written in NSP Documents and shared with researchers who interview such indi-

503 Many research projects on the NSP allowed the research teams to spend one day or less in each research location or community in which they conducted their fieldwork, such as Boesen (2004: 3) who covered 30 communities over 29 days in the field, or Beath et al. (2011: 16) who appear to have spent only one day in each village for each round of their randomized evaluation.

viduals. It can also result in an over-inflated understanding of what the NSP is accomplishing in rural communities in Afghanistan.

Not only is there evidence of positive biases stemming from those who are most commonly asked about the NSP, there is also evidence that even when challenges or problems in the implementation of the NSP have emerged in research, it is often brushed aside by policy-makers and development stakeholders, as it does not align with the positive image they hold of the NSP. This image is generally of a development program that is helping empower rural populations, while instilling democratic values and providing needed infrastructure, as is reflected in the NSP rhetoric.

I contend that these NSP reports and documents don't tell the full story of the failings experienced within the NSP for multiple reasons: (i) many reports rely, at least in part, on NSP Documents, which do not always represent experienced reality, (ii) people who participated in the implementation of the NSP do not always present a realistic picture of what happened for different reasons, and (iii) researchers do not have the time to discover and follow-up on information that is contradictory in nature.

By spending multiple months working in Shah Raheem, I began to see that there were discrepancies between what had been recorded in NSP Documents for Shah Raheem and what community members told me had happened. These differences were not minor in nature, but instead called into question various assumptions, including: (i) the success of the FP in implementing the NSP as a participatory program in the community of Shah Raheem, (ii) transference of knowledge and skills to community members, and (iii) the usefulness of the subprojects implemented. For example, on paper, the community of Shah Raheem implemented three successful infrastructure projects and women actively participated in the NSP process. I learned, however, that one of the subprojects is used only rarely, another subproject is only used by men, and that women did not participate in the program at even the most basic of levels. These and other factors were only revealed over time, and they provide a very different perspective of how the NSP was implemented in the community than the NSP Documents or the first days of research findings.

If I had based my research on the internal documents of the NSP and/or a one day visit to Shah Raheem, it would have been easy to presume that these differences did not exist, that Shah Raheem was a strong example of the NSP having been implemented successfully in accordance with the NSP framework. Shah Raheem reportedly elected twelve CDC members (six male and six female), 301 people reportedly voted on development prioritizations in the community (137 male and 164 female), three subprojects were successfully completed and presumed to be utilized, and all with no reported serious delays. This would also have been supported by the more powerful males in the community that I was able to speak with during my first few days in the community. All in all, the NSP community of Shah Raheem would be considered another beacon of light for democracy and development. Due, however, to the research methodology employed for this research, I became more intimately acquainted with a variety of people in Shah Raheem and their experiences, and it became evident that the NSP Documents did not tell the whole story or even an accurate version of the story. There are many ways in which the implementation of the NSP in Shah Raheem failed to create the inclusion and participation reported in the NSP Documents and in my first interviews. In the end, it has been shown that the NSP's impact on the areas of investigation in the NSP community of Shah Raheem have been incremental at best.[504]

These inaccuracies, and discord, were not reflected in the official statistics stemming from the NSP, nor have they been identified by other research projects that have only spent more limited amounts of time conducting interviews in the district. This leads me to question the accuracy of the NSP produced data and the ability of more time constrained research to accurately reflect the lived experiences of the NSP communities, resulting in an inaccurate portrayal of the NSP as "the" answer to Afghanistan's development problems.

504 I would argue that the indicators tracked in the NSP Documents fall far short of providing enough information to discuss deeper impacts of the program on a community, as they focus strongly on quantitative data regarding the distribution of block grant funds, the completion of subprojects (with no space for utilization information), and tallies of the number of people who voted for CDC members or community development priorities.

8.2 Reviewing the Research Findings

8.2.1 Shah Raheem: Incremental Changes

What do these findings and conclusions really mean for the people of Shah Raheem? The research shows that only small, incremental positive impacts from the NSP were experienced in the community of Shah Raheem. These incremental changes are important to note, however, as they could help lead to greater changes in the future.

The fact that the community was asked about its development priorities was new, as was the idea of the people leading a development project, both of which were raised by the NSP. Even though the CDC members and community did not play a large role in the planning and implementation of the subprojects in Shah Raheem, the idea that they should or could have been involved introduced a new concept into the community, which could affect things from a longer-term perspective. This can only be determined with time.

The foundation of the NSP as a participatory development program that should involve the people was understood by many of the men of the community, bringing forward what can be seen as a new way of conceptualizing development; one that not only involves external development actors. It is possible that this inclusion of the people, even if only in theory, will shift villagers' understandings of the kinds of roles they can play in development. Again, this is something that can only be determined in time and only directly applies to the people who participated in the NSP. It does not appear to have happened as of yet, as seen by the incredibly limited shifts in empowerment revealed through this research.

More men were included in the introductory decision-making process of the NSP, including electing CDC members and prioritizing the development needs of the community. This in itself is noteworthy in a community where only male elders and those with power from the community were previously involved in such activities. Through the NSP more men were included in development meetings, though not all men in the community. However, younger men and others who were traditionally marginalized were included in greater numbers than before. How actively they participated is questionable, but their inclusion in the meetings was itself a small step towards greater inclusion.

As has been shown, the women did not participate in the NSP in Shah Raheem. The fact that a women's CDC was formed, even though only on paper, however, was seen as a small positive step forward for the empowerment of women. Women had not been considered previously when discussing community level discussions, and even though the women were not included in the NSP, the simple fact of a discussion having taken place about the possibility of women being included and the establishment of a women's CDC on paper, were seen as important shifts by FP staff (e.g., KII: FP06, January 14, 2010).

The lack of knowledge about the NSP found in Shah Raheem in general, and especially amongst the women, indicates that there is much greater room to increase the level of participation and knowledge transfer within the NSP. At the same time, however, it does seem that a larger number of people had information about the NSP than about other development projects conducted previously.

Through their participation in group NSP meetings, men from the three villages of Shah Raheem came together and discussed many issues, expanding their interactions and increasing their social connections. This was seen as, simultaneously, positively impacting on social security in the community. While the implementation of the NSP did not increase many other kinds of relationships in the community, such as between females, or between males and females, the increase in social relations and connections between males of the three villages is a step towards greater social security. It is possible that this will play a role in expanding social relationships between other villagers in the future.

The NSP did appear to change community members' attitudes about, or perceptions of, the local government. It increased the positive sentiments that were expressed, touching on bridging social capital. Though, it must be noted, this change in attitude or perceptions was not accompanied by increased interactions between the community members and local government actors.

A larger shift was found in the local government's assessment of NSP communities, along with NGOs and other development stakeholders. Communities with the NSP were seen as being more competent and easier to work with than non-NSP communities, whether this assertion was true or not was not clear.

Overall, the NSP had positive impacts on the community of Shah Raheem, but to label them as anything but incremental would be an exaggeration. These incremental shifts are essential to changing the human security in the community overall and the empowerment of the people, but also indicate that there still remains greater room for further changes.

8.2.2 The NSP and Afghanistan: Raising Concerns

Even though the NSP is one of the most researched development programs in Afghanistan, this research on the program differentiates itself, from an academic perspective, by its commitment to a micro-level perspective. Most research on the NSP, as is the case for most participatory development, has been conducted from a broader perspective, covering a larger geographic area and committing less time to each community. It also either has a strongly quantitative perspective (e.g., Beath, Fotini et al. 2012), is large in scope geographically (e.g., Barakat 2006), or is focused on only one or two specialized topics, such as gender (Azarbiajani-Moghaddam 2010; Echavez 2010, 2012; Echavez and Bagaporo 2012), governance (Nixon 2008a, b), the role of facilitating partners (Altai Consulting 2004; Kakar 2005), the sustainability of the CDCs (Brick 2008), the role of the NSP as a peace building program (Zakhilawal and Thomas 2008), or elite capture (Beath, Christia et al. 2011).

This research has provided a new perspective on a program that has been operating since 2003. It provides a historical analysis of the area of implementation, which acts as a background upon which the implementation of the program can then be explored. It presents an experienced reality that raises questions about what has really happened in NSP communities throughout the country, and highlights ways in which current evaluations of the program might be based upon faulty or misleading data; the NSP Documents. It attempts to be as holistic as was possible given the working environment.

In a country which has experienced decades of political upheaval and conflict; fluctuating degrees of personal, economic, and social security, a lack of empowerment, and development aid that has been concentrated on urban areas and has stemmed from the top-down, it is difficult to imagine that participation in a short-term development program will change the social structure, human securities, or empowerment of a community. This is especially true when this research indicates that communities are at most meeting only one time per month.

Afghanistan is a very challenging environment in which to work, as trust, social structures, infrastructure, and more are complicated and in stages of reemergence or change. It seems logical to assume that a participatory development program would require more implementation time to make significant changes in Afghan communities than the two and a half years committed on average to each NSP communities (MRRD and NSP III 2012: 25). Successful participatory development programs in other countries that have had less tumultuous histories have required longer implementation timeframes, such as Kecamatan in Indonesia and AKRSP of Pakistan.[505] These longer timeframes are often cited as contributing to the positive nature of the programs, and is something that should be contemplated for the NSP.

This research also points to another concern, which is that the positive rhetoric about the NSP is raising expectations that are not being met. I believe this is dangerous and is likely to be detrimental to Afghanistan's development moving forward. The way that the NSP is talked about today, it would be easy to assume that this program has created highly functioning development councils that will take over the burdens of development from the FPs, and will continue their efforts with the help of the MRRD, an assessment that I believe to be overly optimistic.

As each community "graduates" and is handed over to the MRRD, FPs and other stakeholders in the program step away from those communities, with the understanding that the people in the community have been mobilized and are empowered to address development on their own. There are two problems with this: (i) my analysis and that of others (Barakat 2006: 73; Brick 2008; Zakhilawal and Thomas 2008: 21) indicates that CDCs are not generally self-sufficient[506] and that the FPs

505 Kecamatan was started in 1998 and operated for nine years (World Bank 2012) and Aga Khan has been implementing AKRSP for over twenty-five years (AKDN 2012).

506 There is evidence of some very highly functioning CDCs (Affolter, Richter et

have played a much larger role in the implementation of the NSP in communities than was planned for in the NSP framework and (ii) the MRRD at the provincial level is not ready monetarily or in terms of personnel to provide the kind of support the CDCs need.[507] As was shown in the case of Shah Raheem, this necessary support is likely to require a great deal of effort, as the CDC has not actively been seeking out development projects, funding, etc. on its own since the completion of its NSP subprojects. With the MRRD unable to provide direct support to all of the CDCs, it is likely that many of the CDCs established by the NSP will cease to be active, if they have not already done so. This could, in turn, lead to a disregard for the participatory approach in general, as the structures created by the NSP are not able to stand on their own.[508]

This research should be seen as a means of raising well-intentioned questions about the strength of the NSP as a participatory development program in Afghanistan. It is intended to start a conversation about experienced realities in NSP communities, in an attempt to let these stories speak on behalf of the benefits or impacts the program has been delivering, instead of allowing less grounded research and assumptions about causalities to dominate.

8.2.3 Participatory Development in General: Hidden Realities

Research such as this case study is important and should be taken seriously, even at the level of development theory. Development theories and approaches should be informed by fieldwork that actively engages with the people who have experienced the theories and approaches first hand. It is quite possible that further case study research on the NSP and other participatory development programs will support the findings of this research, and support my call for greater honesty from de-

al. 2009: 325; Kakar 2005: 34-37; Nixon 2008a: 34), but from my understanding they are in the minority. It is likely that these CDCs are better equipped to function without the FP and the NSP to support them.

507 The MRRD staff in Balkh Province only numbered nine people in 2009, when it was supposed to start taking on the responsibilities of the NSP communities. Their ability to serve the 844 CDCs that existed in the province back then and the additional CDCs that were created under NSP III, does not seem plausible (cf. EX04, July 29, 2009).

508 This concern that CDCs would disappear or fade was also made by Brick (2008: v) in her unpublished paper on the sustainability of CDCs in Afghanistan.

velopment stakeholders about what is being accomplished, in order to deepen the discussion about what *can* be accomplished.

I believe that it is time to change the way that the NSP and perhaps other participatory and non-participatory development projects are looked at and spoken about. I suggest that it is better for the industry of development, development funders, development participants, and all other development stakeholders to be realistic about what development programs can do, be they participatory development programs or those that are top-down in nature. It is time to stop sweeping issues and concerns under the proverbial rug, and start to bring them out into the open, learning to accept them as a consequence of the complicated nature of development in general.

With the NSP in Afghanistan, too often negative experiences have been hidden from view or ignored. There is no research that provides insights into the failures of NSP projects, despite there being enough failures to establish a "black list" of projects that NSP block grant money cannot support (NSP 2009a: 19-20). No official statistics are given about how many communities experienced corruption in some form related to the NSP process or how many communities actually utilize the infrastructure that was implemented with NSP block grant funds. The possible negative aspects or failures of the program remain unexplored, which I believe is problematic.

Even research that unearths negative aspects of the NSP or ways in which its implementation did not coincide with the goals of the NSP framework (e.g., Barakat 2006; Beath, Fotini et al. 2012; Boesen 2004; Echavez and Bagaporo 2012), is often brushed aside. Those findings are seen as only exceptions to the more positive findings, not important in their own right. Yet, when one looks closely at this research situated in Shah Raheem, which is firmly grounded in the lived experiences of the people of the community, and other qualitative research conducted on the NSP, the cautionary tales, the exclusion of certain sectors of a community, the lack of accurate information in the community, and the limited evidence of empowerment are confirmed and do not appear to be limited to only one or two cases out of the tens of thousands of NSP communities. However, these negative, or not so positive, findings are not what are discussed openly, as the positive expectations of the program are provided as evidence of the program reaching these expectations, with stakeholders highlighting only optimistic research findings, while ignoring evidence of struggles or problems that might contradict this view. In general it seems that because the program is participatory in nature, and participation is assumed to lead to more efficient projects, to empower, and to create or establish greater ownership, that it is presumed this is actually occurring because the NSP is a participatory program.

This type of circular thinking is detrimental to understanding how a development program is working and how it is influencing the lives of those living in the areas where it is implemented. It is also detrimental in that it creates expectations that are not likely to be met, which can increase mistrust and resentment. Additionally, it eliminates the possibility of finding ways to improve the program and its impacts, as the impacts are already presumed to exist, or to ground expectations more solidly in experienced realities. This is a trend that was alluded to in the section on development theories and approaches historically. As each development theory or approach was perceived to have failed to bring about the hoped for changes, a new theory and/or approach was put forward and the old one abandoned without a great deal of analysis or attention paid to understanding the "failure" of the previous theory or approach.

I contend that it is time for the international development community to start asking itself more questions about what can and should be expected from participatory development programs and development programs in general, in order to start an honest dialogue. There is no reason to expect a single development program to provide a perfect answer to the development problems of a country, a district, or even a community. Development is an incredibly complex concept with myriad factors and influences. If development practitioners and beneficiaries could sit down and discuss their expectations, perhaps the veil of rhetoric that surrounds such programs could be lifted and new expectations could be put forward that more closely align with lived realities of what different projects and programs can do. In the case of the NSP in Afghanistan, I would suggest that more incremental impacts and changes should be focused on.

Without this kind of nuanced perspective, the development complex will continue to make false claims of having found the golden ticket or "The Answer." One decade the answer was fiscal responsibility and financial restructuring in the form of the Washington Consensus, and the next decade it was inclusion and participation in the form of community-driven and –based development. Yet none of the different development theories or approaches has managed to make the changes that their proponents promised. Perhaps this has less to do with the programs and more to do with expectations about what the programs can do.

Without a perspective based in experienced realities, development stakeholders will continue to raise expectations to levels that cannot be met, leading to frustrations both among the project participants and the external stakeholders, and a dismissal of development aid efforts. Movements towards a more complicated and nuanced perspective have been supporting the drive towards participatory development, with the advent of the basic needs approach, human security, social capital, etc. expanding the range of development indicators, however when it comes to understanding the impacts of projects a complicated picture continues to be rejected in favor of a more easily told story. For example, the President of Afghanistan, Hamid Karzai praised the NSP for enabling communities to solve problems and develop capacities that will make them self-reliant, at the CDC Conference held in the *Loya Jirga* tent in October 2010 (NSP, MRRD et al. 2011: 10).

What this research shows is that perhaps expectations of social changes and human security enhancements as a result of implementing a single participatory development project or program are too optimistic. The community of Shah Raheem shows us that even in a community which would be seen as a NSP success on paper, a different reality has been experienced. If only quantitative data had been gathered about this community, it would be seen as a community where both genders were able to participate fully, and where projects implemented were being regularly utilized. This would lead, and has, to the assumption that the people have been empowered due to their participation.

Getting to know the people of the community, and not just what they might want to present to a passing outsider, it became clear that not all of these assumptions or documented impacts were quite what they seemed in Shah Raheem. It is possible that this is the case for many communities that have participated in development projects, both of a participatory and a top-down nature. This only feeds into the circular thinking discussed above, which can be dangerous in terms of raising expectations that will not be met, resulting in an approach being rejected *en masse.*

It is time for the expectations and rhetoric about the NSP and other similar participatory development projects to be carefully evaluated and scaled back, based upon further micro-level research. This research is also a call to development stakeholders to start engaging with their development projects from a micro-level perspective that is grounded in experienced realities, to discuss shortcomings, and start to bring expectations down to a more realistic level.

I believe that this research and similar research in the future should be integrated into participatory development projects worldwide. Conducting more intensive and extensive research in a case study community can provide a deeper understanding of the program, of communities in the area, and of the complexities of both. Having insights into how the program is working in one community can help researchers in other communities more fully engage with the people when they are conducting less time-intensive quantitative or qualitative research and possibly help improve implementation. It can also help inform the questions, structure, and methodology of more standardized means of research. If researchers are aware of some of the ways in which lived realities can be misrepresented by some actors and documents, then perhaps the depth of information gained and the accuracy of the research overall can be improved.

The research is also a reminder to avoid Chambers' bias of person (Chambers 1983: 18), speaking with only those who are most easily accessible. In a participatory development program environment in Afghanistan, this can easily result in an overstatement regarding the level of participation experienced across different categorizations of people within the community and can result in an overly optimistic retelling of the implementation process. This is because those who hold power, usually men, do not necessarily represent or understand the experiences of those who are more marginalized in their community. These people with power can also try to present a picture that aligns with what they believe others want to hear about the program.

8.3 Coming to a Close: The NSP in Perspective

The NSP has brought to fruition some needed infrastructure projects to rural communities that it is unlikely would have been reached without a participatory approach, especially in a country where development efforts have been focused mainly on large infrastructure projects conducted in urban areas. For this, the NSP should be credited. I would contend, however, that the praise it is receiving today is raising expectations to unattainable heights based upon assumptions that are not necessarily grounded in community experiences. I do not believe the NSP is the silver bullet that will help Afghanistan to develop, but see it instead as another program of many that can contribute to slow, steady, and incremental shifts towards greater human security and empowerment of the people. By recognizing this and shifting the expectations and rhetoric to more accurately reflect what is happening in the communities of Afghanistan, the small and important steps that are being made can be recognized and appreciated for what they are.

This dissertation is really a call for greater honesty in the development field. It is not intended to be a strike against the NSP or the facilitating partner who implemented this program in the community of Shah Raheem. It is instead a call to reinvigorate the development discourse, to reevaluate the assumptions that currently exist in the development paradigm and to call on development funders, implementers, and beneficiaries to put forward their version of reality for discussion. To continue to invest in development programs without having an open and honest dialogue will perpetuate an environment where everyone is afraid to tell the truth, which can only lead to a continuation of expectations being raised and unmet.

While some would argue that Shah Raheem is a special case and that the findings of a single case study should not be taken as anything more than a solitary experience, I believe that the community of Shah Raheem is representative of some deeper truths about the NSP, about participatory development approaches and efforts, and about the theories of development and the approaches that stem from them.

Annex A: NSP Form 1: Community Profile

NSP Operational Manual, Version IV, November 2007

Islamic Republic of Afghanistan

Ministry of Rural Rehabilitation and Development (MRRD)

National Solidarity Programme

Form 1: Community Profile

Facilitating Partner		District/Community #	
Province/District		Geo codes/GPS	
Community		Date FP first entered the community for NSP	
Villages/ Subvillages Names		#	

Is this a cluster of settlements? ☐ yes ☐ no

Inhabitant profile

This section of the form is to describe the population within the community and asks "How many people are in the community?", "How many families are there?" etc. Try to assess only what number of the community is in each group. This allows the community to see how many children will need schools for instance, or whether the community is ageing. It also allows the community to see how many vulnerable persons there are who may need support.

Number of community inhabitants ☐ Number of community families ☐

No. of inhabitants that are: ☐ # less than 6 yrs ☐ # 6 - 17 yrs ☐ # 18 - 45 yrs ☐ # 46 - 65 yrs ☐ # over 65 yrs

Number of families that are: ☐ IDP ☐ FHF* ☐ DHF* ☐ Returnees

* Female/Disabled Head of Family

Access to transportation

This section provides an assessment of how isolated or accessible the community is. The first question asks by what kind of road or track is available to access the community, please tick the box that best describes access to the community. The second question asks how far the community is from important centres, please enter kilo meters for each place.

Community accessed by: ☐ asphalt road ☐ stone or gravel ☐ graded track ☐ ungraded track ☐ no road

Distances by Km from community to: ☐ km nearest road ☐ km nearest village ☐ km district centre ☐ km provincial centre

Access to water and sanitation

This sections covers the community's access to drinkable water: communities with no water are rare, but please identify how many reservoirs, taps, rivers or wells the community has. In the question about improved toilets, a simple tick will say if all toilets are traditional or what percentage of homes have improved toilets.

Sources of drinking water: ☐ none ☐ reservoir ☐ river ☐ hand dug well ☐ depth to water (metres) ☐ communal tap ☐ household tap ☐ metres to river ☐ handpump

Toilets? ☐ traditional ☐ 0 - 25% ☐ 26 - 50% ☐ 51 - 75% ☐ 76 -100%

Resources and economy

In this section of the form, the community can assess how much productive land is available, how the village gets its income, how the income is distributed and what skills and equipment are available in the community. You can choose only number to put in the number of jeribs.

Total Number of jeribs of land ☐

☐ irrigated, 1 crop/year ☐ irrigated, 2 crops/year ☐ rain-fed ☐ pasture ☐ forest/woodland

☐ regular community maintenance of irrigation (y/n) ☐ Mir ab (y/n)

Agriculture is income from farming or farm labour; Livestock means income from any animals or animal by-products; Trade covers income from selling imported materials or equipment but not agricultural produce; Crafts means income from handiwork and artisans (pots; carpentry, weaving); Services covers transportation, prepared food sales, and household personnel; Manufactures covers welding, tin-smithing, cast iron or aluminium production. You can choose only number of families.

Number of families/men/ women with income from:	agriculture	livestock	trade	crafts	services	manufactures	abroa abroad
Families	#	#	#	#	#	#	#
men	#	#	#	#	#	#	
women	#	#	#	#	#	#	

Ask the community to identify how many families fall in the different wealth categories. Then identify how many skilled trades people there are. How many pieces of equipment does the community have?

# well-off families		# middle income		# poor families	
# landowners		# tenants/sharecroppers		# families - main ethnicity	Pashtun / Tajik / Othe

# of people with skilled trades in community					
	stone masons	bricklayers	carpenters	metal workers	electricians
	rug weavers	mechanics	other		

How many?	tractors	trucks	buses	motor and bicycles	horse/donkeys	oxen (plough)
	#	#	#	#	#	#

Education

In this section, please identify how far the community is from each place of education, and how many people in the community have finished each level of education. You can choose numbers. Please identify the number of teachers there are in the community for each level of education. Please estimate how many of the boys and girls of school-going age attend school.

	primary	secondary	higher		
Distance to nearest school (kms)	km	km	km	# boys in school: #	# girls: #
#/% people finished	#	#	#		
# teachers	#	#	#	Quality of teachers should be assessed later.	

Health

Please show how far you must travel to the nearest clinic or hospital and then identify how many doctors, nurses, midwives etc., you have in the community.

Distance to nearest in kms.	clinic	hospital
	km	km

Number of	doctors	nurses	midwives	pharmacists
	#	#	#	#

Organization

Please indicate whether you have a traditional shura in your community or not, and then identify the main languages spoken in your community.

Traditional shura present?	yes	no	primary language	second language	third language	fourth language

Annex B: NSP Form 2: CDC Election and Composition

NSP Operational Manual, Version IV, November 2007

Islamic Republic of Afghanistan
Ministry of Rural Rehabilitation and Development (MRRD)

National Solidarity Programme

Form 2: CDC Election and Composition

Province	District	Community	Community ID#

Number of community inhabitants ☐ Number of community families ☐

This section is used to describe how the community election was held.
It helps us judge how fair and free the elections were and how representative the CDCs are.

Election Monitoring

Did each voter have only one vote?	Yes	No	*This question allows us to assess whether the number of votes used in the election were fairly and evenly distributed.*
Total number of eligible voters?	Men	women	*This question helps us know how many people could have voted in the election*
Was there any electioneering? (*)	Yes	No	How comment/explain
Was there any voter intimidation? (**)	Yes	No	How comment/explain
Was help given to illiterate voters? (***)	Yes	No	How comment/explain
How many illiterate voters were helped?			
Was the election by secret ballot?	Yes	No	*The secret ballot allows people to vote without fear. If the voting is not secret the CDC may represent only the powerful and not everyone.*
How many eligible voters did vote?	Men	women	*Knowing how many eligible voters have voted helps us know how representative the CDC is (must be over 60% for block grant eligibility).*

* *Electioneering is when people interested in winning the election try to convince you to vote for them*

** *Intimidation or threats of violence or retaliation of some kind are important to register so we can know how free the elections were.*

*** *In this question we need to know how people who cannot sign their name were helped to make sure that their voting was secret.*

Use the following pages to list the CDC [or sub-committees in case of two separate sub-comittees]

Form 2: CDC Election and Composition (continued)

Province	District	Community	Community ID#

* Commander, Mullah/Maulawi, Landowner

* * Female or Disabled Head of Family

Mixed CDC or male sub-committe						CM/M/L * Returnee, IDP, FHF * *, DHF * *				Qaum
No.	Position	Literate		Occupation						
		Yes	No			R	IDP	FHF	DHF	
1	Chair									
2	Vice-Chair									
3	Secretary									
4	Treasurer									
5	Member									
6	Member									
7	Member									
8	Member									
9	Member									
10	Member									
11	Member									
12	Member									
13	Member									
14	Member									
15	Member									
16	Member									
17	Member									
18	Member									
19	Member									

20	Member								
21	Member								
22	Member								
23	Member								
24	Member								
25	Member								
26	Member								
27	Member								
28	Member								
29	Member								
30	Member								

Form filled out by ______ Position ______ Date ______

"I certify that the election was held in compliance with NSP principles and regulations and that >60% of the eligible voters did vote"

Certified by ______ Position ______ Date ______

Date received by RRD/NSP PMU	Name	Title	Signature

Form 2: CDC Election and Composition (continued)

Province	District	Community	Community ID#

* * Female/Disabled Head of Family

Female Sub-Committee

No.	Position	Literate		Occupation		Returnee, IDP, FHF * *, DHF * *				Qaum
		Yes	No			R	IDP	FHF	DHF	
1	Chair									
2	Vice-Chair									
3	Secretary									
4	Treasurer									
5	Member									
6	Member									
7	Member									
8	Member									
9	Member									
10	Member									
11	Member									
12	Member									
13	Member									
14	Member									
15	Member									

Form filled out by ________ Position ________ Date ________

"I certify that the election was held in compliance with NSP principles and regulations"

Certified by ________ Position ________ Date ________

Date received by RRD/NSP PMU	Name	Title	Signature

In case of separate female and male sub-committees, an executive committee will be formed consisting of 2 representatives from each sub-committee.

Executive Committee						Returnee, IDP, FHF * *, DHF * *				Qaum
No.	Position	Literate		Occupation						
		Yes	No			R	IDP	FHF	DHF	
1	Chair									
2	Vice-Chair									
3	Secretary									
4	Treasurer									

Form filled out by ______ Position ______ Date ______

"I certify that the election was held in compliance with NSP principles and regulations"

Certified by ______ Position ______ Date ______

Date received by RRD/NSP PMU	Name	Title	Signature

Annex C: NSP Form 5: Certification of Registration of CDC

Operational Manual, Version IV, November 2007

Islamic Republic of Afghanistan
Ministry of Rural Rehabilitation and Development (MRRD)
National Solidarity Programme
Form 3: CDC Registration Application

Province	District	Community	Community ID#
Number of families	Number of clusters	Number of CDCs	Election date

Villages / Subvillages Names | # |

Members of the Community Development Council

For mixed CDCs, FPs only fill out Page 1 of this form. For separate male and female subcommittees, please use one table for the male and one table for the female sub-comittee. In the case of separate sub-committees, the community will have to elect as well an executive committee.

A) Mixed CDC or Male Sub-Comittee

Full Name	Position	Gender		Age	Signature
		F	M		
	Chair				
	Vice-Chair				
	Secretary				
	Treasurer				
	Member				
	Member				
	Member				
	Member				
	Member				
	Member				
	Member				
	Member				
	Member				
	Member				
	Member				

Chair | Vice-Chair | Secretary | Treasurer

Certified by

CDC (or sub-committee) Chairperson	Name	Signature
FP District Manager	Name	Signature
FP Provincial Manager	Name	Signature

Date received by RRD/NSP PMU	**Name**	**Title**	**Signature**

Form 3: CDC Registration Application (continued)

Province	District	Community	Community ID#

B) Members of the Female Subcommittee

No.	Full Name	Position	Age	Signature
1		Chair		
2		Vice-Chair		
3		Secretary		
4		Treasurer		
5		Member		
6		Member		
7		Member		
8		Member		
9		Member		
10		Member		
11		Member		
12		Member		
13		Member		
14		Member		
15		Member		

Photo	Photo	Photo	Photo
Chair	Vice-Chair	Secretary	Treasurer

Certified by

	Name	Signature
Sub-committee Chairperson		
FP District Manager		
FP Provincial Manager		

Date received by RRD/NSP PMU	Name	Title	Signature

Form 3: CDC Registration Application (continued)

Province	District	Community	Community ID#

C) Members of the Executive Committee

(this only applies in the case of separate sub-committees), 2 should be male, 2 female

No.	Full Name	Position	Sex	Age	Signature
1		Chair			
2		Vice-Chair			
3		Secretary			
4		Treasurer			

Please indicate the type of the CDC by ticking one of the boxes.

Male	Female	Mix	Separate

Annex D: NSP Form 6: Community Development Plan

NSP Operational Manual,Version IV, November 2007

Islamic Republic of Afghanistan
Ministry of Rural Rehabilitation and Development (MRRD)
National Solidarity Programme
Form 6: Community Development Plan

Province	District	Community	Community ID#

Facilitating Partner:

According to the revised gender policy in Operations Manual, a community may have separate male and female sub-committees. In case these 2 sub-committees can not meet and discuss the CDP together, the should hold separate discussions, fill in a CDP each, and then negotiate a common CDP for the whole village, that is then signed by members of both sub-committees.

Community Priorities: Self Help

Describe the priority projects that your community can manage to implement by itself. Rank them in terms of importance

	Problems Identified	Suggested Solutions (using assets)	Estimated Costs (in AFA)
1		1	
		2	
		3	
2		1	
		2	
		3	
3		1	
		2	
		3	
4		1	
		2	
		3	

Community Priorities: Outside assistance needed

Describe the your community's priority projects that need outside help - on the technical front. Rank them in terms of importance

	Problems Identified	Suggested Solutions	Estimated Costs
1		1	1
		2	2
		3	3
2		1	1
		2	2
		3	3
3		1	1
		2	2
		3	3
4		1	1
		2	2
		3	3
5		1	1
		2	2
		3	3

Form 6: Continuation of Community Development Plan

Province	District	Community	Community ID#

Previous Community Projects

Type of project	Who sponsored the project? [NGO, UN, Govt., Self]	Year Started	Year Ended (Actual or Planned	Project costs

Presently managed communal facilities

Type of communal facility	How are contributions made	How often are contributions made

Are there any plans by the Government or other Agency to satisfy community priorities with programs? ☐ *YES* ☐ *NO*

If yes, who and what:

Members of the Community Development Council -
Male Sub-committee (for separate sub-committees) or mixed CDC

No.	Full Name	Position	Father's Name	Signature
1		Chair		
2		Vice-Chair		
3		Secretary		
4		Treasurer		
5		Member		
6		Member		
7		Member		
8		Member		
9		Member		
10		Member		
11		Member		
12		Member		
13		Member		
14		Member		
15		Member		

Form 6: Continuation of Community Development Plan

Province	District	Community	Community ID#

Members of the Community Development Council - Female Sub-committee (for separate sub-committees)

No.	Full Name	Position	Father's Name	Signature
1		Chair		
2		Vice-Chair		
3		Secretary		
4		Treasurer		
5		Member		
6		Member		
7		Member		
8		Member		
9		Member		
10		Member		
11		Member		
12		Member		
13		Member		
14		Member		
15		Member		

Annex E: NSP Form 7: Subproject Proposal

NSP Operational Manual, Version IV, November 2007

Islamic Republic of Afghanistan

Ministry of Rural Rehabilitation and Development (MRRD)

National Solidarity Programme

Form 7: Sub-project Proposal

Province	District	Community	Community ID#

Sub-project approval request

Sub-Project ID code	
Sub-Project title/description (including all component parts and measurements)	

Expected start date	Expected completion date	Total requested AFA

Install-ment	Description	Date expected	Amount (AFA) from NSP	Other Donors	Community Contribution
No. 1	90% of the block grant entitlement on approval of all subproject proposals that together amounts to 100% of the community block grant entitlement				
No. 2	10% of the block grant entitlement on expenditure of 70% of the first installment payment				

See Operations Manual (Financing and Fund Flow section) for further key conditions that MUST be met for installment release

Names and signatures

Community	Facilitating Partner
Signature	Signature
Name/Position	Name/Position
Signature	Signature
Name/Position	Name/Position
Signature	Signature
Name/Position	Name/Position
Signature	Signature
Name/Position	Name/Position

Place office stamp here

Date received by NSP/PMU	Name	Title	Signature

Form 7: Sub-project Proposal (continued)

Province	District	Community	Community ID#

Village Map and Location of Sub-Project

Facilitating Partner Provincial Manager Initials

Form 7: Sub-project Proposal (continued)

Province	District	Community	Community ID#

Engineering design if applicable (Attach extra sheets if necessary)

Facilitating Partner Provincial Manager Initials	

Form 7: Sub-project Proposal (continued)

Province	District	Community	Community ID#

Sub-Project Implementation Plan

No.	Physical Achievements	Anticipated Duration (in Weeks)																									
		1	2	3	4	5	6	7	8	9	10	11	12	13	14	15	16	17	18	19	20	21	22	23	24	25	26
1																											
2																											
3																											
4																											
5																											
6																											
7																											
8																											
9																											
10																											
11																											
12																											
13																											
14																											
15																											
16																											
17																											
18																											
19																											
20																											
* Indicate activities under first and second installments clearly																											

Form 7: Sub-project Proposal (continued)

Province	District	Community	Community ID#

Disbursement Schedule and Budget

Cost Breakdown of Items to be Completed Using the First Installment						
Item (e.g. Materials, Equipment, Labour, Transport, Training, etc.)	Quantity	Unit	Unit Cost	Total Cost	Contributions	
					Community	NSP
TOTAL COST						

Form 7: Sub-project Proposal (continued)

Province	District	Community	Community ID#

Disbursement Schedule and Budget

Cost Breakdown of Items to be Completed Using the Second Installment

Item (e.g. Materials, Equipment, Labour, Transport, Training, etc.)	Quantity	Unit	Unit Cost	Total Cost	Contributions	
					Community	NSP
TOTAL COST						

Form 7: Sub-project Proposal (continued)

Province	District	Community	Community ID#

Disbursement Schedule and Budget

Cost Breakdown of Items Required to Finish the Project						
Item (e.g. Materials, Equipment, Labour, Transport, Training, etc.)	**Quantity**	**Unit**	**Unit Cost**	**Total Cost**	**Contributions**	
					Community	**NSP**
TOTAL COST						

Form 7: Sub-project Proposal (continued)

Province	District	Community	Community ID#

Operations & Maintenance Plan (Attach extra sheets if necessary)

Explain methodology:
Cost Recovery Plan and Calculations:

Facilitating Partner Provincial Manager Initials	

Form 7: Sub-project Proposal (continued)

Province	District	Community	Community ID#

Land

1) **Does the sub-project require land?** ☐ Yes ☐ No

If so, how much land? ☐ Jreeb

How is this land provided? ☐ Donation ☐ Purchase: Willing seller, willing buyer ☐ Acquisition by authorities ☐ Available govt. land without squatters or encroachers

Documentation in case of donation, purchase or use of Govt. land attached? ☐ Yes ☐ No

Is the compensation plan for acquisition (including number of affected families, type of losses, compensation amounts and timing) attached? ☐ Yes ☐ No — Safeguard Forms #2 and/or #3 (see NSP Operational Manual) must be attached

Minorities

2) **Are there minority ethnic groups in the community?** ☐ Yes ☐ No

Are these groups included among the beneficiaries? ☐ Yes ☐ No

Mines

3) **Are mines suspected within the project area?** ☐ Yes ☐ No — If 'Yes', fill in and attach Form #7a; see also Note 1.

Cultural Property

4) **Will the sub-project have any negative impact on Afghan cultural property (e.g. historical monuments, archaeological sites)?** ☐ Yes ☐ No — If 'Yes', the project is not feasible

Poppy Cultivation

5) **Does the project involve irrigation?** ☐ Yes ☐ No

Is poppy cultivation planned within the land to be irrigated? ☐ Yes ☐ No

Are poppies cultivated in the community? ☐ Yes ☐ No

If the answer to either #5.2. or #5.3. of these questions is 'Yes', then the project must be rejected.

Environmental Impact

6) **Will the project have any negative environmental impact?** ☐ Yes ☐ No

If the answer is 'Yes', Safeguard Form #1 (see NSP Operational Manual) must be attached

---

Note 1

Mine Risk - The community must either confirm that there are no mines, or if mines are suspected, the project would either require survey and demining by MACA (or a MACA accredited agency) or abandoned as unfeasible. If the project is implemented by a contractor, mine risk information must be obtained from MACA. The community should be made aware, that it (or the CDC) is solely liable in the case of a mine-related accident, and that neither MRRD or the donors have any legal liability.

Annex F: Household Survey Questionnaire

تاریخ مصاحبه: ____________ (روز ماه سال) Interview Date	قریه: ____________ Village	شماره پرسشننامه: ____________ Questionnaire ID
وقت آغاز: ____________ Start Time	ولسوالی: ____________ District	امضا و یا نشانه مصاحبه شونده: Signature or Mark of Interviewee (نشاندهنده اجازه)
وقت ختم: ____________ End Time		

امنیت در قریه: Security in the Community

جواب	سوال	
	شما و خانواده شما فعلا به کدام تهدید هاي امنيتي روبرو هستيد؟ *لطفا از آنها بخواهيد تا در مورد امنيت فزيكي (شخصي)، امنيت اجتماعي، و امنيت اقتصادي سخن بگويند.* What security threats do you and your household currently face? *Please ask them to talk about Physical security (personal), Social security, and Economic security.*	1.
	تهدید های امروزی با تهدید های که در زمان قدرت طالبان به ان روبرو بودید چه تفاوت دارد؟ (1996-2001) *لطفا از آنها بخواهيد تا در مورد امنيت فزيكي (شخصي)، امنيت اجتماعي، و امنيت اقتصادي سخن بگويند. همچنان از آنها در مورد حكومتداري بپرسيد.* How do today's threats differ from the threats you faced when the Taliban were in power? (1996-2001) *Please ask them to talk about Physical security (personal), Social security, and Economic security. Also ask them about Governance.*	2.
	تهدید های امروزی با تهدید های که در زمان قدرت طالبان به ان روبرو بودید چه تفاوت دارد؟ (1996-2001) *لطفا از آنها بخواهيد تا در مورد امنيت فزيكي (شخصي)، امنيت اجتماعي، و امنيت اقتصادي سخن بگويند. همچنان از آنها در مورد حكومتداري بپرسيد.* What security threats did you face when the Mujahiden were in power? (1989-1996) *Please ask them to talk about Physical security (personal), Social security, and Economic security. Also ask them about Governance.*	3.
	در زمان اقتدار روس ها با کدام تهدیدها مواجه بودید؟ (1989-1996) *لطفا از آنها بخواهيد تا در مورد امنيت فزيكي (شخصي)، امنيت اجتماعي، و امنيت اقتصادي سخن بگويند. همچنان از آنها در مورد حكومتداري بپرسيد.* What security threats did you face when the Russians were in power? (1979-1989) *Please ask them to talk about Physical security (personal), Social security, and Economic security. Also ask them about Governance.*	4.
لطفا از کمترین الی بیشترین دوره امنیت درجه بندی نمائید ، طوريکه (1) نشاندهنده بیشترین امنیت باشد (1-4). _______ دوره اقتدار روس ها Russian Rule _______ دوره اقتدار مجاهدین Mujahideen Rule _______ دوره اقتدار طالبان Taliban Rule _______ دوره بعد از طالبان Post-Taliban Rule	در کدام دوره احساس امنیت کمتر را مینمودید؟ During which time period did you feel the least secure?	5.
	منابع اساسي كشمكش (نزاع) در قريه شما چيست؟ لطفا توضيح نمايد. What are the main sources of conflict in your community? Please provide details.	6.
a. حكومت Government b. پوليس Police c. اعضاي شورای انكشافي محل (CDC) CDC members d. خان (ملك) و يا كدام شخص بانفوذ ديگر Khan or power person e. موسفيدان (بزرگان) محل Community elders *f.* اعضاي فاميل Family members *g.* سايرين: Other ____________	معمولاً كي ها در راستاي حل اين مشكلات همكاري مينمايند؟ Who normally helps to resolve these problems?	7.

	سوال	جواب
8.	لطفا بگویید که پروگرام همبستگی ملی چیست؟ *اگر آنها در مورد پروگرام همبستگی ملی چیزی را نمي دانستند، لطفا آنرا در اینجا یادداشت نماید، اما قبل از اینکه سوالات آتي را مطرح سازید براي آنها پروگرام همبستگی ملی را توضیح نماید.* Can you please explain what NSP is? *If they did not know what NSP is, please note that here, but then explain for them before asking the following questions.*	
9.	پروگرام همبستگی ملی توسط كي تمويل میشود؟ Who provides funding for NSP?	لطفا آنها را بگذارید که پاسخ خود را ارایه نمایند و شما مداخله ننماید. prompt Please do not a. جامعه Community b. موسسه پاملرنه CARE c. وزارت احیأ و انکشاف دهات MRRD d. حکومت Government e. سایرین: Other ____________
10.	پروگرام همبستگی ملی چه تغیرات را براي شما و فامیل شما بمیان آورده است؟ *لطفا از آنها در مورد تغیرات مثبت و منفي بپرسید. مثالها.* What differences has NSP made for **you and your family**? *Please ask them for both positive changes and negative changes. Examples.*	
11.	پروگرام همبستگی ملی برای قریه شما کدام منفعت ها را ببار آورده است؟ What are the benefits NSP has brought to **your community**?	
12.	م همبستگی ملی برای قریه شما کدام هزینه ها و یا مشکلات را ببار اورده است؟ What are the costs or the problems NSP has brought to **your community**?	
13.	آیا شما در انتخابات شورا انکشافی محل رای داده اید؟ Did you vote in the CDC elections?	a. بلی Yes b. نخیر No
14.	چه تعداد از اعضای فامیل شما رای داده اند؟ How many of your family members voted?	________ ذکور و ________ اناث Males Females
15.	ایا کدام یک از اعضای خانواده شما عضو شوراي انکشافی محل است؟ Are any of your household members CDC members?	a. بلی، مشخص نمایید: Yes: Specify ____________ b. نخیر No c. نمیدانم Don't know
16.	سه اولویت مهم انکشافي شما کدام ها بودند؟ What were your top three development concerns?	لطفا از آنهایکه داراي اهمیت بیشترین اند بسوي آنهایکه داراي اهمیت کمتر اند درجه بندی نمایید، در حالیکه (1) از ارزش بیشترین نماینده گي نماید (1-3). ____ آب صحي نوشیدني Clean drinking water ____ آب زراعتي Agricultural water ____ سرك هاي بازسازي شده Fix roads ____ مرکز اجتماعي Community center ____ اعمار يك مكتب Build a school ____ برق Electricity ____ كلينيك صحي Health clinic ____ بيكاري Unemployment ____ سرپناه ها Shelters ____ مالداري Livestock ____ ديوار هاي مانع سيلاب Flood walls ____ سایرین: Other ____________
17.	چه تعداد از اعضای بالغ خانواده شما در جلسه هاي برنامه همبستگی ملی اشتراک نموده اند؟ How many adults from your household participated in NSP meetings?	________ ذکور و ________ اناث Males Females

18.	در نیابت آنعده بالغین خانوادهء شما که در مجالس انکشافی محل اشتراک ننموده اند، لطفا واضح سازید که آنها چرا اشتراک نه کردند؟ For those who did not attend the meetings, please explain why they did not participate.	
19.	كي ها به بهترين شکل به ضروريات انکشافي شما پاسخ ميگويند؟ Who serves your development needs the best?	لطفا آنها را بگذارید که پاسخ خود را ارایه نمایند و شما مداخله ننماید. Please prompt with the following answers a. جامعه Community b. حکومت Government c. موسسات بین المللي International Organizations d. نمیدانم Do not know e. سایرین: Other ________
20.	لطفا واضح سازيد که شما چرا این گزینه را انتخاب نمودید. شما در این چه منفعت ها را مي بینید؟ کمي ها و کاستي هاي دو گزینه دیگر کدام ها اند؟ Please explain why you answered as you did above. What benefits do you see? What are the drawbacks of the other two options?	

مهاجرت: Migration

	سوال	جواب
21.	آیا کدام عضو از فامیل شما تا به حال خلم را ترك گفته است؟ *این در مورد مهاجرت است.* Has anyone from your household ever left Khulm? *This is about migration*	اگر هیچ کسي مهاجرت نکرده باشد، لطفاً موضع را به بخش معلومات احصاییوي در مورد خانواده عوض نمائید. **If no one migrated, please skip to Household Demographics** a. بلي Yes b. نخير No
22.	اگر پاسخ تان بلي است، پس کدام عضو فامیل شما خلم را ترك گفته است؟ If yes, which members of your family left Khulm?	a. همه فاميل شما The whole family b. رئيس فاميل شما (مرد) Male head of household c. ديگر اعضاي مرد فاميل شما Other male household member d. رئيس فاميل شما (زن) Female head of household e. ديگر اعضاي زن فاميل شما Other female household member f. سايرين: Other ________
23.	آنها به کجا رفتند؟ Where did they go?	a. مزار شريف Mazar b. كابل Kabul c. جاي ديگري در افغانستان: Other AF ________ d. پاكستان Pakistan e. ايران Iran f. مملکت خارجي ديگر: Other foreign ________
24.	آنها در کدام سال خلم را ترك گفتند؟ When did they leave Khulm?	سال Year ________
25.	آنها چرا خلم را ترک نمودند؟ Why did they leave Khulm?	
26.	آيا انها براي فاميل شان در خلم پول فرستادند؟ *حواله (فرستادن پول)* Did they send money back to family in Khulm? Remittances	a. بلي، آنها قبل از بازگشت شان پول ميفرستادند Yes, they did before they returned b. بلي، آنها تا به حال پول ميفرستند Yes, and they still do c. نخير، در خلم کسي از فاميل انها باقي نه مانده بود No, there was no family left in Khulm d. نخير No

اگر آنها بر نه گشته اند، در اینجا NA بگذارید و همچنان موضع را به بخش معلومات احصاییوي در مورد خانواده عوض نمائید. **If they did not return, enter NA and skip to Household Demographics** __________ Year سال	آنها در کدام سال به خلم برگشتند؟ When did they return to Khulm?	**27.**
Why did they return to Khulm?	آنها چرا به خلم برگشتند؟	**28.**

Household Demographics **معلومات احصاییوي در مورد خانواده**

سوال	جواب	
هر فردیکه عمرش 18 سال و یا بالا از آن باشد بزرگسال خوانده میشود *Adults are defined as anyone 18 years or older* ______ ذکور و ______ اناث Males Females	تعداد بزرگسالان که در خانواده زنده گی میکنند # of Adults living in the household	**29.**
______ ذکور و ______ اناث Males Females	تعداد بزرگسالان باسواد # of literate adults	**30.**
______ ذکور و ______ اناث Males Females	تعداد اطفال بین سنین 7 و 18 سال # of Children between the ages of 7 and 18	**31.**
______ ذکور و ______ اناث Males Females	تعداد اطفال (بین سنین 7 و 18 سال) که به مکتب میروند. # of Children (7 to 18 years old) who are attending school	**32.**
______ ذکور و ______ اناث Males Females	تعداد اطفال زیر سنین 7 سال # of Children below 7 years old	**33.**
فامیل = یك مرد و همسر (یا همسران) او و اطفالشان. ______ فامیل ها Families	تعداد فامیل های هسته اي که در خانواده زنده گی میکنند # Nuclear families living in the household	**34.**
لطفا از آنها بخواهید که زبانها را درجه بندي نمايند، در حالیکه 1 از زبان مادري و 2 از زبان دوم نماينده گي نمايد و غيره. Please ask them to rank the languages with the mother tongue being 1, the second language being 2, etc. ______ دري Dari ______ پشتو Pashtu ______ ازبکي Uzbeki سایرین: ______	فامیل شما به کدام زبان ها صحبت میکنند؟ What languages does your family speak?	**35.**

Household Livelihood and Poverty: **تنگدستی و تامین معیشت خانواده**

جواب	سوال	
لطفا از آنهایکه داراي اهمیت بیشترین اند بسوي آنهایکه داراي اهمیت کمتر اند درجه بندی نمایید، در حالیکه (1) از ارزش بیشترین نماینده گی نماید (1-4). زراعت ______ Agriculture کارگر روز مزد ______ Daily Labor نسل گیری حیوانات ______ Animal Husbandry دوکاندار ______ Shop Keeper صنایع دستی ______ Handicrafts دیگر مشخص نمایید ______ Other: Specify ______	فعالیت های اقتصادی عمده این خانواده کدام ها اند؟ What are the main economic activities of this household?	**36.**
هر فردیکه عمرش 18 سال و یا بالا از آن باشد بزرگسال خوانده میشود *Adults are defined as anyone 18 years or older* ______ ذکور و ______ اناث Males Females	چه تعداد از بزرگسالان در پیدا کردن درآمد اقتصادي خانواده سهیم اند؟ How many adults contribute to the household economically?	**37.**

	اطفال به آنهاي گفته میشود که عمرشان بین سنین 7 و 18 سال باشد. *Adults are defined as anyone 18 years or younger* _______ ذکور و _________ اناث Males Females	چه تعداد از اطفال در پیدا کردن درآمد اقتصادي خانواده سهیم اند؟ How many children contribute to the household economy?	**38.**
	__________ مرتبه هر ماه per month	بطور اوسط فامیل شما ماهانه چند مرتبه از پیدا نمودن اعاشه عاجز است (اعاشه روزمره)؟ On average, how many times per month is your family unable to find *roozi* (daily subsistence)?	**39.**
	a. خوبتر Better b. یکسان Same c. بدتر Worse	بمقایسه سال گذشته حالت اقتصادی فامیل شما چطور است؟ Compared to last year, how is your family doing economically?	**40.**
________ جریب زمین که ملکیت آنرا دارد و یا هم آنرا به دهقاني و یا هم اجاره گرفته است. jeribs of land owned or rented	a. بلي Yes b. نخیر No c. نخیر، من زمین را به دهقاني و یا هم اجاره میگیرم No, but I rent land	آیا شما درین قریه مالک زمین شخصی هستید؟ اگر بلی، چقدر؟ Do you own land in this community? If yes, how much?	**41.**
	a. نل داخل خانه Tap in the house b. چاه شخصی Personal well c. چاه عامه (قریه) Communal well d. جوی محلی Local stream e. سایرین: مشخص سازید: Other ____________	کدام منبع اب را استفاده میکنید؟ What water source do you use?	**42.**
________ ساعت در روز hours per day	a. برق شهری City power b. جنراتور قریه Community generator c. جنراتور شخصی Personal generator d. سایرین: مشخص سازید: Other ____________ e. هیچکدام None	اگر پاسخ تان بلی باشد، روزانه چند ساعت برق دارید، و از کدام منبع؟ What kind of electricity do you have and for how many hours a day?	**43.**

Description of the House and the Interviewee: **توضیحات در مورد خانه و مصاحبه شونده**

سوال	جواب	
نمبر خانه و یا مشخصات آن. لطفا موقیعت این خانه را در نقشه ایکه شما ساخته اید نشاني نماید. House number or identifying characteristics. Please mark the location of this house on the map you created.		**44.**
a. مرد Male b. زن Female	جنسیت مصاحبه شونده کدام بود؟ What was the gender of the interviewee?	**45.**
__________ Year سال	عمر تخمیني مصاحبه شونده Approximate age of the interviewee	**46.**
a. رئیس فامیل شما (مرد) Male head of household b. دیگر اعضاي مرد فامیل شما Other male household member ____________ c. رئیس فامیل شما (زن) Female head of household d. دیگر اعضاي زن فامیل شما Other female household member ____________ e. سایرین: Other ____________	موقف مصاحبه شونده در فامیل اش Family position of interviewee	**47.**

اسم مصاحبه کننده : ______________________________
Name of Interviewer

امضاي مصاحبه کننده: ______________________________
Signature of Interviewer

Annex G: Household Survey Participants

No.	ID	Date	Village	Gender	Age	Position	No. of Families in Household
1.	CS001f	February 16, 2010	Chahar Sukh	Female	40	Daughter of head of household	1
2.	CS002	February 16, 2010	Chahar Sukh	Male	43	Male head of household	1
3.	CS003	February 17, 2010	Chahar Sukh	Male	23	Son of head of household	1
4.	CS004	February 20, 2010	Chahar Sukh	Male	42	Male head of household	2
5.	CS005	February 20, 2010	Chahar Sukh	Male	60	Male head of household	2
6.	CS006	February 21, 2010	Chahar Sukh	Male	19	Son of head of household	2
7.	SR001	January 20, 2010	Shah Raheem	Male	50	Male head of household	1
8.	SR002	January 26, 2010	Shah Raheem	Male	68	Male head of household	2
9.	SR003	January 27, 2010	Shah Raheem	Male	79	Male head of household	1
10.	SR004	January 27, 2010	Shah Raheem	Male	30	Male head of household	3
11.	SR005	February 2, 2010	Shah Raheem	Male	72	Male head of household	1
12.	SR006	February 2, 2010	Shah Raheem	Male	60	Male head of household	1
13.	SR007	February 3, 2010	Shah Raheem	Male	35	Male head of household	2
14.	SR008	February 7, 2010	Shah Raheem	Male	60	Male head of household	1
15.	SR009	February 7, 2010	Shah Raheem	Male	40	Male head of household	3
16.	SR010	February 9, 2010	Shah Raheem	Male	20	Son of head of household	1
17.	SR011	February 9, 2010	Shah Raheem	Male	30	Male head of household	1
18.	SR012	February 10, 2010	Shah Raheem	Male	44	Male head of household	1
19.	SR013	February 14, 2010	Shah Raheem	Male	22	Son of head of household	2
20.	SR014	February 14, 2010	Shah Raheem	Male	23	Son of head of household	1
21.	SR015f	February 21, 2010	Shah Raheem	Female	34	Female head of household	1
22.	SR016f	February 21, 2010	Shah Raheem	Female	29	Female head of household	1
23.	YB001	February 21, 2010	Yaqubai	Male	50	Male head of household	1

No.	ID	Date	Village	Gender	Age	Position	No. of Families in Household
24.	YB002	February 21, 2010	Yaqubai	Male	79	Male head of household	2
25.	YB003	February 23, 2010	Yaqubai	Male	53	Male head of household	1
26.	YB004	February 24, 2010	Yaqubai	Male	47	Male head of household	1
27.	YB005	February 24, 2010	Yaqubai	Male	65	Male head of household	3
28.	YB006	February 27, 2010	Yaqubai	Male	22	Son of head of household	1
29.	YB007	February 28, 2010	Yaqubai	Male	50	Male head of household	3
30.	YB008	February 28, 2010	Yaqubai	Male	42	Male head of household	4
31.	YB009	February 28, 2010	Yaqubai	Male	50	Male head of household	1
32.	YB010	February 28, 2010	Yaqubai	Male	22	Male head of household	1

Annex H: Market Survey Questionnaire

Survey Date:		Name:

Please walk the main road of Khulm and list each store you see on the LEFT-hand side of the road. You should begin recording this information at the corner of the main road and the road of the governor's office. Start here and walk away from the main road from Mazar to Kabul. *(Please draw a double line to indicate where there are roads).*

Please stop and speak with any shopkeepers who sell the following products and record in the Shopkeepers Questionnaire (one form per shop): **pomegranates, almonds, eggs, wheat, oil, rice, sugar, diesel, and petrol**. Please speak to at least four shopkeepers who sell each of the products above and triangulate the information they provide. If at this point you are getting the same answers for a product you can stop asking about it, if not, try to find out why you are getting different answers by asking a few more shopkeepers.

	Type of Shop	Additional specifics about products sold
1.		
2.		
3.		
4.		
5.		
6.		
7.		
8.		
9.		
10.		
11.		
12.		
13.		
14.		
15.		
16.		

Codes for common types of shops:			
a. Wood seller	**g.** Pharmacy	**l.** Blanket and scarves	**r.** Plastic goods seller
b. Rope seller	**h.** Bread seller	**m.** Jewelry	**s.** Grain sellers
c. Shoes	**i.** Fruit and vegetable seller	**n.** Metal workers	**t.** Gasoline
d. Tailor	**j.** Meat seller	**o.** Bookstore	**u.** Movies/DVDs
e. Telephone vendor	**k.** Handicrafts	**p.** Grocery	**v.** Clothing
f. Photocopy and photos		**q.** Auto repair shop	**w.** Other: ________

Annex I: Shopkeeper Questionnaire

Survey Date:		Name:

Please introduce yourself: Good morning (or afternoon). My name is Waheed. I am helping conduct research on Khulm District and its economy for a PhD in Germany. We are trying to talk to shopkeepers in order to better understand how the market of Khulm works. Do you have ten minutes to answer a few questions for us? At the end of the interview, please remember to thank them for their time and information.

	Question	Answer
1.	*Number and type of shop from sheet above*	
2.	Name of Shopkeeper	
3.	# years in business	

Please circle any products sold in this shop below and fill in the table for those products.

	Product	Suppliers (Name and Village or City)	# of suppliers you purchase from	How much do you buy on average each month?	Price shopkeeper pays the producer or supplier	Price Shopkeeper charges customers	How much sold per month?
1.	Pomegranates						
2.	Almonds						
3.	Eggs						
4.	Wheat						
5.	Oil						
6.	Rice						
7.	Sugar						
8.	Diesel						
9.	Petrol						

13. How has the supply changed for these products over the last thirty years? What caused these changes? *Please try to get them to be specific about the time period they are discussing: Post Taliban, Taliban, Mujahideen, or Soviet.*

14. How have your sales changed over the last 30 years? What caused these changes? *Please try to get them to be specific about the time period they are discussing: Post Taliban, Taliban, Mujahideen, or Soviet.*

15. How has the price of this product changed over the last thirty years? Please provide details, such as hard numbers for increases and decreases.

16. Please place any additional information here. If it is for a specific question, please list that question number fist before writing the answer.

17. Please describe the size of the store and its location from a major landmark.

Annex J: Shopkeeper Questionnaire Participants

No.	ID	Shop Type	Date	Gender	# Years in Business	Products Discussed
1.	SKQ01	Grains	March 3, 2010	Male	20	Pomegranates, almonds, wheat, and rice
2.	SKQ02	Grains	March 3, 2010	Male	15	Almonds, flour, wheat, barley, and rice
3.	SKQ03	Grocery	March 7, 2010	Male	20	Pomegranates, almonds, eggs, oil, rice, and sugar
4.	SKQ04	Grocery	March 7, 2010	Male	20	Eggs, oil, rice, and sugar
5.	SKQ05	Grains and Almonds	March 8, 2010	Male	30	Almonds and rice
6.	SKQ06	Grocery	March 9, 2010	Male	15	Oil, rice, and sugar
7.	SKQ07	Gasoline	March 9, 2010	Male	12	Diesel and petrol
8.	SKQ08	Grains	March 9, 2010	Male	10	Almonds and rice
9.	SKQ09	Grocery	March 10, 2010	Male	11	Oil and sugar
10.	SKQ10	Mechanic	March 10, 2010	Male	10	Diesel and gasoline
11.	SKQ11	Mechanic	March 10, 2010	Male	6	Diesel and gasoline
12.	SKQ12	Grocery	March 10, 2010	Male	15	Eggs, oil, rice, and sugar
13.	SKQ13	Auto parts	March 10, 2010	Male	17	Diesel and gasoline
14.	SKQ14	Grocery	March 14, 2010	Male	27	Almonds, eggs, wheat, oil, rice, and sugar
15.	SKQ15	Grocery and fruits	March 14, 2010	Male	30	Pomegranates, almonds, eggs, wheat, oil, rice, and sugar
16.	SKQ16	Grains	March 15, 2010	Male	25	Wheat and rice
17.	SKQ17	Grocery and fruits	March 15, 2010	Male	35	Pomegranates, almonds, eggs, oil, rice, and sugar
18.	SKQ18	Fruit and vegetables	March 15, 2010	Male	12	Pomegranates
19.	SKQ19	Grocery	March 15, 2010	Male	50	Almonds, eggs, wheat, and rice

Annex K: Facilitating Partner Staff Key Informant Interview Participants

No.	ID	Staff Position	Date	Gender
1.	FP01	CARE Country Director, Kabul	June 30, 2009	Male
2.	FP02	CARE Security Officer, Kabul	June 30, 2009	Male
3.	FP03	CARE NSP Program Manager, Kabul	June 29, 30, 2009 July 1, 2009 January 11, 14, 2010	Male
4.	FP04	CARE Data Management Officer, Kabul	January 12, 2010	Male
5.	FP05	CARE NSP Manager, Balkh Province	July 4, 2009 January 14, 31, 2010	Male
6.	FP06	CARE Provincial Manager, Balkh Province	July 6, 7, 8, 2009 January 14, 2010	Male
7.	FP07	CARE NSP Senior Community Mobilizer, Khulm District	July 12, 2009	Male
8.	FP08	CARE NSP Community Mobilizer, Khulm District	July 9, 14, 15, 16, 20, 22, 28, 2009 January 17, 18, 26, 2010 February 7, 14, 17, 2010	Male
9.	FP09	CARE NSP Technical Facilitator, Khulm District	July 20, 2009	Male
10.	FP10	CARE Acting District Supervisor, Khulm District	July 21, 2009	Male
11.	FP11	CARE Head of Security, Balkh Province	July 23, 2009 January 16, 2010	Male
12.	FP12f	Former CARE NSP Community Mobilizer, Khulm District	August 5, 2009 February 4, 2010	Female
13.	FP13	CARE Technical Facilitator, Balkh Province	July 6, 2009	Male
14.	FP14	CARE NSP Technical Facilitator, Khulm District	July 15, 16, 2009	Male

Annex L: Community Member Key Informant Interview Participants

No.	ID	Position of Title	Date	Gender	Village
1.	Com01f	Housewife	March 6, 2010	Female	Yaqubai
2.	Com02	Male *kalāntar* for Shah Raheem	January 20, 24, 2010 February 2, 2010 March 10, 2010	Male	Shah Raheem
3.	Com03f	Female *kalāntar* of Shah Raheem	January 20, 24, 2010 February 10, 14, 2010	Female	Shah Raheem
4.	Com04	Unemployed	January 26, 2010	Male	Shah Raheem
5.	Com05	Former carpenter, map maker, and elder	January 26, 31, 2010 February 2, 3, 7, 2010	Male	Shah Raheem
6.	Com06	Retired elder	January 27, 2010	Male	Shah Raheem
7.	Com07f	Hairdresser	January 27, 2010	Female	Shah Raheem
8.	Com08	Driver in Mazar-i-Sharif	February 2, 2010	Male	Shah Raheem
9.	Com09f	Head of household	February 3, 2010	Female	Shah Raheem
10.	Com10	Candy producer	February 7, 2010 March 14, 2010	Male	Shah Raheem
11.	Com11	Student	February 9, 2010	Male	Shah Raheem
12.	Com12f	Housewife	February 14, 2010	Female	Shah Raheem
13.	Com13	Brother of *kalāntar*, unemployed	February 14, 2010 March 14, 2010	Male	Shah Raheem
14.	Com14f	Housewife	February 16, 2010	Female	Chahar Sukh
15.	Com15	Landowner, teacher, and CDC Treasurer	February 17, 2010 March 10, 2010	Male	Chahar Sukh
16.	Com16f	Wife of a gardener	February 20, 2010	Female	Shah Raheem
17.	Com17	Brother of candy producer	February 7, 2010	Male	Shah Raheem
18.	Com18	Retired elder	February 23, 2010	Male	Yaqubai
19.	Com19	Landowner, businessman, elder	February 24, 2010	Male	Yaqubai
20.	Com20f	Housewife	February 26, 2010	Female	Chahar Sukh
21.	Com21f	Female *kalāntar* of Yaqubai, elder	February 26, 2010	Female	Yaqubai
22.	Com22f	Housewife	February 27, 2010	Female	Shah Raheem
23.	Com23f	Daughter-in-law of female *Kalāntar*	February 26, 2010	Female	Yaqubai
24.	Com24	Gardener	January 27, 2010	Male	Yaqubai
25.	Com25	Male *kalāntar* of Yaqubai	March 3, 4, 2010	Male	Yaqubai
26.	Com26	Miller	March 3, 7, 2010	Male	Yaqubai
27.	Com27f	Head of the local girls school	March 6, 2010	Female	Yaqubai
28.	Com28f	Mother-in-law of farmer	March 7, 2010	Female	Yaqubai
29.	Com29	Former *kalāntar*	March 7, 2010	Male	Yaqubai
30.	Com30	Gardener	March 7, 2010	Male	Yaqubai
31.	Com31	Retired gardener	March 9, 2010	Male	Yaqubai
32.	Com32	Gardener	March 10, 2010	Male	Chahar Sukh
33.	Com33	CDC member, teacher	March 10, 2010 September 26, 2011	Male	Chahar Sukh
34.	Com34	Imam	March 16, 2010	Male	Yaqubai
35.	Com35	Teacher	March 16, 2010	Male	Yaqubai
36.	Com36	Farmer	March 16, 2010	Male	Shah Raheem
37.	Com37	Son of CDC member	May 31, 2011	Male	Chahar Sukh
38.	Com38	Farmer	August 20, 2011	Male	Chahar Sukh
39.	Com39	Teacher	February 23, 2010	Male	Yaqubai
40.	Com40	Farmer	February 26, 2010	Male	Yaqubai

Annex M: External Key Informant Interview Participants

No.	ID	Organization	Date	Position	Gender
1.	EX01	UNDP - NABDP	July 26, 2009	Provincial Officer	Male
2.	EX02	ACDI/VOCA	July 27, 2009	Project Officer	Male
3.	EX03	Mazar-i-Sharif Hospital	July 28, 2010	Doctor	Male
4.	EX04	Provincial level Directorate of Rural Rehabilitation and Development (RRD), Balkh	July 29, 2009 January 25, 2010	RRD NSP Provincial Manager	Male
5.	EX05	District Education Department, Khulm	August 2, 2009	Directorate of Education	Male
6.	EX06	District Government, Khulm	August 3, 2009 January 18, 2010	District Governor	Male
7.	EX07	District level Directorate of Rural Rehabilitation and Development (RRD), Khulm	August 4, 2009	RRD District Representative	Male
8.	EX08	District Hospital, Khulm	August 3, 2009	Chief of Hospital and Medical Doctor	Male
9.	EX09	Tufts University, Feinstein International Center	January 16, 23, 2010	Former resident of Khulm, Visiting Scholar	Male
10.	EX10	District Government, Khulm	January 18, 2010	Mayor of Tashqurghan	Male
11.	EX11	District Police, Khulm	January 18, 2010	Chief of Police	Male
12.	EX12	Self-employed	January 19, 2010	Head of the Kabul-Mazar Transport Station	Male
13.	EX13	District Agricultural Department, Khulm	August 15, 2009 January 24, 31, 2010 February 7, 2010	Directorate of Agriculture	Male
14.	EX14	Multiple	January 16, 26, 27, 2010 February 17, 2010	Plumber and NGO staff	Male
15.	EX15	ACTED	January 26, 2010	Project Officer	Male
16.	EX16f	HLP	January 31, 2010	Trainer	Female
17.	EX17	Self-employed	February 2, 2010	Taxi Driver	Male
18.	EX18	Self-employed	February 2, 2010	Taxi Driver	Male
19.	EX19	Self-employed	February 2, 2010	Taxi Driver	Male
20.	EX20	Self-employed	February 2, 2010	Taxi Driver	Male
21.	EX21	Mercy Corps	February 4, 2010	Staff member	Male

Annex N: Focus Group Discussions Held

No.	ID	Type	Date	Location	Male	Female
1.	FGD01	Heads of male families from all three villages	January 19, 2010	Yaqubai	12	0
2.	FGD02f	Female youth	January 20, 2010	Shah Raheem	0	12
3.	FGD03	German police trainers (PMT)	January 24, 2010	Regional Command North	4	0
4.	FGD04mix	Female and male elders of one family	January 26, 2010	Shah Raheem	3	3
5.	FGD05f	Female elders	January 27, 2010	Shah Raheem	0	4
6.	FGD06	Male CDC members	February 3, 2010	Shah Raheem	13	0
7.	FGD07f	Female heads of families	February 9, 2010	Shah Raheem	0	18
8.	FGD08f	Females involved in NGO activities	February 9, 2010	Chahar Sukh	0	9
9.	FGD09f	Females involved in Hand in Hand	February 16, 2010	Chahar Sukh	0	12
10.	FGD10f	Females of one family	March 6, 2010	Shah Raheem	0	4
11.	FGD11	District Level FGD	July 10, 2011	Tashqurghan	4	0
				Total	**36**	**62**

Annex O: Map of the Khulm River Basin

Figure 26: The Khulm Watershed, a Dendritic River System Whose Drainage Flows North Towards the Amu Darya River, but Falls Well Short, Draining Into the Turkestan Plain

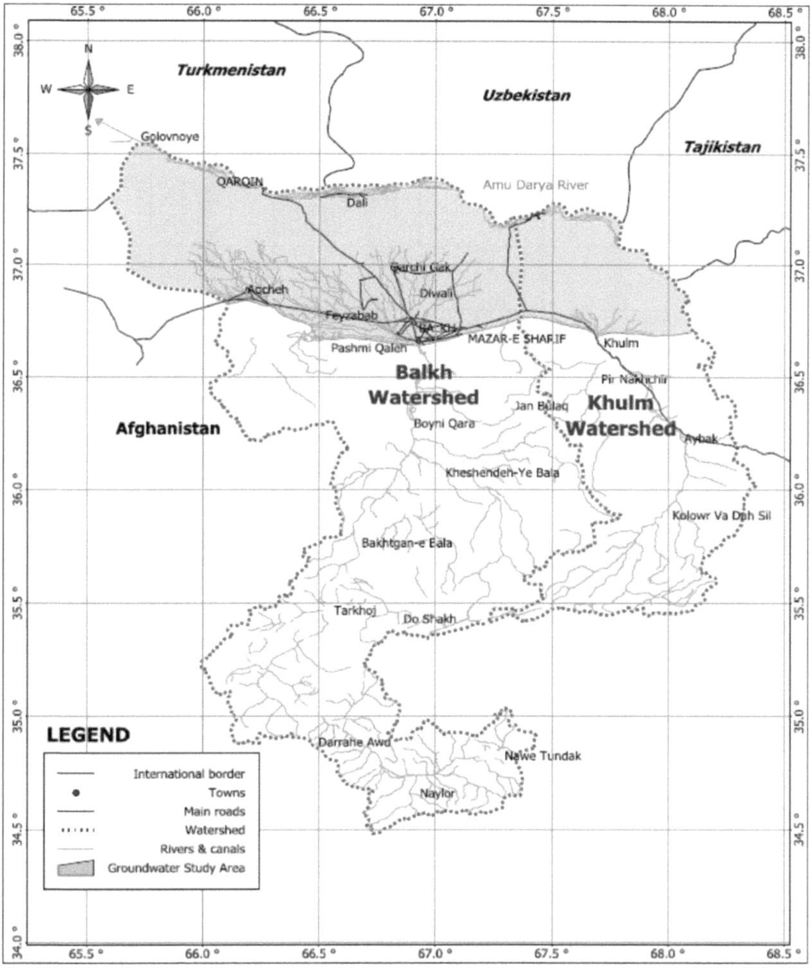

Source: (Ashworth 2005: Figure 1)

Annex P: Transliteration of the Persian Alphabet

Persian Alphabet	Transliteration	Persian Alphabet	Transliteration	Persian Alphabet	Transliteration
ا	a	ر	r	ف	f
ب	b	ز	z	ق	ġ (q)
پ	p	ژ	ž (zh)	ك	k
ت	t	س	S	گ	g
ث	s (th)	ش	š (sh)	ل	l
ج	ğ (j)	ص	S	م	m
چ	č (sh)	ض	z (d)	ن	n
ح	h	ط	t	و	w (v)
خ	x (kh)	ظ	z	ه	h
د	d	ع	‘	ي	i, j
ذ	z (dh)	غ	ġ (gh)		

Glossary of Terms

abāyah	A simple loose robe like garment worn mostly by women that covers the whole of the body except for the face, the hands, and the feet, often written *abaya* or *abiyah*
amir	A title of high office used throughout the Muslim world, it can also be written *emir*
anār	Pomegranate
bādām	Almond
Block grant	The grant provided by the National Solidarity Programme (NSP) to eligible communities to fund approved development subprojects
burqa (borġa')	A full body veil with only a fabric grill from which the wearer can see, known in Dari as a *chaderi*, often written *burqa*
chaderi	Please see *burqa*
Dari	One of the two official languages of Afghanistan, a Persian dialect which is the mother tongue of many northern Afghans
Eid	Eid is the abbreviated term for *Eid al-Fitr* and *Eid al-Adha*, two Muslim holidays, the first of which marks the end of the holy month of *ramazān* and the second which follows the end of the *hāǧǧ* annually
Facilitating	National Solidarity Programme (NSP) partners, usually an
Partner	international or national non-governmental organization (NGO), who assists communities in implementing the program
family	Consists of an adult male and his wife (or wives) and their non-married children. This is the definition used by National Solidarity Programme (NSP) and therefore used to maintain standardization in this research
gozār	Small street, often written *gozar*
habitus	A term used by Pierre Bourdieu to indicate a set of socially learned dispositions, skills and ways of

	acting, that are often taken for granted, and which are acquired through the activities and experiences of everyday life
hāǧǧ	Annual pilgrimage to Mecca that is one of the five requirements of every able-bodies Muslin who can afford it to partake in at least once in their lifetime, often written *haj*
ham-bastegi melli	Name of the National Solidarity Programme in Dari
hašr	Voluntary communal work
HESCO	A collapsible wire mesh and fabric bag that is filled with sand, soil, or rocks and acts as a barrier against bullets, explosions, etc.
Household	All individuals who eat from the same pot
jerib (ǧarib)	Unit of measure used for area (one *ǧarib* is equal to 1,953.64 meters squared or about 0.2 hectares), often written *jerib*
jihad (ǧehād)	Derived from the Arabic root meaning "to struggle" or "to strive," often written *jihad*
jirga	A tribal or deliberative council, a traditional means of conflict resolution in Afghanistan (Pashtu)
kalāntar	Village leader selected by male community members
kār(e) wān-sarāj	A roadside inn where travelers can rest and recover for the night, popular during the time of the Silk Road, often written *caravanserai*
Khalq	A faction of the People's Democratic Party of Afghanistan (PDPA)
Khan (Xān)	A Turkic word for a leader, in Afghanistan it is used to describe someone with power, usually an individual who owns land, often written *Khan*
Khanate	A polity ruled by a *Xān*
kocha	A small street or alley
Koran	The holy book of Islam consisting of the revelations of God to the Prophet Muhammad (PBUH), also written as Qur'an
Loya Jirga	Grand council (Pashtu)
ma'dan	A geographical plain, often written *maidan*

madrasa	A religious school, usually attached to a mosque, where young Muslims study theology, often written *madrasa*
mahr	Bride price, also written as *mehr*
mahram	A male relative who accompanies a women outside of the home
malik	Village headman
masǧed	A mosque, often written *masjid*
Mazar	The city of Mazar-i-Sharif, the capital of Balkh Province
Melli Pejwastan	Name of the National Solidarity Programme in Pashtu, also written *Milli Pawastoon*
mirab (mir'āb)	Person responsible for the distribution of water in a village
Mo'assasāt-i hamkār	Facilitating partner of the National Solidarity Programme (NSP)
moǧāhad	In the Afghan context, it refers to the individuals who fought on behalf of the Mujahideen groups against different central governments based in Kabul, often written *Mujahid*
Mujahideen	The plural of *moǧāhad*, or "one who engages in *ǧehād*." It is often translated as "holy warrior." Different Mujahideen groups tried to rule Afghanistan from 1992 to 1996
mullah (mollā)	Someone who is educated in Islamic theology and sacred law, often the religious leader of a village, who leads Friday prayers in the local mosque, often written *mullah*
Nou-ruz	Persian New Year celebrated on the 21st of March
NSP community	A community in Afghanistan consisting of 25-300 families who have registered with the Ministry of Rural Rehabilitation and Development (MRRD) and where the National Solidarity Programme (NSP) has been implemented
Oversight Consultant	The organization charged with overseeing the management of the National Solidarity Programme (NSP)
Parcham	A faction of the People's Democratic Party of Afghanistan (PDPA)

parde	The practice of keeping strict physical divisions between men and women who are not related, often written *purdah*
Pashtu	One of the two official languages of Afghanistan, which is the mother tongue of many Afghans, especially from the South and the East
ramazān	The ninth month of the lunar Islamic calendar during which Muslims fast, refraining from eating or drinking from the time the sun rises to when it sets
ser	Measure of weight, the most common versions are the Kabul *ser*, which is equal to 7 kg (15.4 lbs) and the Mazar *ser*, which is equal to 14 kg (30.86 lbs)
Shabname (šabnāme)	"Night letters" usually distributed under the anonymity of night and threatening in nature
Shah (Šāh)	Afghan term for King, often written *Shah*
šalwār kameez	Traditional long flowing shirt and pants worn by Afghan men
šarākat	Government industrial businesses
Šari'at	Islamic law seen as derived from the Quran, also written as *Sharia*
shura (šourā)	A traditional village council or group of elders, almost exclusively male in the Afghan context. *Šourā* is a Dari word (*jirga* in Pashtu) that describes traditional councils in Afghanistan that are convened to resolve disputes and address issues that arise at all levels, from the community to the state level
Shura-i-enkeshāfi	Community development council (CDC), a part of the National Solidarity Programme (NSP)
Shura Nazar	A supervisory council that included members from different political parties, ethnic groups, and Mujahideen groups, established by Ahmad Shah Masoud
Taliban	Originally this term meant simply the plural of religious student. The term is now used to reference a group of individuals who took over the government of Afghanistan in 1996 with a strong focus on implementing a strict version of Islamic law from 1996 to 2001

Tang	Narrowness or straightness, often the word used to describe valleys or gorges that emerge out of the mountains
toshak (tošak)	Large cushions commonly used in Afghanistan to sit on the floor, often written as *toshak*
toxm	egg
village	Internally determined local boundaries, usually related to irrigation in the context of rural Afghanistan, this sometimes but does not always align with National Solidarity Programme (NSP) community divisions
xāreğe	Foreigners, also written *kharaji*
zamin-i-pejġule	Term for the necessary rotation of irrigated and planted land utilized in Afghanistan in some locations, also written *zamin-i-paikaly*

Acronyms

ABCD	Afghan Business Capacity Development (NGO)
ACBAR	Agency Coordinating Body for Afghan Relief
ACTED	Agency for Technical Cooperation and Development (NGO)
ADB	Asian Development Bank
AFP	Alliance for Progress and *Agence France-Presse*
AGE	Anti-government elements
AIA	Afghan Interim Authority
AIHRC	Afghanistan Independent Human Rights Commission
AIMS	Afghanistan Information Management Services
AKDN	Aga Khan Development Network (NGO)
AMUL	Anand Milk Union Ltd. of India
ANDS	Afghanistan National Development Strategy
ANP	Afghan National Police
ANSO	Afghanistan NGO Safety Office (NGO)
ARC	Australian Relief Committee (NGO)
AREA	Afghan NGO
AREP	Afghan Rehabilitation & Education Programme (NGO)
AREU	Afghanistan Research and Evaluation Unit, a Kabul-based independent think tank
ARRP	Afghan Rural Rehabilitation Programme
ARTF	Afghanistan Reconstruction Trust Fund
ASAP	Accelerating Sustainable Agriculture Project
AWEC	Afghan Women's Education Center
BCE	Before common era
BDF	Bakhtar Development Fund (NGO)
BRAC	Originally, the Bangladesh Rehabilitation Assistance Committee, but then changed to the Bangladesh Rural Advancement Committee, and finally the organization has decided to use the word BRAC as a name and not an acronym (NGO)
CARE	Cooperative for Assistance and Relief Everywhere (NGO)
CBD	Community-based development
CD	Community development

CDC	Community Development Council, a National Solidarity Programme (NSP) institution of community elected members who are in charge of the planning, management, and monitoring of the NSP subproject(s) in their village, also known locally as a development šourā or šourā-i-enkeshāfi in Dari
CDD	Community-driven development
CDP	Community development plan
cf.	Confer, but also that the information provided in the proceeding sentences also refer to the citation that follows the abbreviation
CFDO	Community Forum Development Organization, a UN participatory development program
CHA	Coordination of Humanitarian Assistance (NGO)
CONCERN	Concern Worldwide (NGO)
CSO	Central Statistics Office (of Afghanistan) and Civil society organization
DACAAR	Danish Committee for Aid to Afghan Refugees (NGO)
DAFA	*Délégation archéoogique française en Afghanistan* (French Archaeological Delegation in Afghanistan)
DED	*Deutscher Entwicklungsdienst* (German Development Service)
DFID	UK Department for International Development
ECLA	Economic Commission for Latin America
FAO	Food and Agriculture Organization of the United Nations
FDD	Focus District Development, a police reform program to increase the operational and institutional capabilities of the Afghan National Police (ANP)
FGD	Focus group discussion
FP	Facilitating Partner of the National Solidarity Programme or mo'assasāt-i hamkār in Dari
GATT	General Agreement on Tariffs and Trade
GDI	Gender Development Index
GDP	Gross domestic product
GEM	Gender Empowerment Index
GIZ	*Deutsche Gesellschaft für Internationale Zusammenarbeit* (German Agency for International Cooperation)
GNP	Gross national product
GoA	Government of Afghanistan

GTZ	*Deutsche Gesellschaft für Technische Zusammenarbeit* (German Technical Cooperation), now a part of GIZ
HALO	Demining organization (NGO)
HAVA	Hilmand Arghandab Valley Authority
HDI	Human Development Index
HHS	Household survey conducted by the author for this research between January and March of 2010
HLP	Horticulture Livestock Program
HPI	Human Poverty Index
IAM	International Assistance Mission (NGO)
ICRC	International Committee of the Red Cross
IDA	International Development Association, the World Bank's fund for the poorest
IDP	Internally displaced people or persons
IED	Improvised explosive device, homemade bombs
ILO	International Labour Organization
IMF	International Monetary Fund
IPACS	Initiative to Promote Afghan Civil Society
IRC	International Rescue Committee (NGO)
IRD	International Relief and Development, Inc.
IRoA	Islamic Republic of Afghanistan
ISAF	International Security Assistance Force
ISI	Import substitution industrialization
JICA	Japan International Cooperation Agency
JSDF	Japanese Self Defense Forces
KDP	Kecamatan Development Programme
KfW	Kreditanstalt Für Wiederaufbau (German Development Bank)
KHAD	Khadamat-i Etela'at-i Dawlati, the main security and intelligence agency of Afghanistan
KII	Key informant interview
LDC	Least Developed Country
MADERA	*Mission d' Aid au Development des Economies Rurales Programme Afghanistan* (Aid Mission to the Development of Rural Economies) (NGO)
MEDICO	German non-governmental organization (NGO)
MoC	Ministry of Commerce

MoE	Ministry of Education
MoI	Ministry of the Interior
MoPH	Ministry of Public Health
MRRD	Ministry of Rural Rehabilitation and Development
MSF	*Médecins Sans Frontières* (Doctors Without Borders) (NGO)
NABDP	National Area-Based Development Programme
NATO	North Atlantic Treaty Organization
NGO	Non-governmental organization
NIC	Newly Industrializing Country
NIEO	New international economic order
NRC	Norwegian Refugee Council (NGO)
NRVA	National Risk and Vulnerability Assessment
NSP	National Solidarity Programme
OC	Oversight Consultant of the National Solidarity Programme (NSP)
OCHA	United Nations Office for the Coordination of Humanitarian Affairs
ODA	Official development assistance
OED	Operations Evaluation Department of the World Bank
OIC	Organization of the Islamic Conference
OM	Operational Manual of the National Solidarity Programme (NSP)
OXFAM	Oxford Committee for Famine Relief (NGO)
PBUH	Peace be upon him
PDPA	People's Democratic Party of Afghanistan, ruled Afghanistan from 1978 to 1992
PEACE Initiative	Poverty Eradication and Community Empowerment Initiative, see PI
PI	PEACE Initiative - umbrella term for five UNDP projects implemented in the 1990s in Afghanistan
PLA	Participatory learning approach
PMT	Police Mentoring Team
PMU	Project Management Unit
PPP	Purchasing power parity
PRA	Participatory Rural Appraisal
PRSP	Poverty reduction strategy paper

PRT	Provincial Reconstruction Team
RI	Relief International (NGO)
RPG	Rocket propelled grenade
RRD	Ministry of Rural Rehabilitation and Development at the provincial level in Afghanistan
SAP	Structural adjustment program
SCA	Swedish Committee for Afghanistan (NGO)
SDES	Socio-demographic economic survey
SDO	Sanayee Development Organization (NGO)
UN	United Nations
UNAMA	United Nations Assistance Mission in Afghanistan
UNCTAD	United Nations Conference on Trade and Development
UNDP	United Nations Development Programme
UNEP	United Nations Environmental Programme
UNESCO	United Nations Education Scientific and Cultural Organization
UN Habitat	United Nations Human Settlement Programme
UNHCR	United Nations Refugee Agency
UNICEF	United Nations International Children's Emergency Fund
UNOPS	United Nations Office for Project Services
UNSMA	United Nations Special Mission to Afghanistan (predecessor of UNAMA)
UNRISD	United Nations Research Institute for Social Development
US	United State of America
USD	United States dollar
USIA	United States Information Agency
USAID	United States Agency for International Development
USSR	Union of Soviet Socialist Republics
VAM	Vulnerability analysis and mapping
VAU	Vulnerability Analysis Unit of MRRD
WFP	World Food Programme of the United Nations
WHO	World Health Organization, a UN agency
WTO	World Trade Organization
ZOA	ZOA Refugee Care for Afghanistan (NGO)

List of Figures

List of Photos

Empirical and Primary Sources Consulted and Utilized

The *Household Survey (HHS)* was conducted in the research case study community of Shah Raheem. Please see Annex F for the household survey template and Annex G for further information about the household survey participants. A total of 32 HHS were conducted between January and March 2010, with 32 individuals.

Key Informant Interviews (KIIs) were conducted with people from the community, district, and province, along with stakeholders located in Kabul. KIIs were mostly conducted in an open format with no questionnaire template. One hundred and forty three different key informant interviews were conducted with 94 individuals. Multiple interviews were conducted with some individuals. Key informant interviews were conducted from 2009 through 2011.

- 56 Community KIIs were conducted with 40 individuals. Please see Annex L for a full listing of participants.
- 38 facilitating partner staff KIIs were conducted with 14 individuals. Please see Annex K for a full listing of Participants and their positions within the FP.
- 30 external KIIs were conducted with a total of 21 individuals. Please see Annex M for a full listing of participants, including their professional positions.
- 19 Shopkeeper KIIs were conducted with 19 individuals. Please see Annex I for the questionnaire template and Annex J for a full listing of participants.

Focus Group Discussions (FGDs) were held with people from the research community and the research district in 2010 and 2011. A total of 11 FGDs were held, with a total of 98 individuals. For further information on the FGDs that took place, including dates, participants, and categorizations, please see Annex N.

NSP Documents for the community of Shah Raheem, were completed by CARE International and stored in the CARE Khulm District Office. These documents are in alignment with the forms required by the NSP under NSP II.

- Form 1: Community Profile
- Form 2: CDC Election and Composition (for the elections in 2006 and 2009)
- Form 3: CDC Registration (for the CDC elected in 2006 and 2009)
- Form 4: Bank Account Opening and Withdrawal Authorization
- Form 5: Certificate of Registration of Community Development Council
- Form 6: Community Development Plan
- Form 7: Subproject Proposal for all three subprojects in Shah Raheem: shallow well, deep well, and community center, including documents for demining, of land transfers, subproject summary, implementation and maintenance, project eligibility appraisal sheet, etc.
- Form 8: Bilateral agreement to implement NSP in Shah Raheem
- Form 9: Block Grant Request
- Less formal contract of Agreement between Shah Raheem and CARE International
- Community list of families by head of household name disaggregated by gender and age points
- List of eligible voters
- Election form for identifying the CDC executive committee members
- Executive council voting tracking
- Training information and attendance sheets
- Work plans for each of the three projects
- Documents addressing the reasons for the original rejection of a subproject proposal
- Oversight consultant engineering monitoring reports
- Procurement documentation, including receipts, and cost evaluation documents
- Blueprints for the wells
- Map of the community
- Certificates of completion for each subproject
- Map of the district created by CARE staff, highlighting NSP projects

NSP Database created for the district of Khulm by MB Wilson, covering all 75 NSP communities in existence as of 2009. NSP Forms 1, 2, and 6 were utilized to create this database.

Additional Primary Sources: Statistics gathered and shared in hard copy, hand-written from the Directorate of Education, the Directorate of Agriculture, and the head of the District Hospital in Khulm.

- Map of the district of Khulm created during the District Development Assembly (June 2012)
- Participatory maps of each of the three villages (Chahar Sukh, Shah Raheem, and Yaqubai), drawn by one community member and confirmed by various community members
- Hand in Hand project documents provided by their staff
- Sofar hen project documents maintained by the female *kalāntar* of the village of Shah Raheem
- List of widows in one village of the NSP community written up by the female *kalāntar* of the village of Shah Raheem
- Posters and documentation from the HLP development program

Secondary Sources Consulted and Cited

Abdullah, Zaheeruddin (1998). "Taliban Closes Afghan Aid Groups." Accessed February 23, 2011 from http://www.afghanistannewscenter.com/ news/1998/july/july20e1998.htm

ACBAR (2010a). "ACBAR Strategic Plan 2010-2012." Kabul: ACBAR.

ACBAR (2010b). "NGO's Are Not a Soft Power." Accessed August 14, 2011 from http://www.afghanaid.org.uk/news.php/16/acbar_press_release_ngos_are_not_a_soft_power

ACSOR (2010). "Findings of the ABC News, BBC, and ARD Survey." Accessed September 8, 2011 from http://news.bbc.co.uk/2/shared/bsp/hi/pdfs/11_01_10_afghanpoll.pdf

Adamec, Ludwig W. (1979). *Mazar-i-Sharif and North-Central Afghanistan.* Austria: Akademische Druck- u. Verlagsanstalt Graz, Austria.

Adamec, Ludwig W. (2002). *Historical Dictionary of Afghanistan.* Lanham, Maryland and London: The Scarecrow Press, Inc.

Affolter, Friedrich W., Katja Richter, Karimullah Afaq, et al. (2009). "Transformative Learning and Mind-Change in Rural Afghanistan" In: *Development in Practice,* 19(3): 311-328.

Afghan Analysts Network (2011). "Non-Governmental and International Humanitarian Organizations Operating in Afghanistan." Accessed February 24, 2011 from http://afghanistan-analyst.org/ngo.aspx

Afghanistan News Center (1998a). "ICRC Regrets Departure of NGOs from Kabul." Accessed May 10, 2011 from http://www.afghanistannewscenter.com/news/1998/july/july23e1998.htm

Afghanistan News Center (1998b). "Kabul Paying High Price for Taliban Expulsion of Aid Groups." Accessed May 9, 2011 from http://www.afghanistannewscenter.com/news/1998/july/july23f1998.htm

AFP (1998a). "Afghanistan's Taliban Finalise Closure of Foreign Aid Groups." Accessed May 10, 2011 from http://www.afghanistannewscenter.com/news/1998/july/july21m1998.htm

AFP (1998b). "Foreign Aid Staff Quit Kabul as Fears for Security Mount." Accessed May 10, 2011 from http://www.afghanistannewscenter.com/news/1998/july/july19a1998.htm

AFP (2010). "German National Killed in Northern Afghanistan." Accessed May 24, 2012 from http://www.google.com/hostednews/afp/article/ALeqM5gPd5r23UaNeOvUnnpuhtTv0wbDsg?docId=CNG.81282d483c9208446f7bbcd961623561.c01

Agarwal, Bina (2001). "Participatory Exclusions, Community Forestry, and

Gender: An Analysis for South Asia and a Conceptual Framework" In: *World Development*, 29(10): 1623-1648.
AgriInfo (2011). "Etawah Project." Accessed September 27, 2011 from http://www.agriinfo.in/?page=topic&superid=7&topicid=518
AIA - MoC (2005). "Index of Laws, Decrees, Regulations, and Charters from Beginning of Interim Administration through 1384/09/27 (2005/12/18)." Afghanistan Interim Administration - Ministry of Commerce.
AKDN (2012). "Rural Development in Pakistan." Accessed October 26, 2012 from http://www.akdn.org/rural_development/pakistan.asp
Akhtar Hameed Khan (1996). *Orangi Pilot Project: Reminiscences and Reflections*. Oxford: Oxford University Press.
Albert, Linette (1974). "Afghanistan: A Perspective." In: Louis Dupree and Linette Albert (eds.), *Afghanistan in the 1970s*. London: Praeger, 249-259.
Alkire, Sabina (2003). "A Conceptual Framework for Human Security." Oxford: Center for Research on Inequality, Human Security and Ethnicity (CRISE).
Altai Consulting (2004). "Assessment of NSP Facilitating Partners: Common Final Report." Kabul: Altai Consulting for the World Bank.
Altai Consulting (2010). "Afghan Media: Eight Years Later: Synthesis Report." Accessed February 18, 2012 from http://www.altaiconsulting.com/docs/media/2010/Afghan%20Media%20in%202010.pdf
Amnesty International (1992). "Document - Afghanistan: Nouvelles Formes de Chatiments Cruels, Inhumains Ou Degradants (9210F): Afghanistan: New Forms of Cruel, Inhuman or Degrading Punishment." Accessed March 27, 2012 from http://www.amnesty.org/fr/library/asset/ASA11/002/1992/fr/b4c880a1-ed92-11dd-95f6-0b268ecef84f/asa110021992en.html
Amstutz, J. Bruce (1986). *Afghanistan: The First Five Years of Soviet Occupation*. Washington, D.C.: NDU Press.
ANSO (2008). "ANSO Quarterly Data Report Q.4 2008." Kabul: The Afghanistan NGO Safety Office. http://www.afgnso.org/2008/2008%20Q4.pdf
ANSO (2010). "ANSO Quarterly Data Report Q.4 2010." Kabul: The Afghanistan NGO Safety Office. http://www.afgnso.org/2010Q/ANSO%20Quarterly%20Data%20Report%20(Q4%202010).pdf
ANSO (2011a). "ANSO Quarterly Data Report Q.4 2011." Kabul: The Afghanistan NGO Safety Office. http://www.ngosafety.org/store/files/ANSO%20Q4%202011.pdf

ANSO (2011b). "ANSO Report: Issue 85 (1-15 November 2011)." Kabul: Afghanistan NGO Security Office.
AP (2011). "Karzai Announces 2nd Round of Sites Where Afghan Forces Will Take Over from Foreign Troops, November 27, 2011." Accessed November 27, 2011 from http://www.washingtonpost.com/world/asia-pacific/nato-says-2-service-members-killed-in-eastern-and-southern-afghanistan/2011/11/27/gIQAXW8o0N_story.html
Arnold, Anthony (1994). "The Ephemeral Elite." In: Myron Weiner and Ali Banuazizi (eds.), *The Politics of Social Transformation in Afghansitan, Iran, and Pakistan.* New York: Syracuse University Press, 35-71.
Arnstein, Sherry (1969). "A Ladder of Citizen Participation" In: *Journal of the American Institute of Planners,* 35(4): 216-249.
Ashworth, J.M. (2005). "Groundwater Assessment of the Downstream Sections of the Balkh and Khulm Watersheds: Inception report, updated 2008." *ADB Loan 1997 AFG: Emergency Infrastructure Rehabilitation and Reconstruction Project (EIRPP).* Kabul: Ministry of Energy and Water. http://www.hydrogeologist.net/Software/Afghanistan/Afghanistan.pdf
Asian Development Bank (2009). "Proposed Multitranche Financing Facility and Administration of Grant and Cofinancing Islamic Republic of Afghanistan: Water Resources Development Investment Program, August 2009." *Report and Recommendation of the President to the Board of Directors, Project Number 42091.* n.p.: Asian Development Bank. http://www.adb.org/Documents/RRPs/AFG/42091-AFG-RRP.pdf
Azarbiajani-Moghaddam, Sippi (2010). "A Study of Gender Equity Through the National Solidarity Programme's Community Development Councils." Kabul: DACAAR. http://www.nspafghanistan.org/files/NSP%20Gender%20Study%20Report.pdf
Baitenmann, Helga (1990). "NGOs and the Afghan War: The Politicisation of Humanitarian Aid" In: *Third World Quarterly,* 12(1): 62-85.
Balassa, Bela (1971). "Trade Policies in Developing Countries: Papers and Proceedings of the Eighty-Third Annual Meeting of the American Economic Association" In: *The American Economic Review,* 61(2): 178-187.
Balassa, Bela (1981). *The Newly Industrializing Countries in the World Economy.* New York: Pergamon Press.
Bank Information Center (2011). "Afghanistan." Accessed June 14, 2011 from http://www.bicusa.org/en/Region.5.aspx
Barakat, Sultan (2006). "Mid-Term Evaluation Report of the National

Solidarity Programme (NSP)." Kabul: Post-War Reconstruction and Development Unit of the University of York. http://www.cmi.no/publications/file/2446-mid-term-evaluation-report-of-the-national.pdf

Barakat, Sultan (2009). "The Failed Promise of Multi-Donor Trust Funds: Aid Financing as an Impediment to Effective State-Building in Post-Conflict Contexts" In: *Policy Studies*, 30(2): 107-126.

Baran, Paul (1957). *The Political Economy of Growth*. New York: Monthly Review Press.

Baran, Paul A. and Paul Marlor Sweezy (1966). *Monopoly Capital: An Essay on the American Economic and Social Order*. New York: Monthly Review Press.

Barfield, Thomas J. (1981). *The Central Asian Arabs of Afghanistan: Pastoral Nomadism in Transition*. Austin: University of Texas Press.

Barfield, Thomas J. (2010). *Afghanistan: A Cultural and Political History*. Princeton: Princeton University Press.

Baron, Stephen, John Field and Tom Schuller, Eds. (2000). *Social Capital: Critical Perspectives*. Oxford: Oxford University Press.

Baxter, Craig (2007). "Historical Setting." In: Alfred Aghajanian (eds.), *Afghanistan: Past and Present*. Los Angeles: IndoEuropean Publishing, 1-53.

BBC News (2010). "Afghans Brave Taliban to Vote in Parliamentary Election." Accessed November 16, 2011 from http://www.bbc.co.uk/news/world-south-asia-11349179

BBC News (2011). "UN Staff Killed During Protest in Northern Afghanistan." Accessed July 24, 2012 from http://www.bbc.co.uk/news/world-south-asia-12940014

Beath, Andrew, Fotini Christia and Ruben Enikolopov (2008). "Randomized Impact Evaluation of Afghanistan's National Solidarity Programme (NSP): Baseline Survey Report." Kabul: The World Bank.

Beath, Andrew, Fotini Christia and Ruben Enikolopov (2011). "Elite Capture of Local Institutions: Evidence from a Field Experiment in Afghanistan." Kabul: The World Bank. http://federation.ens.fr/ydepot/semin/texte1011/ENI2011ELI.pdf

Beath, Andrew, Fotini Christia, Ruben Enikolopov, et al. (2010). "Estimates of Interim Program Impact from First Follow-Up Survey." Kabul: The World Bank. http://www.nsp-ie.org/reports/BCEK-Interim_Estimates_of_Program_Impact_2010_07_25.pdf

Beath, Andrew, Christia Fotini and Ruben Enikolopov (2011). "Winning Hearts and Minds? Evidence from a Field Experiment in Afghanistan." *MIT Working Paper*. Massachusetts Institute of Technology. http://

www.humansecuritygateway.com/documents/MIT_WinningHeartsandMinds.pdf

Beath, Andrew, Christia Fotini and Ruben Enikolopov (2012). "Winning Hearts and Minds through Development: Evidence from a Field Experiment in Afghanistan?" In: *MIT Political Science Department Research Paper No. 2011-14.*

Bebbington, Anthony, Scott Guggenheim, Elizabeth Olson, et al. (2004). "Exploring Social Capital Debates at the World Bank" In: *Journal of Development Studies,* 40: 53-64.

Benen, Steve (2008). "Political Animal: National Solidarity Program." Accessed September 28, 2011 from http://www.washingtonmonthly.com/archives/individual/2008_08/014374.php

Bernander, Bernt (2002). "Community Empowerment in Afghanistan: A Review of the UNDP Experience in the 1990s." UNDP.

Besley, Timothy, Rohini Pande and Vijayendra Rao (2005). "Participatory Democracy in Action: Survey Evidence from South India" In: *Journal of the European Economic Association,* 3(2/3): 648-657.

Bhuvana, Bhatnagar, James Kearns and Debra Sequeira (1996). *The World Bank Participation Sourcebook*. Washington, DC: The World Bank.

Bins, Tony (2002). "Dualistic and Unilinear Concepts of Development." In: Vandana Desai and Robert B. Potter (eds.), *The Companion to Development Studies.* London: Arnold, 75-80.

Binswanger, Hans P. and Swaminathan S. Aiyar (2003). "Scaling Up Community-Driven Development: Theoretical Underpinnings and Program Design Implications." *Policy Research Working Paper 3039.* Washington, DC: The World Bank.

Boesen, Inger W. (2004). "From Subject to Citizens: Local Participation in the National Solidarity Programme." Kabul: Afghan Research and Evaluation Unit (AREU). http://www.areu.org.af/Uploads/EditionPdfs/428E-From%20Subjects%20to%20Citizens-WP-print.pdf

Boserup, Ester (1970). *Woman's Role in Economic Development.* London: George Allen and Unwin.

Boserup, Ester and Christine Liljencrantz (1975). "Integration of Women in Development: Why, When, How." United Nations Development Programme.

Botes, Lucius and Dingie van Rensburg (2000). "Community Participation in Development: Nine Plagues and Twelve Commandments" In: *Community Development Journal,* 35(1): 41-58.

Bourdieu, Pierre (1984). *Distinction: A Social Critique of the Judgement of Taste.* London: Routledge & Kegan Paul.
Bourdieu, Pierre (1989). "Social Space and Symbolic Power" In: *Sociological Theory,* 7(1): 14-25.
Bourdieu, Pierre and James S. Coleman (1991). *Social Theory for Changing Society.* Boulder, Colorado and Oxford: Westview Press.
BRAC (2011). "FAQ." Accessed June 20, 2012 from http://www.brac.net/content/faq-0#.T-GMkheybl4
Brandt, Willy (1980). *North South: A Programme for Survival: Report of the Independent Commission on International Development Issues.* Cambridge, MA: The MIT Press.
Brant, Marvin (1974). "Recent Economic Development." In: Louis Dupree and Linette Albert (eds.), *Afghanistan in the 1970s.* New York and London: Praeger, 91-112.
Brick, Jennifer (2008). Final Report: Investigating the Sustainability of Community Development Councils in Afghanistan. Kabul, Afghanistan Research and Evaluation Unit (AREU) and the Japanese International Cooperation Agency (JICA). Unpublished manuscript.
Brinkerhoff, Derick W. and Rudi Klauss (1985). "Mangerial Roles for Social Development Management" In: *Public Administration and Development,* 5(2): 145-156.
Brown, Frances (2012). "The U.S. Surge and Afghan Local Governance: Lessons for Transition." *Special Report 316.* Washington, DC: United States Institute of Peace (USIP). http://www.usip.org/publications/the-us-surge-and-afghan-local-governance
Brown, Tom (2007). "ASAP India Marketing Mission Report." Kabul, Afghanistan: Chemonics International.
CARE (2009). "NSP Quarterly Project Implementation Progress Report." Kabul, Afghanistan: CARE International.
Carlin, Anne (2003). "Rush to Reengagement in Afghanistan: IFI's Post Conflict Agenda with a Special Focus on the National Solidarity Programme." Washington, DC: Bank Information Center.
Carter, Dr. Lynn and Dr. Kerry Connor (1989). "A Preliminary Investigation of Contemporary Afghan Councils." Peshawar, Pakistan: Agency Coordinating Body for Afghan Relief (ACBAR).
CBS News (2009). "National Solidarity Program." Accessed Oct. 18, 2011 from http://www.cbsnews.com/stories/2008/08/22/politics/animal/main4374953.shtml
Centlivres-Demont, Micheline (1994). "Afghan Women in Peace, War, and

Exile." In: Myron Weiner and Ali Banuazizi (eds.), *The Politics of Social Transformation in Afghansitan, Iran, and Pakistan*. New York: Syracuse University Press, 333-365.
Centlivres, Pierre (1970). Un Bazar d'Asie Centrale: Forme et organisation du bazar de Tâshqurghân Faculté des Lettres, Université de Neuchâtel. PhD thesis.
Central Intelligence Agency (2011). "The World Factbook: Afghanistan." Accessed March 22, 2011 and April 19, 2011 from https://www.cia.gov/library/publications/the-world-factbook/geos/af.html
Central Statistics Office (2004). "CSO 2003-2004 Population Statistics - 388 Districts." Accessed June 14, 2011 from http://www.aims.org.af/ss roots.aspx?seckeyt=387
Centre for Global Negotiations (2012). "The Brandt Equation: 21st Century Blueprint for the New Global Economy." Accessed January 19, 2012 from http://www.brandt21forum.info/About_BrandtCommission.htm
Centre for International Heritage Activities (2011). "Afghanistan, the Bagh-e-Jehan Nama Palace Complex (Archive, April 2009)." Accessed October 11, 2011 from http://www.heritage-activities.nl/Projects/Search%20by%20country/Afghanistan/Activities/The%20renovation%20of%20the%20Bagh-e%20Jehan%20Nama%20Palace%20a-0
Cernea, Michael M. , Ed. (1985). *Putting People First: Sociological Variables in Rural Development*. New York: Oxford University Press and the World Bank.
Chambers, Robert (1981). "Rural Poverty Unperceived: Problems and Remedies" In: *World Development*, 9: 1-19.
Chambers, Robert (1983). *Rural Development: Putting the Last First*. Harlow: Pearson Education Limited.
Chambers, Robert (1994). "Participatory Rural Appraisal (PRA): Analysis of Experience" In: *World Development*, 22(9): 1253-1268.
Chambers, Robert (1997). *Whose Reality Counts? Putting the First Last*. Warwickshire: Practical Action Publishing.
Chambers, Robert (2002). *Participatory Workshops: A Sourcebook of 21 Sets of Ideas and Activities*. London: Earthscan.
Chambers, Robert (2005). *Ideas for Development*. London: Earthscan.
Chapman, Nick and Charlotte Vaillant (2010). "Synthesis of Country Programme Evaluations Conducted in Fragile States." *Evaluation Report EV709*. Department for International Development (DFID).
Charpentier, C.J. (1972). *Bazaar-e Tashqurghan--Ethnographical Studies in*

an Afghan Traditional Bazaar. Uppsala: Institution for allmanvetenskaplega och jamforande etnografi.
Chase, Robert, R. Nording and M. Thongyou (2005). "Thailand Social Capital Evaluation: A Mixed Methods Assessment of the Social Investment Fund's Impact on Village Social Capital." Washington, DC: East Asia Environment and Social Development Unit of the World Bank.
Chemonics International (2008). "Mazar Foods Environmental Scoping Statement, June 22, 2008." Kabul, Afghanistan.
Chemonics International and The Cadmus Group (2008). "Environmental Assessment of the Proposed Mazar Foods Agricultural/Irrigation Development: Balkh Province, Afghanistan Final Draft." USAID.
Christensen, Asger (1995). "Aiding Afghanistan: The Background and Prospects for Reconstruction in a Fragmented Society." *NIAS Report, No. 26.* Copenhagen: NIAS Publications.
Christian, David (2000). "Silk Roads or Steppe Roads? The Silk Roads in World History" In: *Journal of World History,* 11(1): 1-26.
Cleaver, Frances (1999). "Paradoxes of Participation: Questioning Participatory Approaches to Development" In: *Journal of International Development,* 11(4): 597-612.
Cleaver, Frances (2001). "Institutions, Agency, and the Limitation of Participatory Approaches." In: Bill Cooke and Uma Kothari (eds.), *Participation: The New Tyranny?* London and New York: Zed Books, 36-55.
CNN (2010). "Many Afghans Denied Right to Vote, Watchdog Says." Accessed November 16, 2011 from http://articles.cnn.com/2010-09-20/world/afghanistan.elections_1_vote-fraud-independent-election-commission-fair-election-foundation?_s=PM:WORLD
Cogan, Charles G. (1993). "Partners in Time: The CIA and Afghanistan since 1979" In: *World Policy Journal,* 10(2): 73-82.
Cohen, John M. and Norman T. Uphoff (1980). "Participation's Place in Rural Development: Seeking Clarity Through Specificity" In: *World Development,* 8: 213-235.
Coleridge, Peter (2002). "Community Based Rehabilitation in a Complex Emergency: Study of Afghanistan." In: M. Thomas and M.J. Thomas (eds.), *Selected Readings in CBR - Series 2: Disability and Rehabilitation Issues in South Asia.* Bangalore: National Press Printing, 35-49.
Constable, Pamela (2008). "As Crime Increases in Kabul, So Does Nostalgia for Taliban." Accessed June 11, 2012 from http://www.washington post.com/wp-dyn/content/article/2008/09/24/AR2008092403339.html
Conway, Dennis and Nikolas Heynen (2002). "Classical Dependency Theo-

ries: From ECLA to André Gunder Frank." In: Vandana Desai and Robert B. Potter (eds.), *The Companion to Development Studies.* London: Arnold, 97-101.
Cooke, Bill (2001). "The Social Psychologial Limits of Participation?" In: Bill Cooke and Uma Kothari (eds.), *Participation: The New Tyranny?* London and New York: Zed Books, 102-121.
Cooke, Bill and Uma Kothari (2001a). "The Case for Participation as Tyranny." In: Bill Cooke and Uma Kothari (eds.), *Participation: The New Tyranny?* London and New York: Zed Books, 1-15.
Cooke, Bill and Uma Kothari (2001b). *Participation: The New Tyranny?* London and New York: Zed Books.
Corbridge, Stuart (1998). "'Beneath the Pavement Only Soil: The Poverty of Post-development" In: *Journal of Development Studies,* 34(6): 138-148.
Craig, Gary and Marjorie Mayo, Eds. (1995). *Community Empowerment: A Reader.* London: Zed Books.
Creswell, John W. (2009). *Research Design: Qualitative, Quantitative, and Mixed Methods Approaches.* Los Angeles, London, New Delhi, and Singapore: Sage Publications, Inc.
CSO (2011). "Afghanistan Statistical Yearbook 2010-2011." Accessed February 17, 2012 from http://cso.gov.af/en/page/4725
Cullather, Nick (2002). "Damming Afghanistan: Modernization in a Buffer State" In: *The Journal of American History,* 89(2): 512-537.
Dale, Reidar (2004). *Development Planning: Concepts and Tools for Planners and Facilitators.* New York: Palgrave Macmillan.
Dasgupta, Aniruddha and Victoria A. Beard (2007). "Community Driven Development, Collective Action and Elite Capture in Indonesia" In: *Development and Change,* 38(2): 229-249.
Desai, Vandana (2002). "Community Participation in Development." In: Vandana Desai and Robert B. Potter (eds.), *The Companion to Development Studies.* London: Arnold, 117-121.
Desai, Vandana and Robert B. Potter (2002). *The Companion to Development Studies.* London: Arnold.
Devex (2012). "Reconstruction and Upgrading of the Kunduz-Khulm National Highway in Afghanistan, Request for Expression of Interest." Accessed October 16, 2012 from http://www.devex.com/en/projects/reconstruction-and-upgrading-of-the-kunduz-khulm-national-highway-in-afghanistan
Dill, Brian (2009). "The Paradoxes of Community-Based Participation in Dar es Salaam" In: *Development and Change,* 40(4): 717-743.

Dongier, Philippe, Julie Van Domelen, Elinor Ostom, et al. (2003). "Chapter 9: Community-Driven Development." *World Bank Poverty Reduction Strategy Paper 2003.* The World Bank. Volume 1, http://siteresources.worldbank.org/INTPRS1/Resources/383606-1205334112622/5805_chap9.pdf

Doniger, Philip, Julie Van Domelen, Elinor Ostrom, et al. (2001). "Community Driven Development." (eds.), *Poverty Reduction Strategy Paper Sourcebook.* Washington, D.C.: The World Bank. 1, 301-331.

Dörre, Andrei and Tobias Kraudzum (2012). "Persistence and Change in Afghan -- (Soviet) Russian Relations" In: *Central Asian Survey,* 31(4): 425-443.

Dorronsoro, Gilles (2005). *Revolution Unending*. New York: Colombia University Press in association with the Centre d'études et de recherches internationales, Paris.

Dupree, Louis (1980). *Afghanistan.* Princeton, New Jersey: Princeton University Press.

Dupree, Nancy Hatch and Thomas E. Gouttierre (2007). "The Society and Its Environment." In: Alfred Aghajanian (eds.), *Afghanistan: Past and Present.* Los Angeles: IndoEuropean Publishing, 54-133.

Echavez, Chona R. (2010). "Does Women's Participation in the National Solidarity Programme Make a Difference in Their Lives? A Case Study in Parwan Province." *Case Study Series.* Kabul: Afghanistan Research and Evaluation Unit (AREU).

Echavez, Chona R. (2012). "Does Women's Participation in the National Solidarity Programme Make a Difference in thier Lives? A Case Study in Kabul Province." *Case Study Series.* Kabul: Afghanistan Research and Evaluation Unit (AREU). http://www.areu.org.af/Uploads/EditionPdfs/1207E%20Womens%20Participation%20NSP%20Kabul%20CS%202012.pdf

Echavez, Chona R. and Jennefer Lyn L. Bagaporo (2012). "Does Women's Participation in the National Solidarity Programme Make a Difference in Their Lives? A Case Study in Balkh Province." *Case Study Series.* Kabul: Afghanistan Research and Evaluation Unit (AREU). http://www.areu.org.af/Uploads/EditionPdfs/1204E%20Womens%20Participation%20NSP%20Balkh%20CS%202012.pdf

Elliott, Jennifer A. (2002). "Development as Improving Human Welfare and Human Rights." In: Vandana Desai and Robert B. Potter (eds.), *The Companion to Development Studies.* London: Arnold, 45-49.

Enikolopov, Ruben (2011). "Development Programs and Security in

Afghanistan." Accessed November 28, 2011 from http://freepolicy briefs.files.wordpress.com/2011/11/free_policy_brief_enikolopov_28nov.pdf
Enikolopov, Ruben, Fotini Christia and Andrew Beath (2008). "Randomized Impact Evaluation of Afghanistan's National Solidarity Programme (NSP): Report on Election Monitoring." Kabul: The World Bank.
Enikolopov, Ruben, Fotini Christia and Andrew Beath (2009). "Radomized Impact Evaluation of Afghansitan's National Solidarity Programme (NSP): Report on Monitoring of Sub-Project Selection." Kabul: The World Bank. http://www.nsp-ie.org/reports/SPSP-MR.pdf
Escobar, Arturo (1992). "Planning." In: Wolfgang Sachs (eds.), *The Development Dictionary: A Guide to Knowledge as Power.* London and New York: Zed Books, 132-145.
Esteva, Gustavo (1992). "Development." In: Wolfgang Sachs (eds.), *The Development Dictionary: A Guide to Knowledge as Power.* London and New York: Zed Books, 6-25.
Étrangéres, Ministére Des Affaires (2009). "Summary Document." Accessed July 25, 2012 from http://www.diplomatie.gouv.fr/en/spip.php?page=rubrique_imprim&id_rubrique=4006
Ewans, Martin (2002). *Afghanistan: A Short History of Its People and Politics.* New York: HarperCollins Publishers Inc.
Faculty of Oriental Studies (2012). "Balkh Art and Cultural Heritage Project." Accessed July 25, 2012 from http://www.orinst.ox.ac.uk/
Favre, Raphy and Golam Monowar Kamal (2004). "Wastershed Atlas of Afghanistan - First Edition: Working Document for Planners." Kabul: Ministry of Irrigation, Water Resources and Environment (MIWRE).
Ferrier, J. P. and William Jesse (1858). *History of the Afghans.* London: J. Murray.
Ferrier, J.P. (1856). *Caravan Journeys and Wanderings in Persia, Afghanistan, Turkistan, and Beloochistan with Historical Notices of the Countries Living between: Russia and India.* London: William Clowes and Sons.
Finsterbusch, Kurt and Warren A. Van Wicklin III (1989). "Beneficiary Participation in Development Projects: Empirical Tests of Popular Theories" In: *Economic Development and Cultural Change,* 37(3): 573-593.
Fishstein, Paul and Andrew Wilder (2011). "Winning Hearts and Minds? Examing the Relationship between Aid and Security in Afghanistan." Medford, MA: Feinstein International Center of Tufts University.

Fonchingong, Charles C. (2006). "Expanding Horizons: Women's Voices in Community-Driven Development in the Cameroon Grasslands" In: *GeoJournal,* 65: 137-149.
Forět, Philippe and Andreas Kaplony (2008). *The Journey of Maps and Images on the Silk Road.* Leiden, The Netherlands: Brill.
Foschini, Fabrizio (2012). "One Year of Transition: A Look Back." Accessed April 3, 2012 from http://www.aan-afghanistan.org/index.asp?id=2615
Fox, Jonathan and John Gershman (2000). "The World Bank and Social Capital: Lessons from Ten Rural Development Projects in the Philippines and Mexico" In: *Policy Sciences,* 33: 399-419.
Frank, Andre Gunder (1969a). *Latin America: Underdevelopment or Revolution: Essays on the Development of Underdevelopment and the Immediate Enemy.* New York and London: Monthly Review Press.
Frank, Andre Gunder (1969b). *Latin America: Underdevelopment or Revolution?* New York: Monthly Review Press.
Frank, Andre Gunder (1991). "The Underdevelopment of Development" In: *Scandinavian Journal of Development Alternatives,* 10(3): 5-72.
Frank, Andre Gunder (1992). "Latin American Development Theories Revisited: A Participant Review" In: *Latin American Perspectives,* 19(2): 125-139.
Freire, Paulo (1996). *Pedagogy of the Oppressed, 2nd edition.* London: Penguin Books.
Frontier Post (2012). "District Chief of NSP Gunned Down." Accessed October 26, 2012 from http://www.thefrontierpost.com/article/173900/
Fukuyama, Francis (2001). "Social Capital, Civil Society and Development" In: *Third World Quarterly,* 22(1): 7-20.
Fukuyama, Francis (2005). ""Stateness" First" In: *Journal of Democracy,* 16(1): 84-88.
Furtado, C. (1964). *Development and Underdevelopment.* Berkeley and Los Angeles: University of California Press.
Furtado, C. (1965). *Diagnosis of the Brazilian Crisis.* Berkeley and Los Angeles: University of California Press.
Furtado, C. (1969). *Economic Development in Latin America, 2nd edition.* Cambridge: Cambridge Unversity Press.
Furtado, C. (1978). *Accumulation and Development.* Oxford: Martin Robertson.
Gannon, Kathy (2004). "Afghanistan Unbound" In: *Foreign Affairs,* 83(3): 35-46.
Garamone, Jim (2010). "Mullen Discusses Surge Successes in Afghanistan."

Accessed June 14, 2011 from http://www.defense.gov/news/newsarticle.aspx?id=62084
General Assembly of the United Nations (1974). "3201 (S-VI). Declaration on the Establishment of a New International Economic Order." *UN Documents.* United Nations. Accessed A/RES/S-6/3201, http://www.un-documents.net/s6r3201.htm
Generalny Stab. (1984-1986). "Topographic Series." *CD-2, Sheet J_42_111_3*
Gibson, Christopher and Michael Woolcock (2005). "Empowerment and Local Level Conflict Mediation in Indonesia: A Comparative Analysis of Concepts, Measures, and Project Efficacy." *World Bank Policy Research Working Paper 3713.* World Bank. http://www-wds.worldbank.org/external/default/WDSContentServer/WDSP/IB/2005/09/08/000016406_20050908104436/Rendered/PDF/wps3713.pdf
Girardet, Edward (1987). "US Humanitarian Aid Under Fire: Helping Afghan Resistance." Accessed June 14, 2011 from http://www.csmonitor.com/1987/1208/oruss4.html
Giustozzi, Antonio (2008). *Koran, Kalashnikov, and Laptop: The Neo-Taliban Insurgency in Afghanistan.* New York: Colombia University Press.
Giustozzi, Antonio and Niamatullah Ibrahimi (2012). "Thirty Years of Conflict: Drivers of Anti-Government Mobilisation in Afghanistan, 1978-2011." Kabul: Afghanistan Research and Evaluation Unit (AREU).
GIZ (2011). "About GIZ." Accessed April 19, 2011 from http://www.giz.de/en/profile.html
Gizabi, Akram (2006). "Donors Pledge More Aid, but Worry How Kabul Will Spend It." Accessed July 24, 2012 from http://www.jamestown.org/single/?no_cache=1&tx_ttnews%5Btt_news%5D=31349
Glad, Marit (2009). "Knowledge on Fire: Attacks on Education in Afghanistan. Risks and Mitagatory Measures." Kabul: The World Bank.
Gloekler, Aalya and Janet Seeley (2006). "Gender and AKRSP: Mainstreamed or Sidelined?" In: Geof Wood, Abdul Malik and Sumara Sagheer (eds.), *Valleys in Transition.* Oxford: Oxford University Press.
Goodhand, Jonathan and Mark Sedra (2009). "Who Owns the Peace? Aid, Reconstruction, and Peacebuilding in Afghanistan" In: *Disasters,* 34(S1): S78-102.
Goodson, Larry P. (2001). *Afghanistan's Endless War: State Failure, Regional Politics, and the Rise of the Taliban.* Seattle and London: University of Washington Press.
Google Earth (April 4, 2009). "Khulm District Center: 36°41'45.76" N 67°41'59.02" E elev 450m."

Gordon, Stuart (2011). "Winning Hearts and Minds? Examing the Relationship between Aid and Security in Afghanistan's Helmand Province." Medford, MA: Feinstein International Center of Tufts University. http://www.humansecuritygateway.com/documents/FIC-Winning-Hearts-Minds-Aid-Security-Helmand.pdf

Gow, David D. and Jerry Vansant (1983). "Beyond the Rhetoric of Rural Development Participation: How Can It be done?" In: *World Development,* 11(5): 427-446.

Graham-Harrison, Emma (2011). "Afghanistan Will Need $7 Billion a Year Over Next Decade: WB." Accessed December 21, 2011 from http://www.reuters.com/article/2011/11/22/us-afghanistan-money-idUSTRE7AL1TD20111122

Gran, Guy (1983). *Development by People: Citizen Construction of a Just World.* New York, NY: Praeger.

Gregorian, Vartan (1969). *The Emergence of Modern Afghanistan: Politics of Reform and Modernization, 1880-1946.* Standford, California: Standford University Press.

Gregory, Derek, Ron Johnston, Geraldine Pratt, et al., Eds. (2008). *The Dictionary of Human Geography.* Malden, MA: Wiley-Blackwell.

Gronemeyer, Marianne (1992). "Helping." In: Wolfgang Sachs (eds.), *The Development Dictionary: A Guide to Knoweldge as Power.* London and New York: Zed Books, 53-69.

Grötzbach, Erwin (1969). "Junge sozialgeographische Wandlungen im Afghanischen Hindukusch" In: *Mitteilungen der Geographischen Gesellschaft in München,* 54: 115-134.

Grötzbach, Erwin (1972). "Staatliche Agrarpolitik und Bodennutzungsgefüge in Nordost-Afghanistan." *Tagungsbericht und wisssenschaftliche Abhandlungen.* Deutscher Geographentag Erlangen-Nürnberg 1971. Wiesbaden,

Grötzbach, Erwin (1990). *Afghanistan : eine geographische Landeskunde.* Darmstadt: Wissenschaftliche Buchgesellschaft.

Gugerty, Mary Kay and Michael Kremer (2008). "Outside Funding and the Dynamics of Participation in Community Associations" In: *American Journal of Political Science,* 52(3): 585-602.

Guha, Amalendu (1964). "Economic Development of Afghanistan: 1929-1961" In: *International Studies,* 6: 421-439.

Guijt, Irene and Meera Kaul Shah, Eds. (1998a). *The Myth of Community: Gender Issues in Participatory Development.* London: Intermediate Technology Publications.

Guijt, Irene and Meera Kaul Shah (1998b). "Waking Up to Power, Conflict and Process." In: Irene Guijt and Meera Kaul Shah (eds.), *The Myth of Community: Gender Issues in Participatory Development*. London: Intermediate Technology Publications, 1-23.
Guijt, Irene and L. Van Veldhuizen (1998). "What Tools? What Steps?: Comparing PRA and PTD." *Drylands Programme Issue Paper (79): 33.* IIED.
Gupta, Arvind (1992). "Soviet Military Intervention in Afghanistan." In: K. Warikoo and Dawa Norbu (eds.), *Ethnicity and Politics in Central Asia*. New Delhi: South Asian Publishers, 270-286.
Habib, Mina (2012). "Afghan Villagers Rise up Again Taliban." Accessed July 25, 2012 from http://iwpr.net/report-news/afghan-villagers-rise-against-taleban
Hager, Rob (1983). "State, Tribe and Empire in Afghan Inter-Polity Relations." In: Richard Tapper (eds.), *The Conflict of Tribe and State in Iran and Afghanistan*. London & Canberra: Croom Helm, 83-118.
Hailey, John (2001). "Beyond Formulaic: Process and Practice in South Asian NGOs." In: Bill Cooke and Uma Kothari (eds.), *Participation: The New Tyranny?* London and New York: Zed Books, 88-101.
Hamilton, Angus (1906a). *Afghanistan*. London: William Heinemann.
Hamilton, Angus (1906b). *Afghanistan: With a Map and Numerous Illustrations*. London: William Heinemann.
Hanifan, Lyda (1916). "The Rural School Community Center" In: *Annals of the American Academy of Political and Social Science*, 67: 130-138.
Haq, Khadija (1999). "Human Development in South Asia 1999: The Crisis of Governance." Karachi: Oxford University Press.
Haq, Khadija (2002). "Human Development Report in South Asia 2002: Agriculture and Rural Development." Oxford: Oxford University Press.
Haq, Mahbub ul (1995). *Reflections on Human Development: How the Focus of Development Economics Shifted from National Income Accounting to People Centered Policies, Told by One of the Chief Architects of the New Paradigm*. New York: Oxford University Press.
Hardin, Garnett (1968). "The Tragedy of the Commons" In: *Science*, 162(1968): 1243-1248.
Harooni, Mirwais (2012). "Afghanistan "Faces Tougher Times" in Next Security Handover, May 10, 2012." Accessed May 11, 2012 from http://news.yahoo.com/afghanistan-faces-tougher-times-next-security-handover-084133533.html;_ylc=X3oDMTEwcnQ3ZmY5BF9TAzIwMjM4Mjc1MjQEZW1haWxJZAMxMzM2NzAxODI3

Hayami, Yujiro (2009). "Social Capital, Human Capital and the Community Mechanism: Toward a Conceptual Framework for Economists" In: *Journal of Development Studies*, 45(1): 96-123.
Hayward, Chris, Lyn Simpson and Leanne Wood (2004). "Still Left Out in the Cold: Problematising Participatory Research and Development" In: *Sociologia Ruralis*, 44(1): 95-108.
Henkel, Keiko and Roderick Stirrat (2001). "Participation as Spiritual Duty: Empowerment as Secular Subjection." In: Bill Cooke and Uma Kothari (eds.), *Participation: The New Tyranny?* London: Zed Books, 168-184.
Hettne, Björn (2002). "Current Trends and Future Options in Development Studies." In: Vandana Desai and Robert B. Potter (eds.), *The Companion to Development Studies*. London: Arnold, 7-12.
Higashi, Daisaku (2008). "The Challenge of Constructing Legitimacy in Peacebuilding: Case of Afghanistan, October 2008." *CIR Working Paper No. 47*. Centre of International Relations.
Higgott, Richard A. (1983). *Political Development Theory: The Contemporary Debate*. London: Croom Helm.
Hildyard, Nicholas, Pandurang Hegde, Paul Wolvekamp, et al. (2001). "Pluralism, Participation and Power: Joint Forest Management in India." In: Bill Cooke and Uma Kothari (eds.), *Participation: The New Tyranny?* London and New York: Zed Books, 56-71.
Hoddinott, John, Michelle Adator, Timothy Besley, et al. (2000). "Participation and Poverty Reduction: Issues, Theory, and New Evidence from South Africa." Washington, DC: International Food Policy Research Institute, London School of Economics.
Holdcroft, Lane E. (1976). The Rise and Fall of Community Development: 1950-1965, Michigan State University. Unpublished manuscript.
Honda, Michael (2009). "A Different Kind of Surge." Accessed October 18, 2011 from http://www.washingtontimes.com/news/2009/oct/12/a-different-kind-of-surge/print/
Humlum, Johannes (1962). *The Geography of Afghanistan: Translation of La Geographie de l'Afghanistan*. L.G. Hanscom Field, Mass.: Research Library, Air Force Cambridge Research Laboratories, Office of Aerospace Research, U.S. Air Force.
Hunt, Diana (1989). *Economic Theories of Development: An Analysis of Competing Paradigms*. Hertfordshire: Harvester Wheatsheaf.
Huntington, Samuel P. (1969). *Political Order in Changing Societies*. New Haven, Connecticut: Yale University Press.

Huntington, Samuel P. (1993). "The Clash of Civilizations?" In: *Foreign Affairs*, 72(3): 22-49.
ICON-Institute (2009). "Nation Risk and Vulnerability Assessment 2007/8: A Profile of Afghanistan." Kabul: Government of the Islamic Republic of Afghanistan.
Ikbal Ali, Shah (1928). *Afghanistan of the Afghans.* London: Diamond Press.
Illich, Ivan (1992). "Needs." In: Wolfgang Sachs (eds.), *The Development Dictionary: A Guide to Knoweldge as Power.* London and New York: Zed Books, 88-101.
IRoA and MoE (2011). "Response to EFA Global Monitoring Report – 2011: (Afghanistan Section) Hidden Crisis: Conflict and Education." Kabul: Islamic Reupblic of Afghanistan, Ministry of Education.
IRoA, MRRD and NSP (2012a). "Monthly Program Report: Asad 1391 (22nd July to 21st August 2012)." Kabul: Islamic Republic of Afghanistan, Ministry of Rural Rehabilitation and Development, and the National Solidarity Programme. http://www.nspafghanistan.org/Files/Programme%20Monthly%20Report%20August%202012%20(31st%20Asad%201391).pdf
IRoA, MRRD and NSP (2012b). "NSP Quarterly Report: Reporting Period 01st Jadi to 29th Hoot 1390 (22nd December 2011 to 19th March 2012)." Kabul: Islamic Republic of Afghanistan, Ministry of Rural Rehabilitation and Development, and the National Solidarity Programme. http://www.nspafghanistan.org/files/NSP-4th%20Quarterly%20Report%2001st%20Jadi%20to%2029th%20Hoot%201390%20(22nd%20December%202011%20to%2019th%20March%202012).pdf
IRoA, MRRD and NSP (2012c). "NSP Quarterly Report: Reporting Period 01st Mizar to 30th Qaws 1391 (22nd September to 20th December 2012)." Kabul: Islamic Republic of Afghanistan, Ministry of Rural Rehabilitation and Development, and the National Solidarity Programme. http://www.nspafghanistan.org/files/NSP-3rd%20Quarterly%20Report%201st%20Mizan%20to%2030th%20Qaws%201391%20(22nd%20September%20to%2020th%20December%202012).pdf
IRoA and the International Community (2006). "The London Conference on Afghanistan (31 January - 1 February 2006): The Afghanistan Compact." Accessed July 27, 2011 from http://www.nato.int/isaf/docu/epub/pdf/afghanistan_compact.pdf
IRoA MoPH (2009). "Annual Report 1387." Kabul: Islamic Republic of Afghanistan Ministry of Public Health.

ISAF (2011). "International Security Assistance Force - ISAF." Accessed March 9, 2011 from http://www.jfcbs.nato.int/jfcbrunssum/isaf.aspx
Ish-Shalom, Piki (2006). "Theory Gets Real, and the Case for a Normative Ethic: Rostow, Modernization Theory, and the Alliance for Progress" In: *International Studies Quarterly*, 50(2): 287-311.
Isham, Jonathan and Satu Kähkönen (2002). "Institutional Determinants of the Impact of Community-Based Water Services: Evidence from Sri-Lanka and India" In: *Economic Development and Cultural Change*, 50(3): 667-691.
Isham, Jonathan, Deepa Narayan and Lant Pritchett (1995). "Does Participation Improve Performance? Establishing Causality with Subjective Data" In: *The World Bank Economic Review*, 9(2): 175-200.
Islamic Emirate of Afghanistan (2011). "Statement of the Islamic Emirate Regarding the Repeatedly Baseless Accustations of UNAMA." Accessed September 9, 2011 from http://www.shahamat.info/english/index.php?option=com_content&view=article&id=9449:statement-of-the-islamic-emirate-regarding-the-repeatedly-baseless-accusations-of-unama&catid=4:statements&Itemid=4
Ittig, Dr. Annette (n.d.). "Urban Development in Kabul: An Overview of Challenges and Strategies." Accessed September 7, 2011 from http://www.institute-for-afghan-studies.org/Contributions/Projects/Dr-Ittig/UrbanDev.htm
Jackson, Ashley (2011). "High Stakes: Girls' Education in Afghanistan." *Joint Briefing Paper, 24 February, 2011.* Kabul: ACSF, ADA, APDA, AWN, AWSE, AAWU, CARE, CCA, CoAR, CHA, ECW, LCSAWC, Oxfam, SDO, Shuhada, and SCA.
Jedelhauser, Ulrich (2009). "Northern Economic Infrastructure Development: Reconstruction of Economic Infrastructure in Northern Afghanistan." Accessed October 11, 2011 from http://www.auswaertiges-amt.de/cae/servlet/contentblob/355312/publicationFile/4481/factsheet_kfw-infrastruktur.pdf
Jensch, Werner (1973). *Die afghanischen Entwicklungspläne vom ersten bis zum dritten Plan.* Meisenheim am Glan: Verlag Anton Hain KG.
Jimenez, Emmanuel and Yasuyuki Sawada (1999). "Do Community-Managed Schools Work? An Evaluation of El Salvador's EDUCO Program" In: *The World Bank Economic Review*, 13: 415-441.
Johnson, Chris and Jolyon Leslie (2004). *Afghanistan: The Mirage of Peace.* London: Zed Books.

Junker, Heinrich F. J. and Bozorg Alavi (1986). Wörterbuch Persisch-Deutsch. Leipzig, VEB Verlag Enzyklopädie Leipzig.
Junker, Heinrich F. J. and Bozorg Alavi (2002). Persisch Deutsch Wörterbuch. Wisebaden, Harasowitz. 9th edition.
Kakar, Palwasha (2005). "Fine-Tuning the NSP: Discussing Problems and Solutions with Facilitating Partners." *Working Paper Series.* Kabul: Afghanistan Research and Evaluation Unit (AREU).
Kanda, Michio and Kyoko Kuwajima (2006). "The Overview and New Orientation of Technical Cooperation of JICA -- In the Framework of Capacity Development and Human Security" In: *Technology and Development,* No. 19: 36-51.
Kapoor, Ilan (2002). "The Devil's in the Theory: A Critical Assessment of Robert Chambers' Work on Participatory Development" In: *Third World Quarterly,* 23(1): 101-117.
Katz, Travis and Jennifer Sara (1997). "Making Rural Water Supply Sustainable: Recommendations from a Global Study." Washington, DC: UNDP World Bank, Water and Sanitation Program. http://www.its.caltech.edu/~e105/readings/rural_water.pdf
Katzman, Kenneth (2011). "Afghanistan: Post-Taliban Governance, Security, and U.S. Policy." *CRS RL30588.* Congressional Research Service (CRS).
Kenski, Henry C. and Margaret Corgan Kenski (1975). "Teaching Political Development at American Colleges and Universities: A Survey" In: *The Western Political Quarterly,* 28(3): 567-578.
Keough, Noel (1998). "Participatory Development Principles and Practice: Reflections of a Western Development Worker" In: *Community Development Journal,* 33(3): 187-196.
Kerry, John F. (2011). "Evaluating U.S. Foreign Assistnace to Afghanistan: A Majority Staff Report Prepared for the Use of the Committee on Foreign Relations, United States Senate One Hundred Twelfth Congress First Session, June 8, 2011." Washington, DC: Committee on Foreign Relations.
KfW Entwicklungsbank (2011). "Economic Infrastructure in the North." Kabul: Islamic Republic of Afghanistan and the Federal Republic of Germany. http://www.ez-afghanistan.de/fileadmin/content/fact-sheets/englisch/2012-2-Infrastruktur-e_web.pdf
Khwaja, Asim Ijaz (2004). "Is Increasing Community Participation Always a Good Thing?" In: *Journal of the European Economic Association,* 2(2/3): 427-436.

Khwaja, Asim Ijaz (2005). "Measuring Empowerment at the Community Level: An Economist's Perspective." In: D. Narayan (eds.), *Measuring Empowerment: Cross-Disciplinary Perspectives.* Washington, DC: The World Bank, 267-284.

Kiely, Ray (1999). "The Last Refuge of the Noble Savage? A Critical Assessment of Post-Development Theory" In: *The European Journal of Development Research,* 11(1): 30-55.

King, Laura (2011). "Afghans Gradually Take Over Security in Several Areas." Accessed July 26, 2011 from http://www.latimes.com/news/nationworld/world/la-fg-afghanistan-transition-20110725,0,3713779.story

Klak, Thomas (2002). "World-Systems Theory: Centres, Peripheries and Semi-Peripheries." In: Vandana Desai and Robert B. Potter (eds.), *The Companion to Development Studies.* London: Arnold, 107-112.

Kleemeier, Elizabeth (2000). "The Impact of Partcipation on Sustainability: An Analysis of the Malawi Rural Piped Scheme Program" In: *World Development,* 28(5): 929-944.

Knyazev, A.A. (2000). *Istoriya afganskoi voiny 1990kh gg. i prevrashchenie Afganistana v istochnik ugroz dlya Central'noi Azii.* Bishkek: Izdatel'stvo Kyrgyzsko-Rossiiskogo Slavyanskogo Universiteta.

Koehler, Jan and Christoph Zürcher (2007). "Assessing the Contribution of International Actors in Afghanistan." *SFB-Governance Working Paper Series, No. 7, October 2007.* Berlin.

Korten, David C. (1980). "Community Organization and Rural Development: A Learning Process Approach" In: *Public Administration Review,* 40(5): 480-511.

Kothari, Uma (2001). "Power, Knowledge and Social Control." In: Bill Cooke and Uma Kothari (eds.), *Participation: The New Tyranny?* London and New York: Zed Books, 139-152.

Krefetz, Sharon Perlman and Alan E. Goodman (1973). "Participation for What or for Whom? Some Considerations for Research" In: *Administration & Society,* 5(3): 367-380.

Kreutzmann, Hermann (1998). "From Modernization Theory Towards the 'Clash of Civilizations': Directions and Paradigm Shifts in Samuel Huntington's Analysis and Prognosis of Global Development" In: *GeoJournal,* 46: 255-265.

Kreutzmann, Hermann (2008). "Dividing the World: Conflict and Inequality in the Context of Growing Global Tension" In: *Third World Quarterly,* 29(4): 675-689.

Kreutzmann, Hermann and Stefan Schütte (2010). "Persistence and Change of Afghan-German Economic Cooperation: The Case of the New Baghlan Sugar Company" In: *Erdkunde*, 64(H.1): 1-16.
Kurien, V. (1997). "The AMUL Dairy Cooperatives: Putting the Means of Development into the Hands of Small Producers in India." In: Anirudh Krishna, Norman T. Uphoff and Milton J. Esman (eds.), *Reasons for Hope: Instructive Experiences in Rural Development*. West Hartford, Connecticut: Kumarian Press, 105-119.
La Voy, Diane, Chanya Charles, Olivier Carduner, et al. (1998). "Engaging Customer Participation: USAID's Oganizational Change Experience, Prepared for the International Conference on Upscaling and Mainstreaming Participation of Primary Stakeholders, Lessons Learned and Ways Forward, November 19-20." Washington, DC: USAID. http://pdf.usaid.gov/pdf_docs/Pnace360.pdf
Lal, Deepak (1983). *The Poverty of 'Development Economics'*. Lancing: The Institute of Economic Affairs.
Landinfo (2011). "Report, Afghanistan: Marriage." Oslo: Utelendingsforvaltningens fagenhet for landinformasion (Country of Original Information Centre). http://www.landinfo.no/asset/1852/1/1852_1.pdf
Larsson, Katarina (2008). "A Provincial Survey of Balkh, Jowzjan, Samangan and Saripul." Kabul: Swedish International Development Cooperation Agency.
Latham, Michael E. (1998). "Ideology, Social Science, and Destiny: Modernization and the Kennedy-Era Alliance for Progress" In: *Diplomatic History*, 22(2): 199-229.
Lee, Barbara (2011). "Ten Years Later: Is Anyone Paying Attention?" Accessed December 21, 2011 from http://www.huffingtonpost.com/rep-barbara-lee/ten-years-later-is-anyone_b_1024049.html
Lee, J.L. (1996). *The 'Ancient Supremacy:' Bukhara, Afghanistan & the Battle for Balkh 1731-1901*. Leiden, The Netherlands: E. J. Brill.
Leopold, Evelyn (1998). "UN Restricts Aid to Kabul to Emergency Projects." Accessed June 15, 2011 from http://www.afghanistannewscenter.com/news/1998/july/july23c1998.htm
Levi, Scott C. (2002). *The Indian Diaspora in Central Asia and Its Trade, 1550-1900*. Leiden, Boston, and Köln: Brill.
Lewis, J. and P. Walker (1998). *Participation Works! 21 Techniques for Community Participation for the 21st Century*. London: New Economics Foundation.
Leys, Colin (1982). "Samuel Huntington and the End of Classical Moder-

nization Theory." In: Hamza Alavi and Teodor Shanin (eds.), *Introduction to the Sociology of "Developing Societies"*. London: Macmillan Press, 332-349.
Leys, Colin (2005). "The Rise and Fall of Development Theory." In: Marc Edelman and Angelique Haugerud (eds.), *The Anthropology of Development and Globalization: From Classical Political Economcy to Contemporary Neoliberalism*. Malden, Oxford, and Victoria: Blackwell Publishing.
Lieberman, Samuel S. (1980). "Afghanistan: Population and Development in the 'Land of Insolence'" In: *Population and Development Review*, 6(2): 271-298.
Lister, Sarah (2007). "Chapter Eight: The Community Fora Process in Mazar-e-Sharif, Afghansitan." In: James Manor (eds.), *Aid That Works: Successful Development in Fragile States*. Washington, DC: The World Bank.
Liu, Helen (2007). "Transition Strategy and Cycle 2+ Communities: A Study of the NSP." Kabul: Agency for Technical Cooperation and Development (ACTED).
Long, Norman and Ann Long (1992). *Battlefields of Knowledge*. London: Routledge.
Lopez, Ralph (2011). "US Special Inspector Gives Glowing Report on Afghan-Run Development Program, June 2, 2011." Accessed September 22, 2011 from http://www.truth-out.org/us-special-inspector-gives-glowing-report-afghan-run-development-program/1307034509
Lorentz, John H. (1987). "Afghan Aid: The Role of Private Voluntary Organizations" In: *Journal of South Asian and Middle Eastern Studies*, 11(1/2): 62-98.
MacChesney, Robert D. (1991). *Waqf in Central Asia: Four Hundred Years in the History of a Muslim Shrine, 1480-1889*. Princeton, NJ: Princeton University Press.
Magnus, Ralph H. and Eden Naby (2002). *Afghanistan: Mullah, Marx, and Mujahid*. Boulder and Oxford: Westview Press.
Maitland and Talbot (1887). *Captain Maitland's and Captain Talbot's Journeys in Afghanistan*: Blackwell Publishing on behalf of The Royal Geographical Society (with the Institute of British Geographers).
Malleson, G. B. (1878). *History of Afghanistan, from the Earliest Period to the Outbreak of the War of 1878*. London: W.H. Allen & Co.
Mansuri, Ghazala and Vijayendra Rao (2004). "Community-Based and -Driven Development: A Critical Review." *Policy Research Working*

Paper 3209. Washington, DC: The World Bank. http://wbro.oxford journals.org/content/19/1/1.full.pdf+html
Mansuri, Ghazala and Vijayendra Rao (2007). Update Note on Community-Based and -Driven Development. Washington DC, The World Bank. Unpublished manuscript.
Mansuri, Ghazala and Vijayendra Rao (2013). "Localizing Development: Does Participation Work?" Washington, DC: World Bank.
Margesson, Rhoda (2007). "Afghan Refugees: Current Status and Future Prospects." *CRS Report for Congress, Order Code RL 33851.* Washington, DC: Congressional Research Services (CRS). http://www.fas.org/sgp/crs/row/RL33851.pdf
Marshall, Don D. (2002). "The New World Group of Dependency Scholars: Reflections on a Caribbean Avant-Garde Movement." In: Vandana Desai and Robert B. Potter (eds.), *The Companion to Development Studies.* London Arnold, 102-107.
Marx, Karl and Friedrich Engels (2002). *The Communist Manifesto.* London: Penguin.
Massoud, Waheedullah (2010). "Afghanistan Dissolves 172 NGOs for Misconduct." Accessed September 6, 2011 from http://www.google.com/hostednews/afp/article/ALeqM5jyTUIkd7EOQfgCLmUabdR9L_PBWg
Max-Neef, Manfred (1991). *Human Scale Development: Conception, Application and Further Reflections.* New York and London: The Apex Press.
Maynard, Kimberly A. (2007). "The Role of Culture, Islam and Tradition in Community Driven Reconstruction: A Study on the Interantional Rescue Committee´s Approach to Afghanistan´s National Solidarity Program [sic]." International Rescue Committee (IRC).
McCarthy, Jennifer (2006). Spaces of Power and Participatory Development in Afghanistan: A Case Study of the National Solidarity Proramm and (Un)Changing Political Power Structures in Faryab Province. Geography. London, King's College. MA.
McIlwaine, Cathy and Katie Willis (2002). *Challenges and Change in Middle America: Perspectives on Development in Mexico, Central America and the Caribbean.* Essex: Pearson Education Limited.
McNamara, Robert S. (1973). "The Nairobi Speech: Address to the Board of Governors." Kenya, Nairobi: The World Bank Group.
Meadows, D.H., D.L. Meadows, J. Randers, et al. (1972). *The Limits to Growth: A Report for the Club of Rome's Project on the Predicament of Mankind.* London: Pan Books.
Mercer, Claire (2002). "The Discourse of Maendeleo and the Politics of

Women's Participation on Mount Kilimanjaro" In: *Development and Change,* 33(1): 101-127.
Merriam, Sharan B. (1998). *Qualitative Research and Case Study Applications in Education.* San Francisco: Jossey-Bass.
Michener, Victoria J. (1998). "The Participatory Approach: Contradiction and Co-Option in Burkina Faso" In: *World Development,* 26(12): 2105-2118.
Mielke, Katja and Conrad Schetter (2007). ""Where is the Village?" Local Perceptions and Development Approaches in Kunduz Province" In: *ASIEN,* 104(July 2007): 71-87.
Mikkelsen, Britha H. (2005). *Methods for Development Work and Research: A New Guide for Practitioners.* New Delhi: Sage Publications India Pvt.
Mohan, Giles (2001a). "Beyond Participation: Strategies for Deeper Empowerment." In: Bill Cooke and Uma Kothari (eds.), *Participation: The New Tyranny?* London and New York: Zed Books, 153-167.
Mohan, Giles (2001b). "Participatory Development." In: Vandana Desai and Robert B. Potter (eds.), *The Companion to Development Studies.* London: Hodder, 49-54.
Mohan, Giles and Kristian Stokke (2000). "Participatory Development and Empowerment: The Dangers of Localism" In: *Third World Quarterly,* 21(2): 247-268.
Mompati, Tlamelo and Gerard Prinsen (2000). "Ethnicity and Participatory Development Methods in Botswana: Some Participants are to be Seen and Not Heard" In: *Development in Practice,* 10(5): 625-637.
Monsutti, Alessandro (2009). "Local Power and Transnational Resources: An Anthropological Perspective on Rural Rehabilitation in Afghanistan." *Central Asia Initiative Working Paper No. 2.* Los Angeles: UCLA Asia Institute. http://www.international.ucla.edu/media/files/Monsutti_CAI_2.pdf
Monsutti, Alessandro (2012). "Fuzzy Sovereignty: Rural Reconstruction in Afghanistan, between Democracy Promotion and Power Games" In: *Comparative Studies in Society and History,* 54(3): 563-591.
Moorcraft, W. and G. Trebeck (1838). *Travels in the Himalayan Provinces of Hindustan and the Panjab: In Ladakh and Kashmir; In Peshawar, Kabul, Kunduz, and Bokhara by Mr. William Moorcroft and Mr. George Trebeck From 1819 to 1825.* London: J. Murray.
Moseley, William G. and Kavita Pandit (2007). "Introduction." In: William G. Moseley, David A. Lanegran and Kavita Pandit (eds.), *The Introduc-*

tory Reader in Human Geography: Contemporary Debates and Classic Writings. Malden: Wiley-Blackwell, 351-353.
Mosse, David (1997). "The Symbolic Making of a Common Property Resource: History, Ecology, and Locality in a Tank-Irrigated Landscape in South India" In: *Development and Change,* 28(3): 467-504.
Mosse, David (2001). "'People's Knoweldge', Participation and Patronage: Operations and Representations in Rural Development." In: Bill Cooke and Uma Kothari (eds.), *Participation: The New Tyranny?* London and New York: Zed Books, 16-35.
MRRD (2012). "Produce and Broadcast Radio Programmes in Dari and Pushto." Accessed October 8, 2012 from http://mrrd.gov.af/en/jobs/10124
MRRD and AIRD (2006). Gender Mainstreaming in MRRD Programmes: A Preliminary Review of the NSP. Kabul, Islamic Republic of Afghanistan, Ministry of Rural Rehabilitation and Development, and the Afgahnistan Institute for Rural Development.
MRRD and CSO (2005). "The National Risk and Vulnerability Assessment: Afghanistan, June 2007." Kabul: Ministry of Rural Rehabilitation and Development and the Central Statistics Organization. http://docustore.wfp.org/stellent/groups/public/documents/vam/wfp153753.pdf
MRRD and NABDP (2006). "Summary of the District Development Plan: Khulm District, Balkh Province." Kabul: Islamic Republic of Afghanistan, Ministry of Rural Rehabilitation and Development and the National Area Based Development Program. http://www.mrrd-nabdp.org/DDP/Balkh/Khulm%20DDP%20English%20Summary.pdf
MRRD and NSP (2011). "Nation-Building from the Ground Up: NSP/MRRD Disbursements for Village-Level Governance and Socio-Economic Development in Rural Afghanistan Top USD 800 Million." Kabul: Islamic Republic of Afghanistan, Ministry of Rural Rehabilitation and Development and the National Solidarity Programme. http://www.nspafghanistan.org/files/English%20press%20release.pdf
MRRD and NSP III (2012). "Operational Manual Version Six ("OM VI")." Kabul: Ministry of Rural Rehabilitation and Development (MRRD) and National Solidarity Programme Phase Three (NSP III). http://www.nspafghanistan.org/default.aspx?sel=16
Mukhopadhyay, Dipali (2009). "Disguised Warlordism and Combatanhood in Balkh: The Persistence of Informal Power in the Formal Afghan State" In: *Conflict, Security & Development,* 9(4): 535-564.
Murotani, Ryutaro, Eiji Wakamatsu, Tomonori Kikuchi, et al. (2010). "State

Building in Fragile Situations: Japanese Aid Experiences in Cambodia, Afghanistan and Mindanao." Tokyo: JICA Research Institute.

Nagar, Richa and Saraswati Raju (2003). "Women, NGOs and the Contradictions of Empowerment and Disempowerment: A Conversation" In: *Antipode*, 35(1): 1-13.

Nagl, Dr. John A., Andrew M. Exum and Ahmed A. Humayun (2009). " A Pathway to Success in Afghanistan: The National Solidarity Program [sic]." *Policy Brief*. Washington, DC: Center for a New American Security. http://www.cnas.org/files/documents/publications/CNAS%20Policy%20Brief%20-%20Supporting%20Afghanistans%20NSP%20March%202009.pdf

Narayan, Deepa, Ed. (2005). *Measuring Empowerment: Cross-Disciplinary Perspectives*. Washington, DC: The World Bank.

Narayan, Deepa, Robert Chambers, Meera Kaul Shah, et al. (2000). *Voices of the Poor: Crying Out for Change*. Oxford: Oxford University Press for the World Bank.

Narayanan, Pradeep (2003). "Empowerment Through Participation: How Effective is this Approach?" In: *Economic and Political Weekly*, 38(25): 2484-2486.

Nawa, Fariba (2006). "Afghanistan, Inc.: A CorpWatch Investigative Report." Accessed July 19, 2012 from http://s3.amazonaws.com/corpwatch.org/downloads/AfghanistanINCfinalsmall.pdf

Nel, Etienne, Tony Binns and Nicole Motteux (2003). "Community-Based Development, Non-Governmental Organizations and Social Capital in Post-Apartheid South Africa" In: *Geografiska Annaler: Series B, Human Geography*, 83(1): 3-13.

Nelson, Nici and Susan Wright, Eds. (1995). *Power and Participaotry Development: Theory and Practice*. London: Intermediate Technology Publications.

Newell, Richard (1974). "Foreign Relations." In: Louis Dupree and Linette Albert (eds.), *Afghanistan in the 1970s*. New York and London: Praeger, 76-90.

Newell, Richard S. (2007). "Government and Politics." In: Alfred Aghajanian (eds.), *Afghanistan: Past and Present*. Los Angeles: IndoEuropean Publishing, 134-200.

Nicosia, Francis R. (1997). "'Drang Nach Osten' Continued? Germany and Afghanistan During the Weimar Republic" In: *Journal of Contemporary History*, 32(2): 235-257.

Nixon, Hamish (2008a). "The Changing Face of Local Governance? Com-

munity Development Councils in Afghanistan." *Working Paper Series.* Kabul: Afghanistan Research and Evaluation Unit (AREU). http://www.areu.org.af/Uploads/EditionPdfs/802E-Changing%20Face%20of%20local%20Governance-WP-print.pdf.pdf
Nixon, Hamish (2008b). "Subnational State-Building in Afghanistan." *Synthesis Paper Series.* Kabul: Afghanistan Research and Evaluation Unit (AREU). http://www.areu.org.af/Uploads/EditionPdfs/806E-Subnational%20State-Building-SP-print.pdf.pdf
Nkansa, Grace Akukwe and David W. Chapman (2006). "Sustaining Community Participation: What Remains After the Money Ends?" In: *Review of Education,* 52: 509-532.
NSP (2003). "National Solidarity Programme Operations Manual." Kabul: Oversight Consultant.
NSP (2004). "National Solidarity Programme Operations Manual." Kabul: Islamic Republic of Afghanistan Ministry of Rural Rehabilitation and Development.
NSP (2006). "National Solidarity Programme (NSP) Operational Manual, 3rd Version." Kabul: Islamic Republic of Afghanistan Ministry of Rural Rehabilitation and Development.
NSP (2007). "National Solidarity Programme (NSP): Operational Manual, 4th Version." Kabul, Afghanistan: Islamic Republic of Afghanistan Ministry of Rural Rehabilitation and Development.
NSP (2009a). "National Solidarity Programme (NSP) Operational Manual Version V." Kabul: Islamic Repbulic of Afghanistan Ministry of Rural Rehabilitation and Development. http://www.nspafghanistan.org/default.aspx?sel=16
NSP (2009b). "National Solidarity Programme Quarterly Report: Second Quarter 1388 (22nd June 2009 to 22nd September 2009)." Kabul: Islamic Republic of Afghanistan Ministry of Rural Rehabilitation and Development.
NSP (2011a). "Facilitating Partners." Accessed September 23, 2011 from http://www.nspafghanistan.org/default.aspx?sel=17
NSP (2011b). "National Solidarity Programme (NSP) Weekly Status Report: 12th March - 18th March 2011 (21st - 27th Hoot 1389)." Kabul: Islamic Republic of Afghanistan Ministry of Rural Rehabilitation and Development.
NSP (2011c). "National Solidarity Programme: Definitions." Accessed April 19, 2011 from http://www.nspafghanistan.org/default.aspx?sel=15
NSP and MRRD (March 2012). "NSP Coverage by Districts."

NSP, MRRD and GoIRA (2011). "Villages Speak and the Nation Listens: The Third National Consultation Conference of Afghanistan's Community Development Councils: A Report from the National Solidarity Programme (NSP)." National Solidarity Programme, Ministry of Rural Rehabilitation and Development, Government of the Islamic Republic of Afghanistan. http://www.nspafghanistan.org/files/2010%20CDC %20-%20NCC%20III%20Report.pdf

Oakley, Peter (1991). *Products with People: The Practice of Participation in Rural Development.* Geneva: International Labour Office.

OCHA (2009). "Afghanistan Flooding Situation Report #4, May 16, 2009." Kabul: United Nations Office for the Coordination of Humanitarian Affairs.

Operations Evaluation Department (2005). "The Effectiveness of World Bank Support for Community-Based and -Driven Development: An OED Evaluation." Washington, DC: The World Bank.

Owen, Daniel (2006). "Community-Driven Development in the Context of Conflict-Affected Countries: Challenges and Opportunities, Report No. 36425-GLB." Washington, DC: World Bank.

Pain, Adam (2008). "Let them Eat Promises: Closing the Opium Poppy Fields in Balkh and its Consequences." *Water Management, Livestock and the Opium Economy.* Kabul: Afghanistan Research and Evaluation Unit (AREU).

Pain, Rachel and Peter Francis (2003). "Reflections on Participatory Research" In: *Area,* 35(1): 46-54.

Pak Tribune (2012). "Four Taliban Killed in Balkh Gunfight." Accessed June 9, 2012 from http://paktribune.com/news/Four-Taliban-killed-in-Balkh-gunfight-250106.html

Parnwell, Michael J.G. (2002). "Agropolitan and Bottom-Up Development." In: Vandana Desai and Robert B. Potter (eds.), *The Companion to Development Studies.* London: Arnold, 112-116.

Patton, Michael Quinn (1980). *Qualitative Evaluation Methods.* Beverly Hills and London: Sage.

Peet, Richard and Elaine Hartwick (1999). *Theories of Development.* New York: The Guilford Press.

Peet, Richard and Elaine Hartwick (2009). *Theories of Development: Contentions, Arguments, Alternatives.* New York: Guildford Press.

Platteau, Jean-Philippe and Frédéric Gaspart (2003). "The Risk of Resource Misappropriation in Community-Driven Development" In: *World Development,* 31(10): 1687-1703.

Poole, Lydia (January 2011). "Afghanistan: Tracking Major Resource Flows 2002-2010." *Briefing Paper.* Somerset: Global Humanitarian Assistance.
Portes, Alejandro (1998). "Social Capital: Its Origins and Applications in Modern Sociology" In: *Annual Review of Sociology,* 24: 1-24.
Potter, Robert B. (2002). "Theories, Strategies and Ideologies of Development." In: Vandana Desai and Robert B. Potter (eds.), *The Companion to Development Studies.* London: Arnold, 61-65.
Potter, Robert B., Tony Binns, Jennifer A. Elliott, et al. (2008). *Geographies of Development: An Introduction to Development Studies.* Essex: Pearson Education Limited.
Poullada, Leon B. (1973). *Reform and Rebellion in Afghanistan 1919-1929: King Amanullah's Failure to Modernize a Tribal Society.* Ithaca and London: Cornell University Press.
Poullada, Leon B. (1974). "The Search for National Unity." In: Louis Dupree and Linette Albert (eds.), *Afghanistan in the 1970s.* New York and London: Praeger, 34-49.
Poullada, Leon B. (1981). "Afghanistan and the United States: The Crucial Years" In: *Middle East Journal,* 35(2): 178-190.
Poullada, Leon B. (1982). "The Failure of American Diplomacy in Afghanistan" In: *World Affairs,* 145(3 (Winter 1982/1983)): 230-252.
Prebisch, Raúl (1950). *The Economic Development of Latin America and its Principal Problems.* Lake Success, New York: UN Department of Economic Affairs.
PREM and World Bank (2002). "Empowerment and Poverty Reduction: A Sourcebook (Draft)." Washington, DC: World Bank. http://siteresources.worldbank.org/INTEMPOWERMENT/Resources/486312-1095094954594/draft.pdf
Preston, P.W. (2002). *Development Theory: An Introduction.* Oxford and Malden: Blackwell Publishers Inc.
Pretty, Jules (1995). *Regenerating Agriculture - Policies and Practice of Sustainability and Self-Reliance.* London: Earthscan Publications Ltd.
Provost, Claire and Jaz Cummins (2011). "Talk Point: Aid to Afghanistan." Accessed October 19, 2011 from http://www.guardian.co.uk/global-development/2011/oct/17/talk-point-afghanistan-aid
Putnam, Robert D. (1993). *Making Democracy Work: Civic Traditions in Modern Italy.* Princeton, NJ: Princeton University Press.
Putnam, Robert D. (2000). *Bowling Alone: The Collapse and Revival of American Community.* New York: Simon & Schuster Paperbacks.
Quarry, Wendy (2008). "Decision Makers Do Want Communication But

They May Not Want Participation." Accessed March 10, 2010 from http://webzone.k3.mah.se/projects/gt2/viewarticle.aspx?articleID=138&issueID=16

Rahimyar, Mohammad Fahim (1997). *ARIC Bibliography*. Peshawar.

Rahmena, Majid (1992). "Participation." In: Wolfgang Sachs (eds.), *The Development Dictionary: A Guide to Knoweldge as Power*. London and New York: Zed Books, 116-131.

Ralph Bunche Institute for International Studies (2011). "Paul Streeten." Accessed July 27, 2011 from http://www.unhistory.org/CD/Streeten.html

Rao, Vijayendra and Ana María Ibáñez (2005). "The Social Impact of Social Funds in Jamaica: A 'Participatory Econometric' Analysis of Targeting, Collective Action, and Participation in Community-Driven Development" In: *Journal of Development Studies*, 41(5): 788-838.

Rashid, Ahmed (2001). *Taliban: Militant Islam, Oil and Fundamentalism in Central Asia*. New Haven, CT: Yale University Press.

Ravallion, Martin, Shaohua Chen and Prem Sangraula (2009). "Dollar a Day Revisited" In: *The World Bank Economic Review*, 23(2): 163-184.

Resurreccion, Bernadette P., Mary Jane Real and Panadda Pantana (2004). "Officialising Strategies: Participatory Processes and Gender in Thailand's Water Resources Sector" In: *Development in Practice*, 14(4): 521-533.

Reuters (2011). "IMF Renews Afghan Ties, Approves Loan Deal." Accessed November 22, 2011 from http://princetoncouncil.org/index.php?option=com_content&view=article&id=6783:imf-renews-afghan-ties-approves-loan-deal-reuters&catid=58:asia&Itemid=133

Ribot, Jesse C. (1995). "From Exclusion to Participation: Turning Senegal's Forestry Policy Around?" In: *World Development*, 23(9): 1587-1599.

Riechmann, Deb (2012). "NATO: Afghans to Take Lead in Majority of Country." Accessed April 3, 2012 from http://www.google.com/hostednews/ap/article/ALeqM5h65CSTHrpmywUNKOn1HsoD9tjU9g?docId=4343dbc88bb04e74912b79dbb417d512

Rist, Gilbert (2008). *The History of Development: From Western Origins to Global Faith*. London and New York: Zed Books.

Rose, Kalima (1993). *Where Women are Leaders: The SEWA Movement in India*. London: Zed Books.

Rostow, Walt Whitman (1956). "The Take-Off Into Self-Sustained Growth" In: *Economic Journal*, 66(261): 25-48.

Rostow, Walt Whitman (1960). *The Stages of Economic Growth, a Non-Communist Manifesto*. Cambridge: Cambridge University Press.
Rostow, Walt Whitman (1991). *The Stages of Economic Growth: A Non-Communist Manifesto*. Cambridge, New York, and Melbrourne: Cambridge University Press.
Rubin, Alissa J. (2010). "Taliban Slay Elders Over Aid Money." Accessed October 17, 2011 from http://www.nytimes.com/2010/05/26/world/asia/26afghan.html
Rubin, Barnett R. (1994). "Redistribution and the State in Afghanistan: The Red Revolution Turns Green." In: Myron Weiner and Ali Banuazizi (eds.), *The Politics of Social Transformation in Afghanistan, Iran, and Pakistan*. New York: Syracuse University Press, 187-227.
Rubin, Barnett R. (2000). "The Political Economy of War and Peace in Afghanistan" In: *World Development*, 28(10): 1789-1803.
Rubin, Barnett R. (2002). *The Fragmentation of Afghanistan: State Formation and Collapse in the International System*. New Haven and London: Yale Unviersity Press.
Ruttan, Vernon W. (1984). "Integrated Rural Development Programmes: A Historical Perspective" In: *World Development*, 12(4): 393-401.
Sachs, Wolfgang (1992). *The Development Dictionary: A Guide to Knowledge as Power*. London and New York: Zed Books.
Said, Edward W. (1979). *Orientalism*. New York: Vintage Books.
Salahuddin, Sayed (1998). "Aid Agencies' Exit Means Hardship for Kabul." Accessed June 16, 2011 from http://www.afghanistannewscenter.com/news/1998/july/july23d1998.htm
Sarwary, Bilal (2009). "Electricity Transforms Kabul Living." Accessed September 27, 2011 from http://news.bbc.co.uk/2/hi/south_asia/8097865.stm
Sayres, William (1974). "Old Trail and New Paths in Education." In: Louis Dupree and Linette Albert (eds.), *Afghanistan in the 1970s*. n.p.: Praeger, 183-202.
Schuurman, Frans J. (2002). "The Impasse in Development Studies." In: Vandana Desai and Robert B. Potter (eds.), *The Companion to Development Studies*. London: Arnold, 12-15.
Scott, James (1998). *Seeing Like a State: How Certain Schemes to Improve the Human Condition Have Failed*. New Haven: Yale University Press.
Seers, Dudley (1969). "The Meaning of Development, Communication Series No. 44" In: *Institute of Development Studies*, 44: 1-26.
Sengupta, Kim (2012). "Karzai Accused of Endangering Troops by Blocking

Night Raids: NATO Commanders Claim President's Demand has Hampered Vital Operations Against the Taliban." Accessed July 24, 2012 from http://www.independent.co.uk/news/world/asia/karzai-accused-of-endangering-troops-by-blocking-night-raids-7546209.html

Shah, Meera Kaul, Sarah Degnan Kambou, Barbara Monahan, et al. (1999). "Embracing Participation in Development: Worldwide Experience from CARE's Reproductive Health Programs with a Step-by-Step Field Guide to Participatory Tools and Techniques." Atlanta, Georgia: CARE. http://pdf.usaid.gov/pdf_docs/pnack787.pdf

Shah, Saeed (2012, July 23, 2012). "After 30 Years, Pakistan Rolls Up Welcome Mat for Afghan Refugees." Accessed July 25, 2012 from http://www.tri-cityherald.com/2012/07/23/2030149/after-30-years-pakistan-rolls.html

Sharipov, U. Z. (1984). "Ekonomicheskie svyazi Demokraticheskoi Respubliki Afganistan i stran - chlenov SEV." In: J. V. Gankovskii, M. R. Arunova and A. D. Davydov (eds.), *Afganistan. Ekonomika, politika, istoriya.* Moskva: Izdatel'stvo "Nauka", 175-190.

Sherif, Muzafer, O.J. Harvey, B. Jack White, et al. (1961). *Intergroup Conflict and Cooperation: The Robbers Cave Experiment*: Norman Institue of Group Relations, University of Oklahoma, reprinted by Wesleyan University Press 1988.

Shrestha, Ava Darshan (1994). Eating Cucumbers Without Any Teeth: Variations in the Capacities of Rural Women to Participate in Rural Development in Nepal. Bath, University of Bath. PhD thesis.

Sibtain, Sayed N. (1981). "To Build a Village" In: *Afghanistan Journal,* 8(3): 79-99.

Sidaway, James D. (2002). "Post-Development." In: Vandana Desai and Robert B. Potter (eds.), *The Companion to Development Studies.* London: Arnold, 16-20.

SIGAR (2011). "Afghanistan's National Solidarity Program Has Reached Thousands of Afghan Communities, but Faces Challenges that Could Limit Outcomes." Washington, DC: Office of the Special Inspector General for Afghanistan Reconstruction.

Sillitoe, Paul (2000). "Let Them Eat Cake: Indigenous Knowledge, Science and the 'Poorest of the Poor'" In: *Anthropology Today,* 16(6): 3-7.

Simon, David (2002). "Neo-Liberalism, Structural Adjustment and Poverty Reduction Strategies." In: Vandana Desai and Robert B. Potter (eds.), *The Companion to Development Studies.* London: Arnold, 86-92.

Smetherman, Robert M. and Bobbie B. Smetherman (1972). "The Alliance for Progress: Promises Unfulfilled" In: *American Journal of Economics and Sociology*, 31(1): 79-85.

Smith, Tony (1979). "The Underdevelopment of Development Literature: The Case of Dependency Theory" In: *World Politics*, 31(2): 247-288.

Social Development Department (2006). "Community-Driven Development in the Context of Conflict-Affected Countries: Challenges and Opportunities." Washington, DC: World Bank.

Solov'eva, M.R. and A.A. Svetlov (1984). "Tradicii Sovetsko-Afganskoi druzhby." In: J. V. Gankovskii, M. R. Arunova and A. D. Davydov (eds.), *Afganistan. Ekonomika, politika, istoriya.* Moskva: Izdatel'stvo "Nauka", 3-27.

South Centre (2011). "South Commission." Accessed June 6, 2011 from http://www.southcentre.org/index.php?option=com_content&task=view&id=635&Itemid=80

South Commission (1990). "The South Commission Report on Population and Population Policy" In: *Population and Development Review*, 16(4): 795-798.

Stanton, Elizabeth A. (2007). "The Human Development Index: A History." *Working Paper Series.* Amherst: Political Economy Research Institute.

Strand, Arne, Hege Toje, Alf Morten Jerve, et al. (2003). "Community Driven Development in Contexts of Conflict, Concept Paper Commissioned by Environmentally and Socially Sustainable Development (ESSD) Unit of the World Bank." Bergen, Norway: Chr. Michelsen Institute (CMI). http://info.worldbank.org/etools/docs/library/35454/CDDCMI.pdf

Streeten, Paul (1981). *Development Perspectives.* Hong Kong: The Macmillan Press.

Strick van Linschoten, Alex and Felix Kuehn (2010). "Kandahar: Portrait of a City." In: Abdul Salam Zaeef (eds.), *My Life with the Taliban.* New York: Colombia University Press, ix-xxvi.

Suhrke, Astri (2007). "Reconstruction as Modernisation: the 'Post-Conflict' Project in Afghanistan" In: *Third World Quarterly*, 28(7): 1291-1308.

Tariq, Mohammad Osman, Najila Ayoubi and Fazel Rabi Haqbeen (2010). "Afghanistan in 2010: A Survey of the Afghan People." Kabul: The Asia Foundation.

Tariq, Mohammad Osman, Najila Ayoubi and Fazel Rabi Haqbeen (2011). "Afghanistan in 2011: A Survey of the Afghan People." Kabul: The Asia Foundation. http://asiafoundation.org/publications/pdf/989

Teplinskii, L.B. (1958). "Sovetsko-Afganskie otnosheniya za sorok let sushchestvovaniya nezavisimogo Afganistana." In: R. T. Akhramovich (eds.), *Nezavisimyi Afganistan: 40 let nezavisimosti. Sbornik statei.* Moskva: Izdatel'stvo Vostochnoi Literatury, 5-38.

Teplinskii, L.B. (1982). *SSR i Afghanistan 1919-1981.* Moskva: Izdatel'stvo "Nauka".

Teves, Lurli B. (2000). "The Sociocultural Dimension of People's Participation in Community-Based Development: The Role of Patron-Client System in the Philippines" In: *Der Tropenlandwirt,* 68: 147.

The Guardian (2011). "Afghanistan Civilian Casualties: Year by Year, Month by Month." Accessed September 6, 2011 from http://www.guardian.co.uk/news/datablog/2010/aug/10/afghanistan-civilian-casualties-statistics

The Local (2010). "German Aid Worker Killed in Afghanistan." Accessed January 6, 2012 from http://www.thelocal.de/national/20101225-32031.html

Thirlwall, A.P. (2002). "Development as Economic Growth." In: Vandana Desai and Robert B. Potter (eds.), *The Companion to Development Studies.* London: Arnold, 41-44.

Thoams-Slayter, Barbara P. (1994). "Structural Change, Power Politics, and Community Organizations in Africa: Challenging the Patterns, Puzzles and Paradoxes" In: *World Development,* 22(10): 1470-1490.

Thorndike, Edward L. (1920). "A Constant Error in Psychological Readings" In: *Journal of Applied Psychology,* IV: 25-29.

Timsina, Netra Prasad (2003). "Promoting Social Justice and Conserving Montane Forest Environments: A Case Study of Nepal's Community Forestry Programme" In: *Geographical Journal,* 169(3): 236-242.

Todaro, Michael P. (1977). *Economic Development in the Third World: An Introduction to Problems and Policies in a Global Perspective.* London and New York: Longman.

Torabi, Yama (2007). "Assessing the National Solidarity Program [sic]: The Role of Accountability in Reconstruction." Kabul: Integrity Watch Afghanistan and Tiri.

Transparency International (2011). "Corruption Perceptions Index 2011." Berlin: Transparency International. http://cpi.transparency.org/cpi2011/

Truman, Harry S. (1949). "Inaugural Address, January 20, 1949." Accessed June 16, 2011 from www.bartleby.com/124/pres53.html

U.S. Government Federal Research Division (2007). "Country Profile: Afghanistan." In: Alfred Aghajanian (eds.), *Afghanistan: Past and Present.* Los Angeles: IndoEuropean Publishing, 201-227.

UN News Centre (n.d.). "Afghanistan & the United Nations." Accessed February 24, 2011 from http://www.un.org/News/dh/latest/afghan/un-afghan-history.shtml
UN News Service (2011). "Afghan Civilian Deaths Rise, Insurgents Responsible for Most Casualties - UN." Accessed September 28, 2011 from http://www.un.org/apps/news/story.asp?NewsID=39036#
UNAMA (2007). "Balkh: Provincial Profile." Kabul: United Nations Department of Safety and Security (UNDSS) Provincial Assessment.
UNAMA and AIHRC (2011). "Afghanistan: Annual Report 2010: Protection of Civilians in Armed Conflict." Kabul: United Nations Assistance Mission in Afghanistan and the Afghanistan Independent Human Rights Commission. http://unama.unmissions.org/Portals/UNAMA/human%20rights/March%20PoC%20Annual%20Report%20Final.pdf
UNAMA and UNHCR (2012). "Afghanistan: Annual Report 2011: Protection of Civilians in Armed Conflict." Kabul: United Nations Assistance Mission in Afghanistan and the UN Office of the High Commissioner for Human Rights. http://unama.unmissions.org/Portals/UNAMA/Documents/UNAMA%20POC%202011%20Report_Final_Feb %202012.pdf
UNCTAD (2012). "UN Recognition of the Least Developed Countries." Accessed February 16, 2012 from http://www.unctad.org/Templates/Page.asp?intItemID=3618&lang=1
UNDP (1990). "Human Development Report 1990: Concept and Measure of Human Development." New York: Oxford University Press. http://hdr.undp.org/en/reports/global/hdr1990/chapters/
UNDP (1991). "United Nations Human Development Report 1991: Financing Human Development." New York: Oxford University Press.
UNDP (1993). "Human Development Report 1993: People's Participation." New York: Oxford University Press.
UNDP (1994). "Human Development Report 1994: New Dimensions of Human Security." New York: Oxford University Press. http://hdr.undp.org/en/reports/global/hdr1994/chapters/
UNDP (1996). "UNDP Human Development Report 1996: Economic Growth and Human Development." New York and Oxford: Oxford University Press. http://hdr.undp.org/en/media/hdr_1996_en_contents.pdf
UNDP (2002). "Human Development Report 2002: Deepening Democracy in a Fragmented World." New York: Oxford University Press.
UNDP (2010a). "Explanation Note on 2010 HDR Composite Indices." New

York: United Nations Development Programme. http://hdrstats.undp.org/images/explanations/AFG.pdf
UNDP (2010b). "Human Development Report 2010: The Real Wealth of Nations: Pathways to Human Development." New York: Oxford University Press.
UNDP (2011). "Human Development Report 2011: Sustainability and Equity, A Better Future for All." New York: Palgrave Macmillan. http://www.beta.undp.org/content/dam/undp/library/corporate/HDR/2011%20Global%20HDR/English/HDR_2011_EN_Complete.pdf
UNEP (2003). "Post-Conflict Environmental Assessment." Switzerland: United Nations Environment Programme. http://postconflict.unep.ch/publications/afghanistanpcajanuary2003.pdf
UNEP and UNCTAD (1975). "The Cocoyoc Declaration" In: *International Organization*, 29(3): 893-901.
UNFPA (2011). "UNFPA Vacancy Announcement for VA-CSO-2011-03-06." Accessed October 11, 2011 from http://countryoffice.unfpa.org/afghanistan/drive/VA-CSO-2011-03-06RegionalCoordinator.pdf
UNHABITAT (2010). "Inter-Communal Rural Development Project (IRDP)." Accessed June 26, 2012 from http://www.fukuoka.unhabitat.org/projects/afghanistan/detail01_en.html
UNHCR (2011). "2011 UNHCR Country Operations Profile - Afghanistan." Accessed September 7, 2011 from http://www.unhcr.org/cgi-bin/texis/vtx/page?page=49e486eb6#
UNHCR (2012). "2012 UNHCR Country Operations Profile - Afghanistan." Accessed June 12, 2012 from http://www.unhcr.org/cgi-bin/texis/vtx/page?page=49e486eb6
UNIDATA (1991). "Samangan Province: A Socio-Economic Profile." Kabul: UNP/OPS and UNOCA.
UNIDATA (1992). "Balkh Province: A Socio-Economic Profile." United Nations. http://www.nps.edu/programs/ccs/Balkh/Balkh_PDP_Provincial_profile.pdf
Uphoff, Norman T. (1992). *Learning from Gul Oya: Possibilities for Participatory Development and Post-Newtonian Social Science*. Ithaca: Cornell University.
Vision of Humanity (2012). "Global Peace Index: Global Peace Index Rankings." Accessed June 12, 2012 from http://www.visionofhumanity.org/wp-content/uploads/2012/06/2012-Global-Peace-Index-Map-and-Rankings1.pdf

Vivian, Jessica (2008). "NGOs and Sustainable Development in Zimbabwe: No Magic Bullets" In: *Development and Change*, 25(1): 167-193.
Wakil, Abdul (1989). "Afghanistan Solution, 'The First Track'" In: *Orient: Zeitschrift des Deutschen Orient-Intsituts*, 30(3): 359-378.
Wallerstein, Immanuel (1974). *The Modern World-System: Capitalist Agriculture and the Origins of the European World-Economy in the Sixteenth Century*. New York and London: Academic Press Inc.
Warner, Gregory (2007). "The Schools the Taliban Won't Touch: One Ingenious Aid Program is Stabilizing the Toughest Parts of Afghanistan. The U.S. is Cutting Its Funding." Accessed September 22, 2011 from http://www.washingtonmonthly.com/features/2007/0712.warner.html
Werner, Cynthia (2003). "The New Silk Road: Mediators and Tourism Development in Central Asia" In: *Ethnology*, 42(2): 141-159.
WFP (2001) Afghanistan District Vulnerability Map - Combined (updated 7th May 2002). Accessed from http://reliefweb.int/map/afghanistan/afghanistan-district-vulnerability-mapping-combined-4-indicators-landminesuxo-health
WFP and MRRD (2004). "Report on Findings from the 2003 National Risk and Vulnerability Assessment (NRVA) in Rural Afghanistan." Vulnerability Analysis and Mapping (VAM) Unit of the WFP and Vulnerability Analysis Unit (VAU) of MRRD. http://docustore.wfp.org/stellent/groups/public/documents/vam/wfp048582.pdf
Wilber, Donald N. (1962). *Afghanistan: Its People, Its Society, Its Culture*. New Haven: HRAF Press.
Wildman, Jim (2010). "As Fighting Escalates in Kandahar, Nostalgia for Taliban Rule, Islamic Law, September 22, 2010." Accessed June 11, 2012 from http://www.npr.org/blogs/thetwo-way/2010/09/21/130026254/as-fighting-escalates-kandahar-business-student-longs-for-taliban-rule-islamic-law
Williams, Maurice, John Keen and Charles Jenkins (1988). "Retrospective Review of US Assistance to Afghanistan: 1950-1979 [Draft]." Bethesda, MD: USAID. http://www.nzdl.org/gsdlmod?e=d-00000-00---off-0acku--00-0----0-10-0---0---0direct-10---4-------0-1l--11-en-50---20-about---00-0-1-00-0-0-11-1-0utfZz-8-00&cl=CL5.39&d=HASH01a59d5f5426c9abe867a597&x=1
Willis, Katie (2010). *Theories and Practices of Development*. London and New York: Routledge.
Winfield, Nicole (1998). "UN, Red Cross Stays in Afghanistan." Accessed

July 16, 2011 from http://www.afghanistannewscenter.com/news/1998/july/july21a1998.htm
Wood, Geof, Abdul Malik and Sumaira Saheer (2006). *Valleys in Transition: Twenty Years of AKRSP's Experience in Northern Pakistan*. Oxford: Oxford University Press.
Wood, John (1872). *A Journey to the Source of the River Oxus with an Essay on the Geography of the Valley of the Oxus*. London: J. Murray.
Woolcock, Michael (1998). "Social Capital and Economic Development: Toward a Theoretical Synthesis and Policy Framework" In: *Theory and Society*, 27: 151-208.
Woolcock, Michael (2009). "The Next 10 Years in Development Studies: From Modernization to Multiple Modernities, in Theory and Practice" In: *European Journal of Development Research*, 21: 4-9.
Woolcock, Michael and Deepa Narayan (2000). "Social Capital: Implications for Development Theory, Research, and Policy" In: *The World Bank Research Observer*, 15(2): 225-249.
Worden, Scott and Nina Sudhakar (2012). "Learning from Women's Success in the 2010 Afghan Elections." United States Institute of Peace (USIP).
World Bank (2001). "Afghanistan: World Bank Approach Paper November 2001." Washington, DC: World Bank. http://siteresources.worldbank.org/INTAFGHANISTAN/Resources/afgApproach.pdf
World Bank (2002). "World Development Report 2000/2001: Attacking Poverty." World Bank.
World Bank (2010a). "Afghanistan Public Expenditure Review 2010: Second Generation of Public Expenditure Reforms, Report No. 53892-AF, April 2010." World Bank and UK Department for International Development (DFID). http://siteresources.worldbank.org/AFGHANISTANEXTN/Resources/305984-1264608805475/6739619-1276813833351/AfghanistanPERfinal.pdf
World Bank (2010b). "Emergency Project Paper on a Proposed Grant In the Amount of SDR 27.2 Million (US$40 Million Equivalent) to the Islamic Republic of Afghanistan for a Third Emergency National Solidarity Project, June 10, 2010, Report No. 54540-AF." from http://www-wds.worldbank.org/external/default/WDSContentServer/WDSP/IB/2010/06/23/000334955_20100623022629/Rendered/INDEX/545400PJPR0P11101Official0Use0Only1.txt
World Bank (2010c). "Implementation Status and Results: Afghanistan National Solidarity Program III." *Report No. ISR3429*. Washington, DC: World Bank. http://www-wds.worldbank.org/external/default/WDS

ContentServer/WDSP/SAR/2010/12/11/4E4AD41AACA6CD9E852577F60066BBB2/1_0/Rendered/PDF/P1171030ISR0Di021120101292092926999.pdf
World Bank (2010d). "Project Information Document (PID): Appraisal Stage, Report No.: AB5745." Accessed January 3, 2012 from http://www.reliefweb.int/sites/reliefweb.int/files/resources/E57ABE817CCEEEAD852577280072453B-Full_Report.pdf
World Bank (2011a). "Growth in Afghanistan: Afghanistan's Recent Economic Performance." Accessed September 6, 2011 from http://web.worldbank.org/WBSITE/EXTERNAL/COUNTRIES/SOUTHASIAEXT/EXTSARREGTOPMACECOGRO/0,,contentMDK:20592478~menuPK:579404~pagePK:34004173~piPK:34003707~theSitePK:579398,00.html
World Bank (2011b). "National Solidarity Programme Phase Three (NSP III) in Afghanistan: Consultancy Firms for Serving as Third-Party Monitor for NSP in High Risk Areas." Accessed April 19, 2011 from https://www.devex.com/en/projects/national-solidarity-programme-phase-three-nsp-iii-in-afghanistan-consultancy-firms-for-serving-as-third-party-monitor-for-nsp-in-high-risk-areas
World Bank (2011c). "National Solidarity Project: Promoting Community-Based Development in Afghanistan." Accessed April 19, 2011 from http://www.worldbank.org.af/WBSITE/EXTERNAL/COUNTRIES/SOUTHASIAEXT/AFGHANISTANEXTN/0,,print:Y~isCURL:Y~contentMDK:22888068~menuPK:305990~pagePK:2865066~piPK:2865079~theSitePK:305985,00.html
World Bank (2011d). "Transition in Afghanistan: Looking Beyond 2014: Executive Summary." World Bank. http://siteresources.worldbank.org/AFGHANISTANEXTN/Resources/305984-1297184305854/AFTransition.pdf
World Bank (2011e). "What are PRSPs?" Accessed April 5, 2011 from http://web.worldbank.org/WBSITE/EXTERNAL/TOPICS/EXTPOVERTY/EXTPRS/0,,contentMDK:22283891~pagePK:210058~piPK:210062~theSitePK:384201,00.html
World Bank (2012). "Kecamatan Development Program, a Slideshow." Accessed October 25, 2012 from http://digitalmedia.worldbank.org/slideshow/?slideshow_id=231
World Hunger Education Service (2012). "2012 World Hunger and Poverty Facts and Statistics." Accessed July 23, 2012 from http://www.worldhunger.org/articles/Learn/world%20hunger%20facts%202002.htm
Xuangzang (1996). *The Great Tang Dynasty Record of the Western Regions.*

Berkeley, California: Numata Center for Buddhist Translation and Research.

Yate, Major C.E. (1888). *Northern Afghanistan, or Letters from the Afghan Boundary Commission.* Edingurgh and London: William Blackwood and Sons.

Yule, H. (1872). "Papers Connected with the Upper Oxus Region" In: *Journal of the Royal Geographic Society,* 42: 438-481.

Zaeef, Abdul Salam (2010). *My Life with the Taliban.* New York: Columbia University Press.

Zakhilawal, Omar and Jane Murphy Thomas (2008). "Chapter 6: Afghanistan: What Kind of Peace? The Role of Rural Development in Peacebuilding." In: Stephen Baranyi (eds.), *The Paradoxes of Peace-building Post-9/11.* Vancouver: University of British Colombia Press, 147-178.

Zyck, Steven A. (2012). "The Role of China in Afghanistan's Economic Development & Reconstruction." *Afghanistan in Transition.* Civil-Military Fusion Centre. https://www.cimicweb.org/Documents/CFC%20AFG%20Economic%20Stabilization%20Archive/CFC_Afghanistan_China-in-Afghanistan_Mar12.pdf

Studien zum Modernen Orient

SMO 11

Charlotte Joppien

Die türkische Adalet ve Kalkinma Partisi (AKP)

Eine Untersuchung des Programms »Muhafazakar Demokrasi« (Konservative Demokratie)

Berlin 2011. Br. 211 S., 978-3-87997-389-7

SMO 12

Sara Winter

»Ein alter Feind wird nicht zum Freund«

Fremd- und Selbstbild in der aserbaidschanischen Geschichtsschreibung

Berlin 2011. Br. 131 S., 978-3-87997-385-9

SMO 13

Fawzi Habashi

Prisoner of All Generations

My Life in the Homeland Egypt

Berlin 2011. Pb. 292 pp., 978-3-87997-350-7

SMO 15

Irene Weipert-Fenner

Starke Reformer oder schwache Revolutionäre?

Ländliche Notabeln und das ägyptische Parlament in der ʿUrabi-Bewegung, 1866-1882

Berlin 2011. Br. 162 S., 978-3-87997-387-3

SMO 17

Nadine Kreitmeyr

Der Nahostkonflikt durch die Augen Ḥanẓalas

Stereotypische Vorstellungen im Schaffen des Karikaturisten Naǧi al-ʿAli

Berlin 2012. Br. 140 S., 978-3-87997-402-3

KS KLAUS SCHWARZ VERLAG • BERLIN

Studien zum Modernen Orient

SMO 18

Yuriy Malikov

Tsars, Cossacks, and Nomads

The Formation of a Borderland Culture in Northern Kazakhstan in the 18th and 19th Centuries

Berlin 2011. Pb. 321 pp., 978-3-87997-359-8

SMO 19

Claus Schönig/Ramazan Çalık/Hatice Bayraktar (Hg.)

Türkisch-deutsche Beziehungen

Perspektiven aus Vergangenheit und Gegenwart

Berlin 2012. Br. 426 S., 978-3-87997-386-6

SMO 20

Ariela Gross

Reaching *wa'y*

Mobilization & Recruitment in Hizb al-Tahrir al-Islami. A Case Study Conducted in Beirut

Berlin 2012. Pb. 111 pp., 978-3-87997-405-4

SMO 21

Christiane Czygan

Zur Ordnung des Staates

Jungosmanische Intellektuelle und ihre Konzepte in der Zeitung Hürriyet (1868–1870)

Berlin 2012. Br. 328 S., 978-3-87997-407-8

SMO 22

Eliane Ursula Ettmüller

The Construct of Egypt's National Self

In James Sanua's Early Satire & Caricature

Berlin 2012. Pb. 328 pp., 978-3-87997-411-5

KS

Klaus Schwarz Verlag GmbH • Fidicinstr. 29 • D-10965 Berlin

Tel. +30-916 82 749 • +30-916 82 751 • Fax +30-322 51 83

www.klaus-schwarz-verlag.com

info@klaus-schwarz-verlag.com

Bei Fragen zur Produktsicherheit wenden Sie sich bitte an:
If you have any questions regarding product safety,
please contact:

Walter de Gruyter GmbH
Genthiner Straße 13
10785 Berlin
productsafety@degruyterbrill.com